Current Research in Sports Sciences
An International Perspective

Current Research in Sports Sciences

An International Perspective

Edited by

Victor A. Rogozkin
Research Institute of Physical Culture
St. Petersburg, Russia

and

Ron Maughan
University of Aberdeen
Aberdeen, Scotland

Plenum Press • New York and London

Library of Congress Cataloging-in-Publication Data

Current research in sports sciences : an international perspective
  edited by Victor A. Rogozkin and Ron Maughan.
      p.   cm.
    Includes bibliographical references and index.
    ISBN 0-306-45319-3
    1. Sports--Physiological aspects--Congresses.   I. Rogozkin, V. A.
  II. Maughan, Ron J., 1951-
  RC1235.C875 1996
  612'.044--dc20   96-27008
                                                                  CIP

Proceedings of an International Conference on Current Research into Sport Sciences, held in association
with the Goodwill Games, July 27 – 30, 1994, in St. Petersburg, Russia

ISBN 0-306-45319-3

© 1996 Plenum Press, New York
A Division of Plenum Publishing Corporation
233 Spring Street, New York, N. Y. 10013

Printed in the United States of America

PREFACE

There are two main reasons for pursuing research in the Sports Sciences. Firstly, by studying responses to exercise, we learn about the normal function of the tissues and organs whose function allows exercise to be performed. The genetic endowment of elite athletes is a major factor in their success, and they represent one end of the continuum of human performance capability: the study of elite athletes also demonstrates the limits of human adaptation because nowhere else is the body subjected to such levels of intensive exercise on a regular basis. The second reason for studying Sports Science is the intrinsic interest and value of the subject itself. Elite performers set levels to which others can aspire, but even among spectators, sport is an important part of life and society.

Apart from the study of top sport and elite performers, there is also another reason for medical and scientific interest in sport. There is no longer any doubt that lack of physical activity is a major risk factor for many of the diseases that affect people in all countries: such diseases include coronary heart disease, obesity, hypertension, and diabetes. An increased level of recreational physical activity is now an accepted part of the prescription for treatment and prevention of many illnesses, including those with psychological as well as physical causes. An understanding of the normal response to exercise, as well as of the role of exercise in disease prevention, is therefore vital.

Although there is little doubt as to the physical and psychological benefits that can follow from a regular exercise programme, there are also possible negative consequences. The risk of injury associated with the repetitive strain on muscles, joints, and other tissues is obvious. While this may be an acceptable risk in the mature individual, there may be special problems for youngsters who train intensively. The identification of sporting talent in youngsters and its development so that potential is realised without compromising physical and psychological development poses special problems.

To address some of these issues, a major international Conference was held in association with The Goodwill Games in St. Petersburg, Russia, in July 1994. This Conference was unique in many respects, as it was the first in the field of exercise science and sports medicine to bring together scientists and clinicians from Western and Eastern Europe, North America, the Republics of the Former Soviet Union, and from the rest of the world. More than 400 participants from 42 countries contributed to the meeting. For many years, the fiercely competitive nature of international sport had stimulated interest and speculation about the advances made in Sports Science in other countries and had provided an impetus to research in this area. This conference offered an opportunity for a free exchange of information and ideas, and these proceedings are a permanent record of the presentations made.

The programme of the conference focused on three broad areas: (1) the identification and development of sporting talent, (2) reaching and maintaining peak performance, and (3) the relationships between physical activity and wellness. This volume contains selected papers in all of these areas, providing a broad overview of key issues in Sports Science, including much information that has not been available until now.

Thanks are due to many people for the contributions that made a success of the conference and who have allowed the production of these proceedings. Much of the editorial work was accomplished with the help of Susan Shirreffs in Aberdeen. Special thanks are due to Mars Incorporated for their major commitment to support high quality research and education in exercise science and sports nutrition, for their sponsorship of the meeting, and for the organisational skills that helped to ensure its smooth running.

V.A. Rogozkin and R.J. Maughan

CONTENTS

IDENTIFICATION AND DEVELOPMENT OF TALENT IN SPORT

V. K. Balsevich

Russian State Academy of Physical Culture
Sirenevyi St.4 Moscow Russia

The problem of discovery and development of sporting talent is an important one and by satisfactorily solving this problem gifted young athletes can be attracted to the elite sporting activity. The enormous material, financial and spiritual expenditure on their training and health assurance will be fully covered. Sporting talent is the most important prerequisite for a young athlete's progress in a chosen sport. That is why sporting talent can be represented by many aspects of an athlete's motor activity. High sporting achievement is always the result of the many-years, goal-directed training ensuring the attainment of essential physical, functional, coordinational and psychological conditions by an athlete. The highest sporting achievements are determined by the effectiveness of the athlete to realise his potential during the process of his or her sporting development. It is quite clear that the timely diagnosis of such opportunities and the correct prognosis of their successful display in the course of training creates real advantages for athlete's future. The complexity and development of sporting talent can be explained by a number of factors determining the individual's prospects for entering a sporting discipline (3,4,5). The interaction of social and biological factors in man's physical activity development is due to the spontaneous application of environmental influences upon hereditary characteristics (3,6). At the moment, we have no grounds to form a strict conclusion with regard to the biological or social determination of sporting ability. Their origin often becomes shaded in the multitude of other external influences. It is important therefore to recognise these other circumstances which may promote the development of physical potential. The family's sporting traditions, the facilities for activity development in place of residence, demographical factors, elements of sport training and its duration, facilities for sports and physical activity etc can all determine the development of sporting talent. To biological factors determining the sporting talent, we can refer functional and morphological characteristics of the young athlete's apparatus of movements. The significance of inheritance attached to the histological characteristics of muscle system (fast and slow muscle fibres), the level of aerobic and anaerobic working capacity, the body composition and the type of constitution (10) etc is well known. There are many indications that these factors are determined by the influences of genetical origin, or at least they are formed early enough

Current Research in Sports Sciences, edited by Rogozkin and Maughan.
Plenum Press, New York, 1996

1

Table 1. Age standards of sprinting ability's basic sign
according to the support time's data(s) when
running with maximal speed

Age (years)	Males	Females
5–6	0.130 ± 0.010	0.135 ± 0.010
7–8	0.130 ± 0.010	0.130 ± 0.010
9–10	0.125 ± 0.010	0.125 ± 0.010
11–12	0.105 ± 0.010	0.105 ± 0.010
13–14	0.100 ± 0.007	0.105 ± 0.010
15–16	0.095 ± 0.005	0.100 ± 0.005
17–19	0.090 ± 0.005	0.095 ± 0.005
20–25	0.080 ± 0.005	0.090 ± 0.005
25–30	0.080 ± 0.005	0.090 ± 0.005

and before any testing periods of a child's life (6,8,9,10). The indisputable fact is that elite athletes possess a characteristic series of essential, basic signs, which distinguish them from athletes of the lower qualification and among representatives of other sport specializations. For example, the main feature of a sprinter is his ability to contract his muscles during working phases and relax them in the phases of relative rest. These indicators which determine the effectiveness of an athlete's movements, we called the basic or dominant signs of sporting talent (1,4). The importance of discovering these signs and their quantitative evaluation must be regarded as the first priority of sporting talents selection theory. The basic or dominant signs of sprinting ability, for instance, are as follows:

- the duration of ground contact phase when running at maximal speed (Table 1)
- the rhythm of running, characterizing by considerable exceeding of flight phase time by the duration of ground contact phase;
- the absence of tension in movements

Our investigations (1,7) have shown that at least 1% of boys and girls amongst those who never participate in regular sport training sessions demonstrate ground contact times equivalent to those of elite sprinters when running with maximal speed (1,4). Thus, talented young athletes who are capable of running along the track "as on the hot stove", according to J.Owen's remark, may have a chance of being discovered with a help of this dominant sign's measuring. However, it is not enough to evaluate this to realize the potential of the discovered sprinting talent. For this it is necessary to take into consideration some other components of sprinting ability. In this regard the ability for motor learning and training, is the most important factor of a future athlete's sporting development (2). The prerequisites of sporting mastery do not ensure sporting success at an elite level without goal-directed, intensive training over many years. The effect of such training, together with other factors, will depend on the development of the individual. Also, there are reasons to believe that different athletes possess different capabilities of rapidly increasing their physical and technical potential. That is why the second principle of sport talent discovery and evaluation asserts the necessity of taking into consideration the real potential of athlete physical capacities and skills development. Therefore, coaches and other specialists play an important role in the development of the young sprinter. Another important factor in sporting talent development is the state of athlete's health at all stages of his training, and it is necessary to forecast possible changes in the athlete's health in connection with the extreme conditions of training which lie ahead of the young athlete. For these reasons we can formulate the third principle of sporting talent's discovery - the principle of biological reliability. The realization of this depends on the concentration of coaches and doctors to the athlete. It is useful to collect information about hereditary diseases in the family of the athlete. The athlete's psychological status may also determine the effectiveness of training and success in his sporting career. The athlete's psychological stability in connection with the hard conditions of competition and his readiness for prolonged training with a high load are important. An important element of an athlete's psychological characteristics is his interest in the many years training and the aspiration for success at the highest level. The necessity to realize all these demands are prerequisites of successful sporting activity.

As made obvious from above, the evaluation of a young athlete's potential in sporting improvement can not be limited by a single test. It is obvious that the discovery of talent is a long process involving a detailed study of the athlete from all angles. This process can be made up of several stages. At the first stage is the evaluation of sporting ability with the help of the basic signs measurement, and examination of the athlete's health. If both these criteria are positive, we move up the second stage of sport talent selection. At the second stage of sport talent's discovery the basic training takes place during 3 or 4 years. The indices of progress in physical and motor qualities and skills are taken into consideration at this period. Simultaneously, the profound psychological, medical and biological study of the athlete must be carried out to evaluate his biological and psychological reliability. The study of development of sporting ability is realized at the third stage of sport talent's discovery during 3 or 4 years. During this period the main information about real opportunities for the athlete's sporting potential realization are accumulated. It is the most crucial time for both the coach and athlete. A coach must possess a vast knowledge and sensitivity so as not to be carried away by the rapid progress of the athlete. It necessary to evaluate exactly the strong and weak points of the athlete, to determine the main strategic directions to progress to the highest achievements. The demands on athletes who had achieved up to the end of this period the results close to the maximum expected are considered for selection to the national team. Their stability and reliability during competitive activity, their ability to achieve high parameters during competition, the capacities to recover after emotional and physical stress, the adaptive resources for different climatic, geographic and time areas, the adaptation to unusual spectators' reactions and to the strong competition of rivals should also be taken into consideration at this stage.

REFERENCES

1. Balsevich, V. K., 1969, "Zur sportlicher Talente" Theorie und Praxis der Korperkultur 6:17–21.
2. Balsevich, V. K., 1970, Wann soll die Ausbildung im Lauf einsetzen? - Die Leibeserziehung, 19,Bell Lehrhifte :77–79.
3. Balsevich, V. K., 1975, The biological rhythms in development of human locomotions in ontogenesis. V-th International Congress of Biomechanics, Abstracts, Jyvaskyla :13.
4. Balsevich, V. K., 1978, The value of selected biomechanical characteristics for diagnosis of motor function in sprinting. International Congress of Sport Sciences, Edmonton, Canada :37.
5. Balsevich, V. K., 1982, The dominant indication of sporting abilities and its use for top level sprinters selection, International Congress of Sports Sciences, Patiala, India :1–2.
6. Balsevich, V.K., 1984, Biomechanical aspects of sport for children and youth, In:Biomechanics-Kinathropometry and Sports Medicine, Exercise Science, Scientific programme abstract of the Olympic Scientific Congress, July 19–25, University of Oregon,Eugene,USA :17–18.
7. Balsevich, V. K., Karpeev, A. G., Martin, E. E., 1981, Hereditary and envirinmental determination of biomechanical characteristics in hyman motion ontogenesis. VIII-th International Congress of Biomechanics. Nagoya :275.

8. Hoshikawa, T., Amano, Y., Kito, N., Matsui, H., 1983, Kinetic analysis of walking and running in twins, Biomechanic VIII-0A (H.Matsui, K.Kobayashi - eds.) - Human Kinetics Publiishers, Campaign :498–502.
9. Murase, Y., Hoshikawa, T., Amano, Y. et al., 1979, Biomechanical analysis of sprint running in twins, VII-th International Congress of Biomechanics :179–180.
10. Vbrova, G. 1979, Influence of activity on some characteristic properties of slow and fast mammalian muscles. Exercise and Sport Sciences Reviews, The Franclin Institute Press. 7:181–213.

EXPERT COACHES' STRATEGIES FOR THE DEVELOPMENT OF EXPERT ATHLETES

John H. Salmela

School of Human Kinetics
University of Ottawa, Canada

ABSTRACT

Discussion on the role that natural talent plays or environmentally based variables play in the development of expertise in sport has seen a gradual swing from the importance of innate abilities to that of environmental determinism. Ericsson and colleagues (1993) have defended a theoretical framework on the development of expert performance in sport and other achievement domains based on the environmental notion of long-term adaptations of performers to deliberate practice. The present paper advances a complementary dimension to Ericsson's view of deliberate practice in terms of the role expert coaches play in the organization and supervision of training and competition. In-depth interviews were conducted with 21 expert Canadian coaches of basketball, volleyball, ice hockey and field hockey whose average age and coaching experience was 45.5 and 18.1 yrs, respectively. All coaches were interviewed in regards to their personal athletic history, evolving visions of coaching, approaches to training and competition. Inductive analyses revealed that the complex primary tasks of expert coaches were designed to plan maximal deliberate practice. The coaches directed their personal resources to the organization of practice and group processes and towards the resolution of the effort and motivational constraints. Motivational and effort constraints were both reduced in the training process by such means as creating a vision for the team and setting the goals, teaching the skills, maintaining a work ethic, and training the physical and mental systems. Competition was an arena for evaluating the effectiveness of previous training where the lessons learned were blended into future practices. Ericsson's model of expertise development is highly compatible with the main reported beliefs and behaviors of these expert coaches.

INTRODUCTION

Guy Lafleur [a National Hockey League superstar] was interesting, he was just such a great athlete. I can always remember, he didn't train all summer, he used to ride a bicycle and he used to

Current Research in Sports Sciences, edited by Rogozkin and Maughan.
Plenum Press, New York, 1996

5

run *down hills*, [italics added] and he said he used to chop wood. The day he turned 40 years old we had two one and a half training periods, the first one he was terrible and the second one he was the best guy on the ice. He is a great natural athlete -- smoked 2 or 3 packs of cigarettes a day, but that was Guy Lafleur, probably one of the greatest talents the game has ever seen. (Former professional ice hockey coach)

During the 1970s, I figured out that sports wasn't talent, sports was mind, and heart was the bottom line. Talent was important, but the bottom line was mental; your mental make up, your mental training, without discounting the importance of technical things and tactical things and physical things in sports. (Former Canadian Olympic basketball coach)

In summary, our review has uncovered essentially no support for fixed innate characteristics that would correspond to general or specific natural ability and, in fact, has undiscovered findings inconsistent with such models. (Ericsson, Krampe & Tesch-Römer, 1993, p. 399)

If an expert coach was asked which citation was the most accurate description of the development of athletic achievement, many would respond as the first citation of the hockey coach by attributing a central role to "the gift", or native talent. Others might suggest that native talent along with various proportions of learned personal characteristics, often called character, would be most important. Although the basketball coach stressed the importance of hard work and directed practice which make up character, few would deny the importance of bringing some natural gift or talent to the gymnasium.

The purpose of this initial section is to trace the development of expert performance in the literature from the first genetically based perspective which relied primarily on innate abilities to one that centers on environmental variables of nurturing and training.

Research on Talent Identification

The notion of the gift or innate talent was the predominant view in the 1970s in elite Canadian sport programs for two reasons. The first was the optimism accompanying the growth of the sport sciences and their positivist research doctrines. At this time sport talent was predicted or identified through isolated and accurate measurements of the appropriate variables. In relation to the Guy Lafleur example, the question would be: "What are those variables which made him an exceptional performer and how do you measure them?" The second reason for pursuing talent identification programs in Canada was the outstanding success of the East German sport program and their reliance on sport science and talent identification procedures.

It is within this context that my associates and I spent over 15 years examining the combination of innate and trainable variables in ralation to predicting performace in the sport of gymnastics. The Testing for National Talent (TNT) project was based on the assessment of all male, national level, Canadian gymnasts. The analysis was based on the assumption that performance outcome could be predicted if sufficient, multidisciplinary variables could be assessed and trained in an applied research program. These variables were sensitive both to the demands of the sport and to developmental changes in athletes, We attempted to determine which areas or combination of innate and trainable abilities were necessary to predict gymnastic performance.

The project was indeed successful both in terms of scholarly production and theory development (Petiot, Salmela, & Hoshizaki, 1987; Régnier, Salmela & Russell, 1993). Figure 1 indicates that it was possible to explain 100% of performance variance within most age groups using a fairly small number of age specific, multidiciplinary variables.

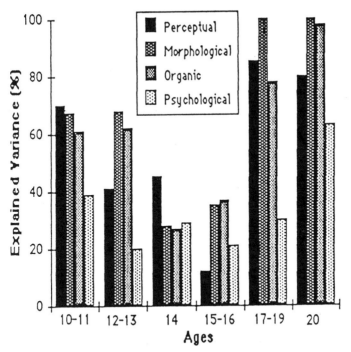

Figure 1. Explained variance of gymnastic performance using all possible families of sport science determinants. (Reproduced with permission of Sport Psyche Editions).

The relative importance of each family of variables was also shown to change across age groups. However, these variables were derived from over 100 potential demographic, psychological, perceptual, organic, physiological and morphological variables elaborated in conjunction with expert gymnastic coaches and sport scientists.

Despite this, TNT failed from a practical standpoint since the time and resource allocation necessary for such an enterprise proved to be unwieldy in the Canadian context: So many measurements, from so many scientists, on so few male gymnasts in such a large country. In addition, doubts began to arise on moral and ethical issues regarding the role of third parties, such as sport scientists, making decisions or career determination (Salmela, 1987). The Canadian situation was a contrast with the former East German context, where talent identification could be carried out in such a geographically limited and politically controlled environment. From our perspective, talent identification was an interesting, but sterile exercise.

Talent Identification versus Talent Development

Bloom (1985) and his colleagues studied the developmental patterns of expert artists, scientists and athletes along with the roles that family and coaches/mentors played in their careers. He presented a parsimonious overview on how the social context played a determining role in shaping these young individuals across the early, middle and late stages of their careers. While not having direct evidence to this effect, he suggested that the situational context, and the role of family members and mentors overrode the natural ability of the performer.

Bloom's seminal work on developing talent in youth also profoundly influenced the thinking behind our research efforts and the important social and pedagogical influences which were demonstrated so clearly in these knowledge domains. Thus the term talent development was added to that of talent identification (Régnier, Salmela, & Russell, 1993). Based upon Bloom's arguments, it appeared that the effects of an appropriate nurturing environment could play a more significant role than any effects of the initial talent base.

Talent Development as Deliberate Practice

A more radical position was the one espoused by Ericsson and colleagues (Ericsson, Krampe & Tesch-Römer, 1993) in a provocative article in which they convincingly argued that the unique determinant of expert performance was deliberate practice. More precisely, deliberate practice was characterized as being "those activities that have been found most effective in improving performance" (p. 367), and was distinct from the activities of play or work.

Ericsson et al. examined the development of expertise in a number of domains and created a theoretical framework based on the "monotonic benefits assumption" or that "the amount of time an individual is engaged in deliberate practice activities is monotonically related to that individual's acquired performance" (p.368). They further relied on a metric for deteremining the amount of time required to develop expertise, invoking the "10 year rule" of Simon and Chase (1973). More precisely, a minimum of 10 years of deliberate practice must be accumulated, while overcoming the constraints of resources, motivation and effort. They argued that pure talent only played a modest role in the early years of an athlete's career. Bloom (1985) termed this period "the early years" and noted that an athlete's motivation could be increased because they were perceived as "special", i.e., being tall and agile in basketball.

However, the main determining factor was that these individuals were placed within a context in which they were able to spend at least 10,000 hours of deliberate practice. Ericsson et al. also argued that deliberate practice was not in itself rewarding and often required supervision, since play rather than deliberate practice usually resulted when an athlete was placed in a spontaneous choice situation.

The work of Ericsson et al. (1993) was interesting and forced us to re-examine earlier research on talent detection (Régnier, Salmela & Russell, 1993). In preliminary analyses of the data set, the single demograhic variable of "number of hours of practice per week" explained at least 80% of the performance variance for each age group of the gymnastic population. However, this variable was mistakenly eliminated from the analysis as a "nuisance" variable which prevented more "scientific" dimensions to emerge (Figure 1). Unfortunately, time, lost data files and geographic relocation of both Régnier and myself have relegated this potential fact to that of an anecdotal recollection of analyses conducted 15 years ago! This is indeed unfortunate in that this national data set would have been a fine test of Ericsson's theoretical framework.

Coaches and Talent Development

Ericsson and colleagues' (1993, 1994) primary arguments on the development of expertise or talent were focused upon the effects of prolonged adaptations of the performer to deliberate practice and to constraints which inhibited the learning process. Whereas Bloom (1985) stressed the role of significant others in talent development, Ericsson et al. only referrred to teachers in regards to the instructional process:

To assure effective learning, subjects should ideally be given explicit instructions about the best method and be supervised by a teacher to allow individualized diagnosis of errors, information feedback, and remedial part training. The instructor has to organize the sequence of appropriate training tasks and monitor improvement to decide when transitions to more complex and challenging tasks are appropriate. Although it is possible to generate curricula and use group instruction, it is generally recognized that individual supervision by a teacher is superior. (Ericsson et al.,1993, p. 367)

The above citation may be appropriate to the teaching of static field tasks, such as chess, or more complex laboratory tasks, such as digit recall. It seems that instruction alone is a necessary but not a sufficient condition to deal with the complex tasks involved in coaching in sport. In the latter context, the environment is both complex and dynamic, and there are added elements of fitness, danger and team work.

Research on Coaches

Until recently, there have been few attempts at conceptualizing what coaches do in their work. In fact, Gould, Giannini, Krane, and Hodge (1990) found it "disconcerting" that only 46% of elite American coaches believed that "...there exists a well defined set of concepts and principles for coaches to use." (p. 337) In fact, the two most important knowledge sources which helped them develop their coaching style were their own "experience" and that of "other successful coaches". Gould, Hodge, Peterson, and Petlichkoff (1987) also surveyed college coaches and the authors recommended that these approaches "...be supplemented in future investigations by actual observation of coaches in practices and competitions, and by in-depth interviews that allow for the acquisition and interpretation of rich qualitative data." (p. 307) These findings suggested that a reality based, 'bottom-up' approach might be useful in investigating such complex domains as high level coaching and that their experiential base should be tapped in order to build up a conceptual base of the coaching process.

Recently, Côté, Salmela, Trudel, Baria, and Russell (1995) proposed a theoretical model of the coaching process which was grounded in the reported practices of expert gymnastic coaches (Figure 2). The model had three central components: organization, training, and competition. In the organizational component the coaches created a vision, set and monitored goals over the training season . Physical and mental training were then combined with team building, skillful teaching and preparation for competition. Competition was not an end in itself, but was used to evaluate the progress of the training program. Following competition, lessons were learned and the coach then organized new operations to reduce the difference between the actual state of the team versus its desired state. The three other factors in the Coaching Model were the respective characteristics of the coach and the athlete as well as the contextual factors of the specific environment, such as the pre-season or the Olympic Games.

In the present paper, the three constraints of resources, effort and motivation of Ericsson et al.'s (1993) theoretical framework on the development of expertise will be considered both in light the inductively derived categories of crafting the training environment (Figure 3) of the knowledge base of expert coaches of team sports (Salmela, 1994) and within the context of the Coaching Model (Côté, Salmela, et al.,1995). However, in that the exhaustive analysis of the full data set on the expert team coaches is still in progress, the present article will be used as a heuristic device to plumb possible avenues of analysis of these three viewpoints on expertise of both performers and their coaches.

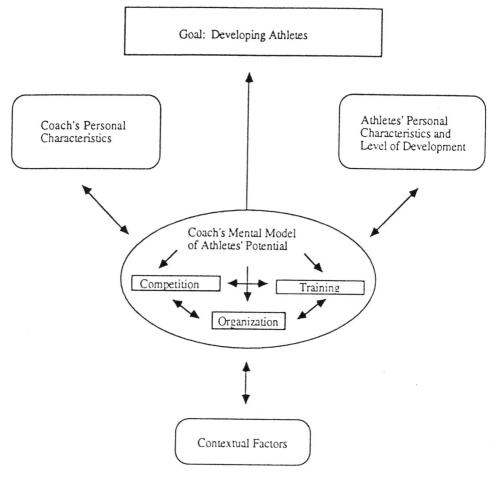

Figure 2. The Coaching Model (Reproduced with permission of Human Kinetics Publishers).

METHODS AND PROCEDURES

Twenty-one expert basketball (BB), field hockey (FH), ice hockey (IH) and volley-ball (VB) coaches were identified by their National Sport Organizations as being the most knowledgeable respected coaches in Canada, having consistently produced athletes at the national and international levels over an extended period of time. The average age of the coaches was 45.5 years and they had coached for an average of 18.1 years, which classi-fied them as experts (Rutt-Leas & Chi, 1993).

All coaches were interviewed in regards to their own personal athletic history, their evolving visions of coaching, their approaches to both the training and the competition proc-esses and their reflections on improving future coach education (Côté, Salmela, & Russell (1995a. 1995b). The intent of these interviews was not only to elicit information on the vari-ous knowledge domains, but also to consider the nature of the transitions experienced during the development of their expert coaching careers (Salmela, Draper & Desjardins, 1994).

The interviews were transcribed verbatim and resulted in over 700 pages of single spaced text files. The interviews were inductively analyzed using Paradox for Windows software following the procedures outlined by Côté, Salmela, Baria, and Russell (1993).

Crafting the Training Environment

Setting the Direction

Vision for the Team
Long-term Goals
Daily Practice Goals
Buying Into Goals
Monitoring the Goals

Building the Team

Knowing the Athletes
Team Building
Long-term Well Being of Athletes
Selecting and Cutting Athletes

Training the Systems

Fitness
Mental Training

Teaching the Skills

Practice Planning
Teaching Styles
Work Ethic
Rules and Discipline
Using Assistants and Support Staff

Figure 3. Components of the training process of expert coaches of team sports.

RESULTS

The three constraints of resources, effort and motivation of Ericsson et al's (1993) model will be used as a framework for the organization of the functional roles used by expert coaches in the organization and implementatation of deliberate practice. Perhaps the most encompassing and least defined dimension of Ericsson et al.'s model was the role of the teacher / coach as a provider of both physical and human resources. Ericsson et al believed that deliberate practice was an effortful activity and did not occur spontaneously . If this is the case, the coaches' and teachers' visions, work ethic and teaching skills are even more important than previously thought. The coach essentially acts as a provider of both physical and human resources which permits the athletes to overcome the constraints of effort and motivation.

Setting the Direction and Deliberate Practice

One of the most important human resources provided by expert coaches was that of setting the program direction. The long-term vision of the coaches must be progressively translated into specific, attainable short-term goals. At the molecular level, the plan becomes implemented into a daily training plan which includes infinitely detailed consideration of how the vision will be shaped each day to meet the needs and characteristics of each athlete:

All coaches had in mind a vision for their team. The process for operationalizing this vision often began through building a road map for the realization of the individualized goals of the group. A sense of group cohesion was a necessary part of building a successful team, and many different situations served as laboratories for the team building process.

Côté, Salmela,Trudel et al. (1995) stated that organizational tasks were a central component in the mental model of the coaching process of developing elite athletes. Organizational tasks were complex, metacognitive initiatives related to the planning of practice and competitive situations which considered diverse factors related to human relations.

The coaches' vision of the entire training process provided the athletes with a stimulating environment and a feeling of assurance that they were on the right track. This, in turn, encouraged them to greater training intensity. To put this within Ericsson et al's (1993) terminology, effective and realistic goal-setting procedures provided exciting, motivational dimensions to the practice environment which in turn helped the athletes overcome the arduous effort constraints which accompanied intensive training:

> I want them to be sensitive to other people around them, to always come to the rink thinking, "What a great job I got going here! Have I ever got it made! Somebody's paying me to play hockey!" Do you say "I'm going out to work hockey today?" No- you say "I'm going to go out and play hockey today", You're getting paid to play. Can you believe how lucky we are? We're getting paid to play. Most guys get paid to work. We get paid to play. (IH4)

> For the most part they stay and I think that is a very positive reflection on kids that when it gets tough most of them don't quit as long as they know where they are going. I have found if a kid has direction and they have standards and they are getting support for what they are doing, they will run through walls, they will suffer pain and they will suffer hardship. (BB3)

> The sacrifices they make here are unbelievable. I don't know how we keep talking them into doing it. But we do. We have other information. We have had different Canadians make six figures at different times in their career. Normal salary I guess would be the $50,000 to $75,000 range for an 8 month season, then they come back and are paid for $750 a month for the next four months. (VB5)

Buying into Goals

The vision of the coach acts as the basis for all other activities in the training and competition program. However, this vision must be translated into operational goals through formal goal-setting procedures. One important part of this process for the coach involves getting the athletes to comply to these goals. It is thus necessary that the athletes "buy in" to these goals. This process necessitates that the coaches give up some control over the ways and means of implementing these goals, a process termed "athlete empowerment":

> I have used goal-setting extensively. I had a certain vision of going somewhere but I had to get everybody on board. People have to buy into the program. If I said I am leading, sure they are going to follow me because I am the coach, but somewhere along the way I am going to lose them. (VB2)

> Now I am to the point where, and it has taken me a while to get this way, I think I am more empowering as a coach than I used to be. It is very difficult to do because you have to give up some aspects of control, but I think it is more effective to do that now, and I think that is just a maturation process that you go through as a coach. (FH3)

Monitoring the Goals

Once goals are agreed upon and the athletes have bought into the coach's vision, one important task remains, that of monitoring the goals. This is one of the most difficult tasks for the coach since keeping the athletes focused on training and goal commitment often results in conflicts between the coach and the athlete. The following citation reflects how one coach monitors, perhaps to an extreme degree, the individual goals of her athletes:

> When we begin the season everybody sets their individual goal, I have meetings with all individuals, then we set individual goals and team goals. You know constantly we are looking at those, I do not let them go too far. I carry them around, I have all these little pieces of paper, I carry them in my briefcase with me. So that I am never too far away from them, and I am constantly re-evaluating where each individual is at. I will confront them: "I thought this is what you wanted to do, it doesn't look like that to me. It is your piece of paper? You wrote it, do you want to change it?" (FH2)

The goals were not only related to the outcome of the games, e.g., the number of goals or assists, but also to the process or the means of achieving the goals during practice. In this case, it was the necessary level of sustained effort during each training:

> We will call everybody in and we always hold hands at the end of practice, we call everybody in and we remind them, "If you saved anything you are wasting your time, don't save anything, don't anticipate what is going to happen because one of our things is to give everything you have to every little thing". (BB1)

The coach placed very high standards of effort in training and sometimes paid a personal price in terms of being liked by the athletes. This also showed the nature of compassion for the individual, which invariably arose in the daily routines of the athletes:

> I am too hard on the athletes a lot of times. I make comments about them with regard to their field hockey ability and they know that I am not making a comment about them personally. I make that distinction very clear. I will tell athletes at the beginning, "I am here to coach you. The best thing for me would be to be your friend, I want to be friends with you but the bottom line is I am paid to coach, not to be your friend. My job is to make you the best field hockey player that I can, and I am going to do what I think it is going to take to get you there." (FH5)

It is clear from the above citations that the organizational tasks related to goal-setting played a central role in keeping the athletes on task and commited to providing the necessary effort required for the development expert performance (Ericsson et al.,1993). However, it does not imply that the external influence of either the coach or parent is the sole reason why athletes are able to overcome the necessary effortful hurdles required for achieving excellence in sport. Orlick's (1992) model of excellence indicated that commitment to goals was a core performance component which had an internal locus of control. It is clear that even exceptional performers can also benefit from coaching influences to facilitate this inner drive.

However, Reilly (1995) recently outlined the self-directed practice habits of golfing prodigy Tiger Woods. This article clearly stated that the athlete was responsible for the committment required for deliberatal practice rather than the coach or the parent:

> When dads drag their seven year olds up to Wayne Gretzky and say "Wayne, will you tell him he's got to practice," Gretzky always says, "Nobody ever told me to practice." (p. 68)

This self-directed commitment to practicing may be the very reason that exceptional athletes perform at such levels. On the other hand, there are no athletes of note who have excelled at the highest level without coaching.

Teaching the Skills

Once the organizational task of setting the direction for the training of the athletes, the complex training task of teaching the physical skills within this training regime becomes foremost (Côté, Salmela, Trudel, et al.,1995). Once into a specific practice session, expert coaches are pivotal in providing athletes with to the physical resources of sport, that is, the facilities, sport equipment, but most importantly access to the practice or playing environment required for athletes to develop their skills. More precisely, it is the coach who controls the selection of the team and their amount of playing time:

> This is not recreation, this is an elite program and you have to do certain things to be in an elite program. If you don't subscribe to that idea then that is OK, it doesn't make you a lesser person but guess what, you can't be part of the program because we can't afford it. (BB3)

> Although I do make all of the important decisions, certain decisions like selections make me extremely important. Once selected, I should be less important, however, the amount of playing time also makes me extremely important. (BB4)

> We run a mile and a half, timed, and the players hate it but they love it. They know at tryout camp they are going to have to run a mile and a half, and basically I put it to them. I say, "I don't really care what your time is, but if you can't run a mile and a half you sure as heck can't train five days a week in my gym. You are going to die". (VB2)

> When they deserve it they will play, but I don't believe that people should play when others have put in more more time and effort to improve their games. They should earn their playing time. (BB5)

The complex task of teaching team sports requires a flexible and creative outlook in order to provide training situations which simulate the demands of the sport, which are progressive in nature and mix technique with tactics and strategy. Variations were found in terms of how fast one should progress as well as how much leeway should be given to self-discovery versus formal teaching:

> My favorite month of coaching in university was October. I used to love it. You have these kids who come out of high school and don't know piss from paint. You get them into a hot house and surround them in a good environment. You see them make quantum leaps. Nobody knows whether they are playing, going to be the starter and they are all full of piss and vinegar and I am at my glory because I like to teach them. (BB4)

> It doesn't matter if there are mistakes, I always encourage athletes to make mistakes, because that is the only way you learn. The more mistakes you make the better off you are going to be, except, of course, in a final game. You hope by the time you get there you have made them all. Of course that is almost impossible to do. The more mistakes you can make during the course of the practices the better. This is quite often a problem with young athletes, they are afraid to make mistakes. (FH3)

> I found it a very interesting experience working with one coach because she liked to make all those decisions, and you can't do that. She is a very very successful coach, and I couldn't figure out if she makes all the decisions how she could be as successful as she has been. That is the fun of coaching, there are a hundred ways to do things, and none of them is particularly the right way. It is just that it happens to be the right chemistry with the right athletes, the right group. (FH3)

Work Ethic

At this high level in sport, the work ethic is central to the achievement of the established goals. This does not mean that the elements of enjoyment are not carefully blended with hard work. In fact, one of the most persistent sources of satisfaction for these coaches is seeing the realization of individual progress by the athletes.

> I think I am personable yet demanding. Kids want to play for me. That is what they have said that to me. I think I motivate kids to go beyond whatever the hell they think they can go to and to get the best out of them. I do it in a way that I stay relaxed. I really believe that you have to put the athlete in environments where they feel comfortable that they can fail. You don't want them to fail, but if they do fail it is not a major crime. That is only a step, and you fail for a reason,. I think that I am pretty successful at that, I think I motivate really well, I think that is one of my strengths. (BB2)

> The national team athletes, they work so hard at practice, you just want to work hard because they are working so hard. It is just fun to be involved at that level. I really enjoy it, I get a lot of personal satisfaction from it. (FH3)

Coaches have certain non-negotiable rules regarding conduct and discipline from their athletes:

> Discipline is not self imposed. On the court discipline is coach imposed. I discipline someone for some action or behavior depending on what I think it deserves, and what is best for the team and for the person. (BB5)

> In terms of discipline, the new athletes coming into the university environment would know not to step out of line. I think there has been maybe 2 or 3 stories from the early years when I was coaching that are now legend that they wouldn't even dare try anything. I think I am very de-

manding but I wouldn't be in it if I didn't enjoy it, and I like to have fun. We have a lot of fun at practice, the athletes also work really hard. (FH3)

Successful teaching is partially attrributed to the creation of exciting learning environments. These coaches provided highly organized situations which permitted the efforts required for deliberate practice to be tempered by periods of recuperation.

The most important thing to practice is having a really organized plan that you have spent time on. You have intensity plus you have some soft time, everybody is organized. You are going in feeling organized, and that is probably when it goes the best. (BB5)

I do a practice plan. In fact, I have my practice plans from 1970 in a book. That is something I always did. I thought that was a small thing that gives me an edge over some other coaches. I have always written it down, so we don't do a lot of shake the sleeve and see what falls out. (VB2)

In practice, we don't tell everybody what we are going to do. I know that a lot of things I have read say it is really good for athletes. I think our athletes are familiar as to the tone of the practice, but as to what we are specifically going to do, I think a little bit of mystery is nice, they know they are going to get something new once in a while, a little wrinkle, they know also that they are going to have a lot of input into it. If it is uncomfortable we are going to listen to you about it. (BB1)

In general agreement with Ericsson et al. (1993), the expert coaches recognized that the effort constraints of sustained practice had to be modulated for maximum effect. Training sessions could not be carried out at full pace and some form of relaxed training was scheduled in order to provide both physical and mental relief:

Every so often, I have practices I call "no brain" practices- and every so often you have to have one of those. You've gone through a lot of games, and the guys are tired, they're mentally fatigued- If I go out there and try to run through a tedious practice, it isn't going to work. So I'll say- "Today (or tomorrow) will be a no-brainer". (IH4)

Always going from a to b to c, and picking up the intensity, finally getting to fatigue and falling off at the end of practice. Sometimes I thought, gees I ought to open up some night with, take them to exhaustion at the start of practice, just to see if one of the things was people who fight their way through fatigue. (BB4)

Simulation

Ericsson et al., (1993) did not fully develop lthe transition from the practice to the competition mode. In the sport psychology literature, however, mental preparation for competition has been shown to be a complex phenomenon and a number of useful performance strategies have been developed (Orlick, 1992). Further consideration in this vein to the notion of deliberate practice may be an important contribution from sport psychology.

Probably one of the greatest skills I have as a coach is getting the guys in practice to concentrate. Whether it is factual or not, they think this is really going to apply to playing that next opponent. That is what I always do to reinforce, this is what you are going to do in a game situation. This is game hockey- this is exactly what happens in a game. (IH4)

I think that if athletes understand what it is they are doing it, what the opposition is going to do to us, this is why we have to play against it in this way. It is much easier for them to do it. If the kids that are not as strong from a coachability perspective, do not understand, they won't buy in. You won't have everybody buying in to what you want to do. (FH2)

Competition is the end result of what we have done in practice, we would like to make the players confident enough that we have enough in our arsenal to win every game. Competition is not a lot different to the training. My biggest goal as a coach is to make practices very similar in a lot of ways to what the competition will be. You try and simulate a competition situation, we have competition all the way through practice in different ways. Making sure that all of a sudden they are not going to experience new fatigue in competition, that they have experienced it in practice. (BB1)

Once the practice has been implemented, some of the expert coaches also monitored the effectiveness of the practice, so that they could ensure that future deliberate practice was maximized. This attention to detail assured that each practice session was coherent with the overall vision of the coach in relation to their athletes:

On the back of our sheets we have a practice plan. We have a thing we go through for 10 minutes at the end of practice. We just put down, what didn't work, what was successful and what to pick up the next day. So the next day when we go back we just flip it and say, "OK", let us see what we said last night". In this column might be: keep an eye on Sue, looks like she is very tired, check with somebody about that exam. So we reinforce that point and it will go on the top of our practice. (FH5)

The Next Practice

The practice after a competition was an important learning component which tied competition to training. The delay in providing feedback until the next day permitted a rational analysis to be made. Discussion on the next practice were made on the results of the previous competition in regards to the appropriateness of the plan, the effectiveness of the individual strategies, and the necessary steps that are required to correct any deficiencies:

At the meetings after a game, 90% of the time, I'll go "Well done guys, good job. Practice is 10 o'clock tomorrow." I don't talk about the game. They know they are not going to get analyzed right after a game, it is too emotional, especially if they lose. You can't communicate with someone who is emotional, it is a waste of time, so the first thing is to give them time to reflect (BB3)

The next day it is usually a matter of revealing the game plan: "This is what we said we were going to do. Why couldn't we do it?" Or "What worked and what didn't work?" We get input from them, because at this point, it is happening to them, and this is where you have to hand it over, so they are going to bring the information back. (FH5)

Concluding Remarks on Talent Development and Coaching

While Ericsson et al.'s (1993) theoretical framework on the development of expertise has yet to be empirically tested within the sport context, the general implications cannot be ignored. By centering upon the environmental dimensions of practice rather than on innate abilities, both aspiring coaches and athletes gain a large measure of personal control over the development of expertise.

This framework also provides a heuristic tool for the examination of current practices of expert coaches in sport. For example, few coaches would disagree with the statement that one of their primary tasks as coaches is to maximize the quantity of deliberate practice for their athletes each day, by "training smarter".

There are also a number of areas where the current data base on expert coaches might make important contributions to a more robust understanding of the phenomena related to deliberate practice. It is clear that the tasks in which expert coaches are involved on a daily basis extend well beyond the instructional management tasks to which Ericsson et al. alluded. For example, the complex organizational tasks of setting, complying and - monitoring goals appear to be central element within which effective practice conditions can occur. In addition, expert coaches often possess many effective pedagogical tools which can not only be used to teach physical skills, but also to push or pull, threaten or nurture, or release and control athletes at appropriate moments within a specific context so that they perform consistently at the highest levels.

Finally, Ericsson and colleagues have yet to consider the task of attaining maximum performance at a given moment in a competitive setting, although this issue has been a central concern within the field of sport psychology. While deliberate practice is undoubtedly a necessary condition for the development of expertise, it is clear that it is not a sufficient one. The rich and complex performance strategies that Orlick and Partington (1988) elicted from Canada's Olympic medalists indicated that "training smart" often required a different set of skills than "competing smart" at crucial Olympic moments.

In sum, Ericsson and colleagues' notion of deliberate practice has extended the research in the area of expertise development. In the sport realm, for example, this paper has attempted to demonstrate the central role of expert coaches in the organization, training and competition of elite athletes. With continued research in this area, individuals may soon come to realize that innate talent may play a lessor role than previously believed and that the focus should be turned to the nature of practice conditions and the facilitatory role that coaches play in its development.

ACKNOWLEDGMENTS

This research has been supported in part by research grants from Sport Canada's Applied Research Program, the Coaching Association of Canada and by the Social Sciences and Humanities Research Council of Canada.

REFERENCES

Côté, J., Salmela, J. H., Baria, A., & Russell, S. J. (1993). Organizing and interpreting unstructured qualitative data. The Sport Psychologist, 2 , 127–137.

Côté, J., Salmela, J. H., & Russell, S. J. (1995). The knowledge of high-performance gymnastic coaches: Competition and training considerations. The Sport Psychologist,,9, 76–95.

Côté, J., Salmela, J. H., & Russell, S. J. (1995). The knowledge of high-performance gymnastic coaches: Methodological framework. The Sport Psychologist,,9, 65–75.

Côté, J., Salmela, J. H., Trudel, Baria, A., & Russell, S. J. (1995). The Coaching Model: A grounded assessment of expert gymnastic coaches' knowledge. Journal of Sport and Exercise Psychology, 17, 1–17.

Ericsson, K. A., Krampe, R. T., & Tesch-Römer, C. (!993). The role of deliberate practice in the acquisition of expert performance. Psychological Review, 3, 363–406

Gould, D., Giannini, J., Krane, V., & Hodge, K. (1990). Educational needs of elite U.S. national team, Pan American, and Olympic coaches. Journal of Teaching in Physical Education, 9, 332–344.

Gould, D., Hodge, K., Peterson, K., & Petlichkoff, L. (1987). Psychological foundations of coaching: Similarities and differences among intercollegiate wrestling coaches. The Sport Psychologist, 1, 293–308.

Orlick, T. (1992). The psychology of personal excellence. Contemorary Thought on Performance Enhancement, 1, 109–122.

Petiot, B., Salmela, J. H., & Hoshizaki, T. B. (1987). World identification systems for gymnastic talent. Montreal: Sport Psyche.

Régnier, G., Salmela, J. H., & Russell, S. J. (1993).Talent detection and development in sport. In R. N. Singer, M. Murphey, & L. K. Tennant (Eds.). Neew York: Macmillan.

Reilly, R. (!995). Goodness gracious, he's a great ball of fire. Sports Illustrated, March 27, 62–72.

Russell, S. J. & Salmela, J. H., (1992). Quantifying expert athlete knowledge. Journal of Applied Sport Psychology, 4, 10–26.

Rutt-Leas, R. R., & Chi, M. T. H. (1993). Analyzing diagnostic expertise of competitive swimming expertise. In J. Starkes & F. Allard (Eds.). Cognitive issues in motor expertise. Amsterdam: North Holland.

Salmela, J. H. (1994). Learning from expert coaches . Journal of Coaching and Sport Science,1, 1–11.

3

SELECTION OF CHILDREN FOR SPORTS

W. Starosta

Institute of Sport
Trylogii Street 2/16, 01–892 Warsaw, Poland

INTRODUCTION

The selection of competitors was and still is one of the main problems of every sports discipline. The right selection ensures the growth of the discipline and the success of the sportsmen. The increasing level of sporting achievements needs the participation of more and more universally skilled sportsmen. Despite the large number of publications concerning different aspects of the selection of children for sports, there is still no ideal solution.

- Is some countries systems of selection are implemented. Some of these were constructed on a "Children for sports" basis, that is treating children as objects. As a result the training process was dehumanized and the system of values was changed. Sporting success became the main measure of effective training and the health, physical and psychological growth of the young person was pushed aside.
- In a few countries the basis for selection was "Sport for children," but the realization of this slogan did not occur. The principles for selecting children for sports must therefore lie between these two principles. I will concentrate on the following aspects:

  - Model for selection for sports;
  - The age to commence systematic training;
  - The manner of achieving early specialization.

Model for Selection for Sports

Solving the problem of selection gives several advantages: it saves spending money on the wrong individuals for many years of training; it saves the competitors from wasting expense, lack of self-realization, loss of health due to over-training and does not put them off practising sport for health: it saves the trainers time; it provides valuable support for representations at different levels. The selection procedure should involve a young person who wants to practice sport. It should not contain elements, which may harm his/her pride

Current Research in Sports Sciences, edited by Rogozkin and Maughan.
Plenum Press, New York, 1996

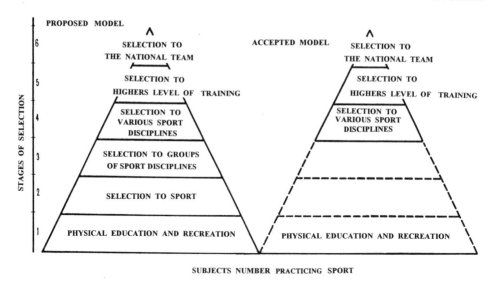

Figure 1. Models of selection of children for sport and for higher level of training.

and deter them from sport. The problems of selection involves elements of a biological, medical, social, pedagogical, psychological and moral nature.

The proposed model (Fig.1) is divided into three phases. The first, Primary selection, is concerned with defining the level of fitness of all healthy children and parameters of body structure. The second, Main selection, provide only the most talented sportsmen with different groups of sport disciplines according to their abilities. The third phase, Special selection, should determine specific abilities within the given discipline. Comparing the abilities of prospective sportsmen with those of established athletes may not be the best selection procedure. This is because the different athletes may have different beneficial features and may compensate the lack of one feature with the growth of another one. The proposed model (Starosta, Handelsman, 1990) has been put into practice in some countries (e.g. Swedish table tennis). In Poland it has been partly implemented in some sporting disciplines or clubs. The current selection system (accepted model, see Fig.1) is not rational, since it lacks the first phase, and starts with the second one ie candidates are put straight into the chosen discipline. From the systematic point of view we may call it a serious organizational-methodological mistake. It means that each discipline looks for candidates suitable only for itself and it is probable that candidates suitable for other disciplines are eliminated. For each disciplines point of view this system might seem rational, when the best are chosen from hundreds and thousand of candidates, but for the sport as a whole, this is an ineffective method. As a result, a large number of skilled children who did not find the right discipline were lost; eliminated from one discipline they never tried another one. If initial training starts with specialization, it should be treated as temporary, and participation in general sport lessons should be continued. Always there should be the possibility of changing the discipline.

THE AGE TO COMMENCE SYSTEMATIC TRAINING

This starts selection and early specialization. The organization of championships for younger and younger participants (including World Championship and Championship of Europe for juniors, young juniors and even for children) forces trainers to start the intensive training at a very early age. This raises a number of questions: Does the tendency to start

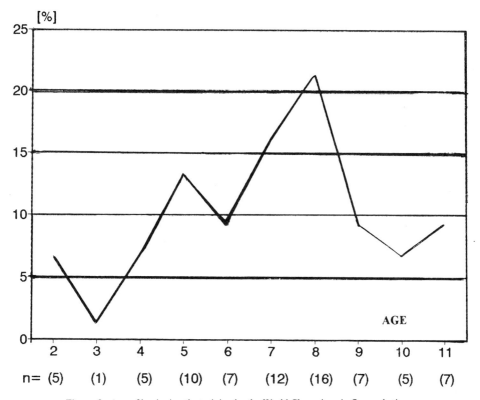

Figure 2. Age of beginning the training by the World Champions in figure skating.

training in a selected sporting discipline come from the young athletes interests? Does it favour his/her development? Many questions of this kind arise. With this objective treatment of children they had no input, but now to save the children from this negative phenomenon of contemporary sport, answering these questions is essential for everyone participating in the training of children. It seems that the contemporary development of sport forces intensive training at early age. Some countries try to win this race but do the results of research confirm this to be necessary? Tumanjan (1984) analyzed the age at which 20000 successful fighting athletes commenced training. He showed, that persons starting their career when slightly older were more successful. The majority of wrestlers both in classical and free style started training at the age of 13 or older. This seems to confirm the following: It is not favourable to start training in wrestling in early age and it is possible that beginning training at later age is a specific biological protection for the competitors, enabling them to more effectively function at higher levels of training. Similar results were obtained in our own research of 40 elite polish wrestlers. We were also interested in how many years was needed in the ex-Soviet Union to bring the novice wrestler into the master class? We found it to be different in different years, but the average duration was declining constantly to reach 6 years in 1940–76 for both classic and freestyle wrestlers. Depending on different factors, it could have taken from 1 year to 12 in classical and to over 19 in freestyle. Because of these differences, the mathematical averages does not to have any practical value. In this context, the average time needed to reach the master class should not be treated as fixed, that is, we should approximate ranges covering the majority of cases. This has been confirmed by researches on wrestlers and by our own research (Puni, Starosta,1979). We analyzed the age at which training started and average time needed to reach the World Championship in figure ice skating (Fig.2)

Ice-skaters started their training at different ages (over 50% at the age of 5, 7 and 8 years) and they needed different times to complete their career from beginner to World Champion; the differences were as big as 10 years! Very interested results were obtained in research on the age of Polish champions in artistic skating (Drozdowski, 1979); reaching the master class in this discipline needed average of 3 years, and the optimal age to begin the training was 11. Girls starting at the age of 8 needed 4–5 years to reach first class, while at the age of 10, 3–4 years were needed, at 11, 1–3 years, at 12, 1–3 and at 14, 2–3 years. This shows that the training should be started at the earliest age but at the optimal time. The optimal ages for commencing different sporting disciplines are slowly being set (Starosta, Handelsman, 1990). The data presented here giving very important conclusion for theory of training. Statistically processed data from many different sporting disciplines does not confirm the necessity of training at a very early age. Neither do bio-social factors confirm this tendency. It is contrary to the "Sport for children" idea and is an invention of people looking for success for any price. Such tendencies should be condemned, because nobody has a right to expose the child to unreliable experiments.

THE MANNER OF ACHIEVING OF EARLY SPECIALIZATION

Early specialization should be treated as the introductory phase of training started at the optimal age. This phase should not be based on adult training, since it has different aims and purposes. The most important factor should be to create a wide base for future specialization. The strongest element of early specialization should be motoric preparation

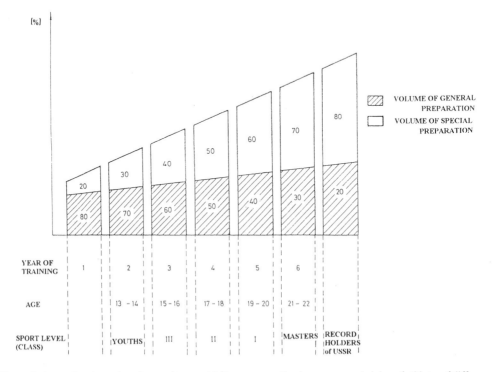

Figure 3. Approximative ratios of general to special fitness preparation in many years training of athletes of different sport disciplines according to years of training, age and sport class.

(Starosta, 1984, 1990). The most important element of specialization is the suitable balance between general and special training (Fig.3).

This should correspond with the ancient Roman rule "Hurry up slowly." It should act as a mainframe in structuring the aims and purposes for different stages of training according to the competitors age (both physical and psychomotoric). They should take place without harming the competitors health. Reaching the top result is possible only in the higher phases.

CONCLUSIONS

1. The model of selection used in many countries includes organizational and methodological mistakes because each discipline is looking for the candidates suitable only for its own particular discipline. As a result, among the eliminated candidates there are some, who might have been highly successful in other disciplines.
2. Practising one particular discipline should be begun at the optimal age. A tendency to lower this age is not confirmed advisable by research results or by bio-social conditions.
3. Early specialization should be realized according to ancient Roman rule "Hurry up slowly," stressing universal motoric development including movement coordination and the suitable proportion between general and special motor training.
4. Children participating in sport training need protection and should not be objects misused by adults for their success. High-level sport is too strong to be used in a way that might be dangerous for children's health.

REFERENCES

Drozdowski, Z.,1979, Antropologia sportowe. Skrypty AWF w Poznaniu, Warszawa-Poznan, PWN: 62–67, 189–192.

Puni, A. C., Starosta, W. 1979, Psychologiczne przygotowanie w sportach niewymiernych, Warszawa: Sport i Turystyka: 29–38.

Starosta, W. 1984, Movement coordination as a sport selection element, In: Genetics of psychomotoric traits in Man. Warsaw: Polish Academy of Sciences - Int.Soc. Sport Genetics and Somatology: 247–272.

Starosta, W., Handelsman, A. B., 1990, Biospoleczne uwarunkowania treningu sportowego dzieci i mlodziezy. Warszawa: Resortowe Centrum Metodyczno-Szkoleniowe Kultury Fizycznej i Sportu: 398–434.

Starosta, W., 1990, Bewegungskoordination im Sport, Warszawa-Gorzow Wlkp: Internationale Geselschaft fur Sportmotorik: 11–27, 60–66, 226–233.

Tumanjan, G. S., 1984, Sportivnaja barba: otbor i planirovanje, Moskwa: Fizkultura i Sport: 122–134.

4

CONTROL OF MOTOR FUNCTION AND ITS ROLE IN SELECTION AND ORIENTATION OF YOUNG SPORTSMEN

O. M. Shelkov and G. A. Hrisanfov

Research Institute of Physical Culture
Dynamo Ave. 2, 197042, St. Petersburg, Russia

INTRODUCTION

The examination of some recent studies in this field testifies that the leading researchers, who have been dealing with the problems of mastering sporting skills (Nabatnikova & Hordin,1982; Balsevich & Zaporozanov, 1987) link the future progress in the training of sports persons with an attempt to individualize them on the basis of selection, orientation and prediction of sports endowments.

It has been shown that the individual motor-psychical regulation of motor activities appears to be the basis for moving abilities. It should be noted as well that the other typical feature of motor-psychical reactions is that despite being partly genetically predetermined, they can be developed by purposeful training (Teng, 1988; Malina, 1989; Pate & Shephard, 1989).

METHODS

The investigation was carried out according to the methods, established out by the RIPC (Research Institute of Physical Culture) in St. Petersburg, Russia (Rogozkin,et al.,1988). The main components of this complex are the indices of provision of physical motion according to the temporal and spatial parameters of efforts. In all, 43 indices were analysed. The indices are grouped into blocks, with one common index.

The first block is set up by the indices of height, weight and maximum hand strength (the physical development and strength block).The second block (subjective evaluation of a sportsman's state) includes self-appraisal of general state, mood, desire for training, goal setting, confidence in its attainment, satisfaction in the process of training itself, and readiness for the highest result indices. The third block (mental state) puts together such indices as situational anxiety, motivation for going in for sport, a vegetative component of

Current Research in Sports Sciences, edited by Rogozkin and Maughan.
Plenum Press, New York, 1996

27

inclination to strength expenditure and unsteadiness of the nervous system. The forth block (the exactness of movement operation) is made up by the following indices: the exactness of perception and reproduction of time intervals, movement speed, efforts, spatial quantity and the exactness of reaction to a moving object. The fifth block (the quickness of movements) is composed of the following: reaction time to a quick movement, the implementation time of the whole body moving activity, the time of movement with small amplitudes. The sixth block (motional readiness) consists of the following kinds of indices: activity in speed and force training, the vegetative component of inclination to strength expenditure and self-appraisal of the desire to compete. And the final and seventh block (asymmetry in movements) has only one index of asymmetry in speed and force.

The value of each index has been calculated according to the method of relative coefficients. This helped to analyse all the indices by using a common scale irrespective of the units of measure.

This investigation was carried out in normal conditions during a number of important contests from 1985 to 1990. Also, the residents of some sports boarding-schools and Olympic training centers of Russia were also examined in the years of 1989–1994.

Thus, the study contains the data on more than 4500 young sports persons, men and women, aged from 10 to 21 years ranking from the first class level up to the degree of the international class master and specializing in 26 different kinds of sports.

RESULTS

The results of the research on peculiarities of motor-psychological manifestation and physical development of young sportsmen, specializing in the kinds of sports, which are connected with the motion activity of a cyclic character and which require a demonstration of coordination are presented here. The data was classified according to the gender and age of the examinees. Every year of their physical development from 13 up 20 has been analysed separately.

By examining the 1st block of the overall physical development indices it was revealed that the maximum increase in value occurs in the age range of 13–17 both for men and women; that accounts for 80% of men and 40% of women. Also, it was established that after this critical period (the age of 17) stabilization in the above mentioned indices occurs without any substantial increase taking place. It should be also be noted, that in sports with a cyclic character, the absolute values are more strongly marked in comparison to the sports which rely on the coordination of movements. The complex evaluation of the subjective mental state and motion activity does not reveal any evident age dynamics. However in the age range between 13 and 18 years we observe a tendency towards lowering of the relative indices that testify that having improved their craftsmanship sportsmen can estimate their state more objectively. This is typical of both of men and women as well as of different sporting groups. Analysis of the 4th block (the indices of exactness of movement) according to temporal, spatial and effort parameters (fig.1) shows that these characteristics improve within the age, reaching the greatest values by the age of 18.

It was also shown that in the kinds of sports with a high level of coordinated movements, the complex index of exactness is higher for women than for men and at the ages of 17–18, these distinctions between the two groups are significant. For cyclic sports there is an inverse relationship: the index at the beginning of training (at the age of 13–14) is considerably higher in sportsmen, who take part in high level coordination sports. There are different oriented age changes observed in connection with the indices of the 7th block

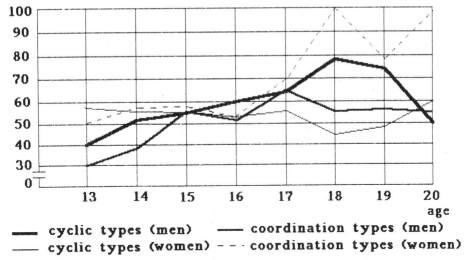

Figure 1. Dynamics of the complex index, characterizing the exactness of movements.

(asymmetry in movements). However, improving skills in the cyclic sports leads to the reduction of asymmetry and the most important characteristic that considerably affects success in sport is the complex index quickness (fig.2).

It was established that within the age range the given index demonstrates an increase in the analysed types of activity. However at the beginning of training (at the age of 13–14) as well as during the process of skill improvement this index is considerably higher in men, and by the age of 17, these differences acquire a less marked character. As to the kinds of sports with high level coordination, the indices of quickness at all stages of training (ages 13–18) are higher than with cyclic sport. The present research made it possible to use the integral evaluation that includes 32 indices that describe motor-physical

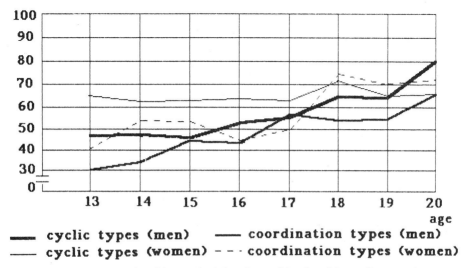

Figure 2. Dynamics of the complex index, characterizing the quickness of movements.

regulative function of young sportsmen and reflect indirectly the level of their fitness. The analysis revealed that within the age range, the motor regulational functions improve, so that the integral index with men at all age groups is higher. At the same time the distinction between different sports groups are of a less marked character.

CONCLUSION

The results of the investigation demonstrate that the most marked differences of the motor-psychological manifestations according to the block, that characterize different moving qualities (quickness, strength, coordination), depend on age both with men and women and reach their utmost values by the age of 17–18. However their dynamics in different periods of age has an irregular character i.e. has its peculiarities (the largest increase in the indices is typical for the age 13–15). It has been shown that the dynamics of motor-psychical reactions depends largely on the character of the moving activity. The analysis of the exactness block (the 4th) in particular shows that high level coordination in cyclic sports gives indices for higher values with women age from 13 to 18 years. The results of this research show, that the indices of physical development, motor regulation, mental state components, being unified in blocks characterizing motor-psychological regulational functions, can be used for evaluation of moving quality development, practical skills and habits according to the temporal, spatial and accuracy parameters depending on the types of activity, gender and age of sportsmen.

REFERENCES

Balsevich, V. K., Zaporozanov, V. A., 1987, Physical Activity of a Men, Kiev, Zdorovia.

Malina, R. M., 1989, Growth and Maturation: Normal Variation and Effect of Training, In: Youth, Exercise and Sport, Ed. Gisolfi, C. V., Lamb, D. R., Benchamark Press Inc., Indiana: 223–261.

Nabatnikova, M. Y., Hordin, A. V., 1982, Outlook on Investigation Young People Sports Problems. Theory and Practice of Physical Training, 8 :30–32.

Pate, R. R., Shephard, R. J., 1989, Characteristics of Physical Fitness in Youth. In: Youth, Exercise and Sport, Ed. Gisolfi, C. V., Lamb, D. R., Benchmark Press.Inc., Indiana: 3–40.

Rogozkin, V. A., Volnov, V. A., Bulkin, V. A., 1988, Unified Complex Controle for Sportsmen Mass Examinations: Methodological Recommendations, Moscow.

Teng, F., 1988, The Effects of Identifing the Stages of Rapid Growth Direction in Selecting Child Athletes, Seoul Olimpic Scientific Congress, Part 2:1101–1102.

5

NEUROLOGICAL AND PATHOPSYCHOLOGICAL CRITERIA FOR SPORT SELECTION

T. P. Koroleva

Kuban State Academy of Physical Education
Budennogo St.161, 350670, Krasnodar, Russia

INTRODUCTION

The methodological basis of sport selection is the psychographical analysis of a person's activity in the relevant sport, i.e. the psychological profile. When analysing and classifying psychological qualities, the identification of criteria which give information that can be the basis of a long term prognosis is of great interest. We proposed that these criteria are important factors that characterise the individual neurodynamic profile in relation to physical activity. From this perspective we have examined the guarding or defence technique specific to the demands of hand-to-hand fighting. Guarding demands high levels of concentration, good ability to read non-verbal signals from the opponent, a highly developed skill at tracking moving objects, high attention distribution and switching, capacious operative memory, a marked capacity for analysis and synthesis and a fast response time. Elite performers are characterised by the following: force, mobility and an unbalanced character of the nervous system according to its excitation state, i.e. choleric temperament. Guarding also places many other demands on the individual. Hunger deadens the attention function and accordingly all the physical and psychological processes that are organized by it. Sleepiness is accompanied by an increased susceptibility to many suggestive influences, especially non-verbal ones. The sensation of hunger can, however, simultaneously cause excessive excitation. The attention stability of the individual is therefore of great importance, together with self-sufficiency, and the ability to relax many distractions are present. Here the neurodynamic preconditions are different: a reserved phlegmatic temperament is ideal. In spite of the obvious importance of these areas for sports research, the individual characterlogical level and psychological profiles of Elite performers in combat sports, especially in respect of guarding behaviour, has received little attention.

Current Research in Sports Sciences, edited by Rogozkin and Maughan.
Plenum Press, New York, 1996

METHODS

The need for checking the possible interconnection of neuro-and patho-psychological indices with traditional methods used for sport selection has defined the methodological approaches used in this study, and both the complex approach and the longitudinal approach have been used. Young athletes from the school of hand-to-hand fighting and guarding activity took part in the investigation during the initial selection procedure and at the end of the school year. Control measurements were carried out in the next year's new enrolment group and also in the group of grown up school students. The following complex of methods was used: quickness of information search in Shulte's and Shulte-Gorbov's tables; visual memory and recognition; duration of movement poses; reaction to a moving object and complex reaction to two stimuli; maximum speed of movement; dynamometric threshold of muscle sensibility. Non-verbal intellect was measured according to the culture-free Kettella test. Formation of the cortex was measured by touch localization and the premotor zones by the synkinesia test. Nervous system activation was defined by use of a galvanometer. The anatomical-morphological signs of psychological and personality qualities were defined with the help of an individual word portrait test.

RESULTS

40 juveniles were examined when enroling at the school of combat and guarding activity. 24 of these were selected to take part in the study. After one year only 17 subjects remained in the group, i.e. 71 per cent of the initial group. The characteristics which distinguished those who persisted from those who were not accepted and stopped their classes were: the psychomotor "portrait" which consisted of a nervous system which was in the middle of the normal range according to its excitation force. They also had a greater total mobility, quicker complex selection reaction to two stimuli, more precise anticipation, good attention stability, and a lower threshold of muscle sensibility. The most important aspect, however, was a two or three times better ability to lower the nervous system activation when relaxing and to increase it when mobilizing. The data collected from the individual tests allowed the construction of a unique picture of school pupils representative of more than 80 per cent of the whole group: an oval skull and oval face of pale complexion without freckles, a wide inter brow distance, a high brow and widely spaced eye sockets. The face is smooth, without prominent cheekbones and also without interbrow vertical wrinkles. There is no diastema. The ears are small and flat against the head. They are of medium to low stature, and the pelvis is narrow. The forefinger is shorter than the fourth one. These anatomical and morphological signs are characterized by the following peculiarities according to the classification of Sheldon and Krechmer.

- Cerebrotonic: asthenic type of constitution, and schizoid character: 35 per cent of pupils;
- Somatotype: athletic type, epileptoid character: 30 per cent of pupils
- Mixed asthenic-athletic type, i.e. schizoid-epileptoid character: 35per cent of pupils.

These three characteristics groups are largely conditioned by hereditary factors. The social environment can strengthen or weaken them but it is very important that the natural characteristics are present.

These three groups were further analysed according to these constitutional types of character, as these young performers are sensitive to factors which may alter the character accentuation. These three groups were found to differ in other respects. The athletic personalities have less will power than others (2.6 times) and the asthenics have a higher capacity for relaxation (6 times) and non-verbal intellect (1.4 times). The mixed type personalities, i.e. those with an asthenic-athletic constitution and schizoid-epileptoid character development showed the most favourable signs of professional fitness: a higher capacity for attention switching and long term auditory memory (1.5 time), a high ability to activate the nervous system when aroused in comparison with the relaxed state (3 times), and finally they have a 1.5 fold weaker nerve system, i.e. it is actually in the middle of the spectrum of strength-weakness of excitation. Comparison of the neurological and pathopsychological results shows that the asthenics have a more marked absence of synkenesia (1.4 times in comparison with the mixed type and 1.13 times in comparison with the athletic personalities).

Statistical analysis has shown than the following parameters correlate with a localization exactness, at the 1–10 percent significance level: absence of synkinesia, maximum speed of movement, and faster complex reaction times. A greater accuracy of response to a moving object is associated with the absence of synkinesia.

The following improvements were observed to occur during the training year: attention stability increased by 50%, attention volume by 12%, maximum speed of movement by 7% and memory of poses of movement by 8%. The response to a moving object was slowed by 34%, i.e. the balance of nerve processes shifted to the side of inhibition. At the same time the force of nervous system expressed as its capacity for excitation increased by 3%. These changes are typical of boys during this phase of maturation. After puberty some of these individuals will show a marked strengthening of the force and instability of the CNS. Phlegmatic/sanguine individuals belong to the object guarding type (strategy - +), whereas choleric-reactive melancholic individuals belong to the subject guarding type (strategy ++) according to the neurodynamic preconditions of their professional fitness.

In the group enrolled for the following year, it was found that 17 pupils out of 27 had mixed asthenic-athletic type personalities. The comparison of the results of the speed of information search using Shulte's and Shulte-Gorbov's table showed that asthenic-athletic types were better in simple attention and visual search tasks and in memory volume tests as well as in information search tasks that required attention switching and speed of visual recognition. This and other additional information gained from this group allowed a more precise definition of the characteristics of the ideal competitor: he is more often lap-eared and his ear lobe is more protuberant, he has sharpened canines, his middle finger is right-angled and the big toe is longer than the second one.

CONCLUSION

These results allow the construction of general behavioural and psychological models of persons competing at high levels in self-defence and hand-to-hand fighting.

Behavioural and Psychological Model of Asthenic-Athletic Type (schizoid-epileptoid character)

Cognitive Sphere

1. Information — Self contained.
2. Problem Decision type — It is built from literature and when organizing the work of other people.

| 3. Thinking peculiarities | It is under the range from wide objective thinking to concentrated subjective one. Rigid logic with argumentation and taking into consideration the opinion of other persons is typical of such people. |
| 4. Decision making up . | Rigidity and persistence of his opinion till the very end are typical of such persons |

Emotional-Requirement Sphere

1. Actual emotional	They need seclusion or action experience under hard conditions
2. Emotional experience	Emotional experience reactions are prognosis stable and predictable.
3. Emotional experience	Reserved character of feelings with manifestation possible emotional lacking self-control and inclination to risk are typical persons.
4. Vocal peculiarities	Fluctuations of loudness and expressiveness take place
5. Socialization	Inclinations to reasoning and dominating are typical of such persons, fear before social contacts can result in brave behaviour

Outward Manifestations

1. Walking	Fluctuations from clumsiness and limpness to take place
2. Pose	It is from constraint, angularity to confidence in carriage
3. Gesticulation	It is restrained and decisive.
4. Mimicry	Fluctuations from impenetrability (high self-control) to expressiveness, artisticity take place
5. Behaviour	Quick responsibility, nonstandard actions take place, nonprediction of purposes is possible
6. Phobias (fears)	Closed rooms and opened space can be not accepted to equal degree
7. Sensibleness to pain	Fluctuations from Spartan indifference to extreme sensibility take prognosis experience

CORRELATION OF PHYSICAL WORKING-CAPACITY WITH MORPHOFUNCTIONAL CHARACTERISTICS IN BOYS AGED 9–15

A. G. Falaleev, V. A. Kobzev, and S. V. Cherenina

Research Institute of Physical Culture
Dynamo Ave. 2, 197042, St. Petersburg, Russia

INTRODUCTION

In the past there have been a number of publications which examined the results of studying the interrelations between physical work-capacity and functional characteristics of athletes. The purpose of studying the correlations was on the scientific grounds of the mechanisms of co-ordination of motor and vegetative functions by the stability of intra- and inter-systemic correlations, as well as revealing the most informative functional characteristics for the evaluation and prognosis of physical work-capacity, fitness and functional readiness. In this paper we show the existence of reliable correlations between the frequency of movements, the power of physical work and heart rate, the frequency of respiratory movements, oxygen consumption, body mass and length, arterial blood pressure, the amplitude characteristics of EEG rhythms and indices reflecting the functional state of muscles (Falaleev, 1988, 1991; Falaleev at al., 1989, 1990).

Data was collected from adult athletes and therefore, any information about age peculiarities of the correlations between the indices of physical work-capacity and morphofunctional characteristics in children and juveniles were not shown. Therefore, the task of the present investigation was to study the age dynamics of correlations between the indices of work-capacity and morphofunctional characteristics in children and juveniles between the ages of 9–15 and the peculiarities of their alteration under the influence of middle-distance running training.

MATERIALS AND METHODS

In a preparatory period of training (PPT) and in a competitive period of training (CPT) we examined 105 boys between the ages of 9 and 15 (15 persons of each age) who attend a special sports school in a group tending towards training in middle-distance run-

Current Research in Sports Sciences, edited by Rogozkin and Maughan.
Plenum Press, New York, 1996

35

ning. Also, we examined 105 boys of the same age who did not take part in sport. In each subject, the following morphofunctional characteristics were measured and used for subsequent processing: biological index of age (BIA) evaluated by secondary sexual characteristics (Tanner, 1979), speed of treadmill running (SRT), time of running on treadmill (TRT) "to refusal", body mass (BM) and body length (BL), chest girth at deep breath (CG deep breath) and at exhalation (CG exhalation), width of chest (WC), pulmonary capacity (PC), dynamometry of right hand (DRH), dynamometry of left hand (DLH), maximum oxygen uptake (VO_2max, l/min.and VO_2max, ml/min/kg.), lactate fraction of O_2 debt (LF O_2debt, l,ml/kg), alactate fraction of O_2 debt (AF O_2debt, l), heart rate at initial state (HR init.), at the level of VO_2max. (HR VO_2max), in restorative period (HR rest.), amplitude (mm) and temporal (s)characteristics of P and T waves at rest (Pmm.init, Tmm.init, Ps.init,Ts.init) at the level of VO_2max. (Pmm.VO_2max., Tmm.VO_2max., Ps.VO_2max, Ts.VO_2max.) and in the period of restoration (Pmm.rest, Tmm.rest, Ps.rest, Ts.rest).

The analysis of the findings consisted in the calculation of linear correlations between the indices of SRT, TRT and the above morphofunctional characteristics for each of the 7 age groups among the pupils of the sports school (in PPT and CPT) and the inactive boys, as well as in studying the dynamics of the tightness of correlation in the age aspect and under the influence of training in middle-distance running.

Table 1. Correlations between the speed of running on treadmill and morphofunctional characteristics in school children who do not take part in sport

Morphofunctional characteristics	Age						
	9	10	11	12	13	14	15
BIA	0.30	0.38	0.45	0.48	0.50	0.51	0.52
BM	0.30	0.39	0.41	0.47	0.49	0.50	0.51
BL	0.20	0.25	0.26	0.30	0.35	0.49	0.50
CG deep breath	0.35	0.38	0.41	0.47	0.49	0.49	0.51
CG exhalation	0.41	0.48	0.49	0.49	0.50	0.51	0.52
WC	0.10	0.20	0.25	0.30	0.35	0.41	0.45
PC	0.40	0.41	0.46	0.47	0.48	0.49	0.50
DRH	0.09	0.10	0.20	0.25	0.38	0.49	0.50
DLH	0.25	0.30	0.31	0.32	0.41	0.47	0.50
VO_2. l/min	0.31	0.37	0.41	0.46	0.49	0.50	0.51
VO_2. ml/min/kg	0.35	0.40	0.42	0.44	0.45	0.49	0.52
LF O_2-debt. l	0.20	0.25	0.31	0.35	0.45	0.48	0.49
AF O_2-debt. l	0.20	0.25	0.26	0.31	0.36	0.37	0.41
O_2-debt. l	0.20	0.25	0.26	0.31	0.35	0.45	0.47
HR init.	−0.35	−0.38	−0.41	−0.45	−0.49	−0.50	−0.51
HR VO_2-max	−0.36	−0.39	−0.41	−0.46	−0.48	−0.49	−0.50
HR rest	−0.35	−0.41	−0.45	−0.47	−0.49	−0.50	−0.50
Pmm init.	−0.30	−0.35	−0.41	−0.45	−0.49	−0.49	−0.50
Ps init.	−0.35	−0.38	−0.41	−0.45	−0.47	−0.49	−0.49
Tmm init.	0.35	0.41	0.45	0.49	0.49	0.50	0.50
Ts init.	0.36	0.41	0.45	0.47	0.49	0.49	0.51
Pmm VO_2-max	−0.41	−0.45	−0.46	−0.47	−0.48	−0.49	−0.49
Ps VO_2-max	−0.45	−0.46	−0.47	−0.48	−0.48	−0.49	−0.50
Tmm VO_2-max	−0.30	−0.35	−0.38	−0.41	−0.45	−0.47	−0.50
Ts VO_2-max	−0.41	−0.45	−0.48	−0.49	−0.49	−0.50	−0.51
Pmm rest.	−0.30	−0.35	−0.38	−0.40	−0.41	−0.45	−0.49
Ps rest.	−0.35	−0.38	−0.41	−0.45	−0.47	−0.49	−0.50
Tmm rest.	−0.41	−0.46	−0.48	−0.49	−0.49	−0.50	−0.50
Ts rest.	−0.45	−0.47	−0.49	−0.49	−0.49	−0.49	−0.49

RESULTS

The analysis of the correlations between the indices of physical work-capacity and morphofunctional characteristics at each age group both in the inactive boys and those who train in middle-distance running (in PPT and in CPT) enabled us to reveal the appropriate increase in the tightness of correlations with increasing age from 9 to 15 years old (Table 1).

In this case the greatest values of correlation were shown by the inactive boys at the age of 10, at ages 12 and 14–15 for those boys who trained in running in a PPT and by boys at age 12 who trained in a CPT.

The comparison of the values of correlation in the boys between the ages 9–15 who do not practice sport with those who train in middle-distance running showed that in the latter both in a PPT and in a CPT the tightness of correlations in each age group was higher (Table 2).

In this case in CPT (Table 3) the tightness of correlations was higher than in the PPT. The number of significant correlations in each age group showed that in the inactive boys the significant correlations ($> \pm 0.50$) appeared at age 13 years and it prevailed over the number of insignificant one at 15 years of age. In the young athletes the number of significant correlations prevailed from the age of 9–10 years.

Table 2. Correlations between the speed of running on treadmill and morphofunctional characteristics in young athletes in a preparatory period of training

Morphofunctional characteristics	Age						
	9	10	11	12	13	14	15
BIA	0.51	0.56	0.58	0.60	0.61	0.64	0.65
BM	0.49	0.49	0.51	0.53	0.58	0.60	0.63
BL	0.50	0.51	0.54	0.56	0.57	0.60	0.63
CG deep breath	0.64	0.67	0.68	0.70	0.74	0.78	0.79
CG exhalation	0.59	0.60	0.61	0.63	0.65	0.67	0.68
WC	0.55	0.56	0.58	0.61	0.63	0.66	0.70
PC	0.65	0.69	0.70	0.72	0.74	0.79	0.81
DRH	0.55	0.56	0.53	0.59	0.61	0.63	0.68
DLH	0.50	0.51	0.53	0.56	0.58	0.60	0.63
VO_2. l/min	0.49	0.50	0.53	0.54	0.58	0.60	0.61
VO_2. ml/min/kg	0.50	0.51	0.54	0.56	0.60	0.61	0.64
LF O_2-debt. l	0.50	0.52	0.54	0.56	0.58	0.61	0.66
AF O_2-debt. l	0.47	0.49	0.51	0.53	0.54	0.56	0.58
O_2-debt. l	0.51	0.54	0.56	0.58	0.59	0.61	0.63
HR init.	−0.50	−0.51	−0.53	−0.54	−0.56	−0.58	−0.60
HR VO_2-max	−0.53	−0.54	−0.56	−0.58	−0.61	−0.67	−0.68
HR rest.	−0.54	−0.56	−0.58	−0.61	−0.64	−0.68	−0.70
Pmm init.	−0.51	−0.53	−0.57	−0.59	−0.63	−0.67	−0.68
Ps init.	−0.50	−0.52	−0.55	−0.58	−0.61	−0.64	−0.69
Tmm init.	0.50	0.51	0.54	0.59	0.63	0.64	0.69
Ts init.	0.49	0.52	0.55	0.57	0.59	0.61	0.64
Pmm VO_2-max	−0.53	−0.54	−0.56	−0.61	−0.64	−0.66	−0.69
Ps VO_2-max	−0.52	−0.54	−0.58	−0.64	−0.67	−0.69	−0.71
Tmm VO_2-max	−0.56	−0.58	−0.60	−0.65	−0.69	−0.71	−0.74
Ts VO_2-max	−0.54	−0.56	−0.61	−0.66	−0.70	−0.74	−0.76
Pmm rest.	−0.51	−0.54	−0.58	−0.60	−0.64	−0.69	−0.73
Ps rest.	−0.50	−0.52	−0.59	−0.63	−0.67	−0.69	−0.74
Tmm rest.	−0.52	−0.54	−0.60	−0.65	−0.68	−0.72	−0.72
Ts rest.	−0.51	−0.57	−0.61	−0.67	−0.70	−0.73	−0.74

Table 3. Correlations between the speed of running on treadmill and morphofunctional characteristics in young athletes in a competitive period of training

Morphofunctional characteristics	Age						
	9	10	11	12	13	14	15
BIA	0.60	0.61	0.65	0.68	0.70	0.71	0.75
BM	0.50	0.51	0.54	0.56	0.58	0.64	0.68
BL	0.54	0.56	0.58	0.64	0.68	0.70	0.72
CG deep breath	0.69	0.71	0.73	0.79	0.81	0.84	0.83
CG exhalation	0.63	0.65	0.69	0.73	0.75	0.76	0.79
WC	0.60	0.61	0.66	0.70	0.71	0.73	0.76
PC	0.69	0.73	0.75	0.79	0.80	0.83	0.85
DRH	0.58	0.60	0.64	0.69	0.70	0.76	0.80
DLH	0.55	0.58	0.61	0.64	0.69	0.71	0.75
VO_2. l/min	0.54	0.56	0.61	0.64	0.69	0.71	0.76
VO_2. ml/min/kg	0.56	0.58	0.63	0.69	0.71	0.74	0.78
LF O_2-debt. l	0.54	0.56	0.58	0.60	0.64	0.68	0.70
AF O_2-debt. l	0.50	0.53	0.56	0.58	0.61	0.64	0.67
O_2-debt. l	0.54	0.56	0.58	0.63	0.67	0.69	0.73
HR init.	−0.53	−0.56	−0.58	−0.59	−0.61	−0.64	−0.66
HR VO_2-max	−0.56	−0.58	−0.61	−0.64	−0.67	−0.70	−0.71
HR rest.	−0.57	−0.60	−0.63	−0.65	−0.69	−0.73	−0.74
Pmm init.	−0.54	−0.58	−0.63	−0.65	−0.71	−0.74	−0.76
Ps init.	−0.54	−0.56	−0.58	−0.61	−0.66	−0.70	−0.71
Tmm init.	0.54	0.58	0.61	0.65	0.67	0.70	0.73
Ts init.	0.55	0.57	0.59	0.64	0.66	0.69	0.70
Pmm VO_2-max	−0.56	−0.59	−0.61	−0.66	−0.69	−0.71	−0.74
Ps VO_2-max	−0.54	−0.61	−0.67	−0.71	−0.74	−0.76	−0.79
Tmm VO_2-max	−0.56	−0.61	−0.64	−0.71	−0.74	−0.78	−0.80
Ts VO_2-max	−0.57	−0.59	−0.63	−0.69	−0.74	−0.79	−0.83
Pmm rest.	−0.54	−0.58	−0.61	−0.64	−0.70	−0.73	−0.76
Ps rest.	−0.53	−0.56	−0.63	−0.71	−0.76	−0.81	−0.84
Tmm rest.	−0.56	−0.58	−0.64	−0.69	−0.71	−0.76	−0.78
Ts rest.	−0.55	−0.61	−0.64	−0.71	−0.73	−0.76	−0.74

The close correlations ($> \pm 0.70$) in the boys who do not practice sport were absent. In the young athletes in PPT, the close correlations were observed only with separate morphofunctional indices (CG deep breath, WC, PC, HR rest., Ps.VO_2-max, Ts.VO_2-max., Ps.rest.,Ts.rest., Tmm.VO_2-max., Pmm.VO_2-max., Pmm.rest., Tmm.rest.) from 11 years of age and even at the age of 15 the number of close correlations did not prevail.

In the CPT close correlations were observed from 10 years of age (BIA, BL, CG deep breath, CG exhalation, WC, PC, DRH, DLH, VO_2 l/m,VO_2 ml/min/kg, O_2-debt, HR VO_2-max, HR rest, Ps.init, Ps.VO_2-max.,Ps.rest, Ts.init, Ts.VO_2-max, Ts.rest, Pmm.init, Pmm.VO_2-max,Pmm.rest, Tmm.init, Tmm.VO_2-max, Tmm.rest). In this group (CPT) the number of close correlations from 14 years of age prevail.

The same regularities were also revealed by the analysis of correlations between the time of running on the treadmill "to refusal" and morphofunctional characteristics.

DISCUSSION

The investigations carried out and the analysis of the findings enabled to trace the age peculiarities (from 9 to 15 years old) of the correlation between the indices of work-capacity and morphofunctional characteristics in boys with various levels of adaptation to middle-distance running.

They revealed a general pattern of an increase in the tightness of correlation as the child is growing irrespective of the fact of whether he is subjected to systematic physical loading or not.

There are theoretical propositions (Zimkin, 1963;Klimova-Tcherkasova, 1972) which state that the correlation of peripheral systems reflect various combinations of the interaction of the nervous centres they regulate.

Basing upon data in the literature and our findings on the dynamics of at different ages and fitness, one can affirm that when an organism grows the perfection of central and peripheral mechanisms of the motor and vegetative systems occur.

Regular training in middle-distance running intensify and accelerate the process of mastering intercentral and interfunctional interactions. Under the influence of training the phenomena of synchronose are seen at an earlier age.

REFERENCES

Falaleev, A., 1988, Correlation of motor and vegetative functions by physical exercises, Physiology of man 14: 263–271.

Falaleev, A., 1991, General regularities of the reconstruction of intrafunctional correlations in human organism by the adaptation to physical exercises. In: Intrafunctional interrelations by the adaptation of an organism to sporting activity. L., Res. Inst. of Phys. Cul. :4–11.

Falaleev, A., Maslova, I., Tcherenina, S., Drychkin, A., 1989, General conformity with a law interfunctional relationship between locomotory and cardiorespiratory parameters in during of bicycle ergometers tests, Abstracts of XXXIth International Congress of physiological sciences, Helsinki, Finland, July 9–14 : 431.

Falaleev, A., Tcherenina, S., Maslova, I., Kuzemsky, V., 1990, The relationship of physiological characteristics of maximum short-term and prolonged of physical loads at rowing, Abstracts of the XXIV First World Congress of sports medicine, Amsterdam, May 21 - June 1 : 166.

Klimova-Tcherkasova, V., 1972, Interstructural tonic interaction in regulation. In: Evolution, ecology and brain, Edited by N.Vasilevsky, L., Medicina : 121–138.

Tanner, Dj., 1979, Growth and constitution of man. In: Human biology, M., Mir : 366–471.

Zymkin, N., 1963, On some physiological mechanisms of motor skills in sport, In: Sensomotory and motor skill in sport, L.: 5–26.

7

NUTRITION FOR YOUNG ATHLETES

R. J. Maughan and S. M. Shirreffs

University Medical School
Foresterhill, Aberdeen AB9 2ZD, Scotland

INTRODUCTION

The nutritional demands of preparing for and participating in sport at the elite level have been extensively studied in recent years, and much information is now available on both the nutritional requirements and nutritional habits of top level performers in many different sports (Burke and Deakin, 1994). Most of this information, however, relates to the major locomotor sports, such as running, cycling and swimming, where performance is easily quantified, and also applies only to adult performers in these sports. In some sports, there is little available information, either on the nutritional demands of training and competing, although the world's best performers in some of these sports are adolescents. Women's gymnastics is a good example of a sport where the elite performers begin intensive training before puberty and reach their peak while still adolescent.

Where the nutritional implications of sports performance by youngsters has been considered, there has often been no attempt to identify nutritional needs, but rather to rely on extrapolations from adult participants. The nutritional preparation of the elite young athlete, however, raises special problems for the nutritionist and dietitian. These physically gifted youngsters are often highly motivated, and undergo prolonged strenuous exercise in training on a daily basis. This period of exercise stress often coincides with a period of rapid growth, so there are some real difficulties in making simple extrapolations.

As well as being highly motivated, young athletes are often easily impressed by their heroes and will seek to emulate not only their training programmes but also their dietary habits. This can lead to extreme behaviour.

THE ELITE YOUNG ATHLETE

Several studies have described the physical characteristics of elite junior performers in different sports as well as in non-elite and non-athletic children (eg. Buckler et al, 1977; Bloomfield et al, 1990; McMiken, 1975). The "nature versus nurture" debate continues, but it is clear that success in sport is achieved by those who have inherited certain physical (and most probably also mental) characteristics. In most events where large size and

Current Research in Sports Sciences, edited by Rogozkin and Maughan.
Plenum Press, New York, 1996

strength favour success, the early-maturing child will succeed, and this effect will be most marked in those sports where age-group competition exists as those children born in the early part of the year have a distinct advantage (Brewer et al, 1992).

The spectrum of sporting events is, however, such that there are also events (the most obvious and most frequently studied example is again women's gymnastics) where small stature and late maturation confer a distinct advantage. It is well recognised that young female gymnasts are shorter and lighter than sedentary girls of the same age: they also show signs of late maturation, as evidenced by late menarche (Lindholm et al, 1994; Malina, 1994). Although there is clear evidence of an effect of intensive training on the hormones of the hypothalamic-pituitary axis, it is not clear that this effect can in itself account for what appears to be a loss of growth potential in elite young athletes. The parents of female gymnasts have been shown to be of less than average height (Theintz et al, 1989), confirming the suggestion that the short stature of the elite gymnast is determined largely by genetic rather than training and nutritional factors: in this study, Theintz et al found no evidence that the predicted adult height for a group of elite young female gymnasts who had already been training intensively for a period of 5 years was less than the target height.

It thus appears that, although restricted energy intake can delay maturation and sexual development, the late development which is observed in some sports is more related to inherited characteristics than to the effects of intensive training of nutritional inadequacy.

ENERGY DEMANDS

Unlike the mature performer, where energy intake should normally equal energy expenditure and where an excess of intake over expenditure will generally result in an increased fat mass, the energy intake of the growing child, whether athlete or not, must exceed energy expenditure if growth is not to be compromised.

There are many studies which report estimates of the energy intake of young athletes, and these studies have the same shortcomings as all investigations which rely on self-reports of intake. Data from a number of these studies have recently been collated by O'Connor (1994), who observed that there was good agreement between studies in the estimates of energy intake for active children: with data from a wide range of different sports, and a mean age range from 7–22 y included in the same comparison, a wider variability might have been expected. Expressed relative to body weight, the energy intake of males (mean values from about 210–360 kJ/kg) was found to be generally higher than that of females (130–230 kJ/kg).

Although there is much concern regarding the potentially negative effects of restricted energy intake in growing children participating in weight category sports or in those where low body fat is deemed desirable, these concerns may not be justified in the case of most participants. There is, however, no doubt that deficiencies of specific nutrients, especially iron (Nickerson et al, 1985; Rowland et al, 1987) and calcium (Warren et al, 1986), secondary to low energy intakes may occur in some performers. Again, these are problems of greatest concern for, but by no means exclusive to, female participants.

PROTEIN INTAKE

The increased needs for protein intake by active adolescents relative to their sedentary peers are a consequence of the increased protein demand imposed by exercise combined with the additional requirement for growth and maturation. The protein requirement of children is greater than that of adults, and this is reflected in the recommendation of higher intakes in children and adolescents compared with adults. Recommended Dietary Intake (RDI) values for the adult population vary widely between countries (from 0.8–1.2 g/kg/d): where separate values are established for adolescents, they are generally in the region of 1 g/kg/d (Lemon, 1992). Bar-Or and Unnithan (1994) have recently proposed, in non-athletic children, an increase from the adult value of 0.8 g/kg/d to 1.2 g/kg/d for boys and girls aged 7–10 y, with a figure of 1.0 g/kg/d for youngsters aged 11–14 y. After this age, the recommended intake is not different from that for adults.

Both strength training and endurance activities have been shown, in studies of adult athletes, to result in an increased protein requirement, resulting from the enhanced rate of protein oxidation during exercise, and also from increased requirements for tissue repair processes. Lemon (1992) recommended intakes, during periods of intensive training or competition, of 1.2–1.7 g/kg/d for strength athletes and 1.2–1.4 g/kg/d for endurance athletes. There is no reason to suppose that the effect of exercise on the protein metabolism of young athletes is fundamentally different from that in adults, but the training loads are unlikely to approach the extremes considered by Lemon (1992) in formulating his guidelines. It seems likely, therefore, that the protein requirement, and therefore the recommended intake, during periods of hard training or competition in youngsters will not exceed the range of 1.2–1.7 g/kg/d set for adult athletes.

Dietary protein intake in intimately linked to total energy intake as, if the composition of the diet remains constant, the protein content of the diet is determined by the total amount of food consumed. Where the energy intake is sufficient to meet the increased needs imposed by both growth and physical activity, it is difficult to select foods that will fail to provide an adequate protein intake. Published studies suggest that most young athletes achieve a dietary protein intake of about 1.6 g/kg/d (O'Connor, 1994), and that even in those sports where energy intake is commonly restricted there is not good evidence of an inadequate intake. It is equally clear, however, that there is a limited amount of information available on the protein requirements of the growing athlete who is engaged in intensive training: it is possible that there are substantial increases in the requirement during periods of rapid growth, as during the adolescent growth spurt. Special attention must be paid to the dietary intake of athletes who are training hard at this time, especially where restriction of energy intake is commonly practised. Where energy, and especially carbohydrate, intake is inadequate, protein oxidation by the exercising muscles is increased (Lemon, 1992), so the adequacy of protein intake cannot be considered independently of the overall diet.

CARBOHYDRATE

The major fuels used by muscles during exercise are carbohydrate, either in the form of muscle glycogen or blood glucose, which is derived from the liver glycogen, and fatty acids which may come from the adipose triglyceride via the plasma fatty acids or from the intramuscular triglyceride depots. The proportions in which these fuels contribute to energy production are determined primarily by the exercise intensity, with the contribution of carbohydrate increasing as the exercise intensity increases. Effective training, in all sports, requires a relatively high volume of exercise at high intensity, thus placing large demands on the body's limited carbohydrate stores (Coyle, 1992).

There is a limited amount of information on substrate utilisation during exercise in children, and even less on the adaptations that occur in children in response to training. This shortage of information is in large part a reflection of the invasive procedures, including muscle biopsy and tracer infusion, that are necessary to elicit this information. It is, however, clearly established that the skeletal muscles of pre-pubertal boys and girls have a limited capacity for anaerobic metabolism: this is generally ascribed to a low activity of the enzymes of the glycolytic pathway (Eriksson et al, 1973). The same authors, however, have shown that the activity of succinate dehydrogenase, which is used as a marker of oxidative capacity of muscle, is not different between boys aged 11–13 y and adults (Eriksson, 1973). The consequences of this metabolic profile are a limited capacity for high intensity exercise, which relies heavily on anaerobic glycolysis for energy provision, and a reduced blood and muscle lactate concentration at maximum exercise Inbar and Bar-Or, 1986). There is likely also to be a greater reliance on fat as a metabolic substrate during prolonged exercise at more moderate intensity, although this seems to have received little attention.

In spite of the evidence for a decreased reliance on carbohydrate for energy provision in children, however, there seems little reason to alter the recommendation made to adults that, for both health reasons and exercise performance (Costill, 1988) the diet should be relatively high in carbohydrate and low in fat. The available evidence suggests that the dietary carbohydrate intake of physically active children is not different form that of the active adult population, giving a carbohydrate intake corresponding to about 50% of total energy intake (O'Connor, 1994).

An increase in the dietary carbohydrate intake to about 55% of total energy intake with a corresponding reduction in the intake of fat might be recommended. This message is consistent with the recommendations for sustaining consistent training, but would also help to establish an eating pattern that is appropriate from a health perspective and could be carried through to adult life.

FLUID BALANCE AND WATER REQUIREMENTS

Hyperthermia and dehydration are potentially serious problems for all athletes engaged in strenuous exercise lasting more than about 30 min: in hot weather, possibility of these problems is greatly increased. The dangers of exercise in the heat are further magnified for young athletes. Compared with adults exercising at the same intensity, pre-pubescent children have a reduced rate of sweat secretion, and a lower sensitivity of the sweat glands to elevated core temperature (Bar-Or, 1989). The larger surface area to mass ratio of children is an advantage in cool conditions, allowing an increased convective heat loss. When ambient temperature exceeds skin temperature, however, heat will be gained by this route, imposing an increased thermal load on the exercising child, and raising the very real possibility that the limited sweating capacity will fail to cope. The rather frequent reports of dizziness, syncope and nausea in children exercising in the heat probably reflect the extensive peripheral vasodilatation that occurs in an attempt to promote convective heat loss but which results in peripheral pooling of blood and a fall in central venous pressure which leads in turn to a fall in arterial pressure.

Because of these potential dangers, several organisations have issued guidelines which suggest that the participation of children in endurance exercise in the heat requires careful supervision, and should in some situations be curtailed (America College of Pediatrics, 1982).

A suitable fluid replacement regimen during exercise can minimise the dangers of heat stress. Even though the sweating capacity of children is low, and the sweat electrolyte content is low relative to that of adults (Meyer et al, 1992), the need for fluid and electrolyte replacement is no less important than in adults. Indeed, in view of the evidence that core temperature increases to a greater extent in children than in adults at a given level of dehydration, the need for fluid replacement may well be greater in children (Bar-Or, 1989). As with adults, however, children seldom consume sufficient fluid at times of high rates of loss to meet their requirements (Bar-Or et al, 1980).

Dilute glucose-electrolyte drinks, containing carbohydrate at a concentration of about 2–8% and sodium at a concentration of 20–50 mmol/l, and which are isotonic or moderately hypotonic have been shown to be most effective at replacing fluid loss and at enhancing exercise performance in adults (Maughan, 1994). There is insufficient evidence in this area in relation to performance effects in children to make anything other than the same recommendations for children. The primary requirement is to ensure an adequate volume intake, and this requires that the fluid is palatable: even so, children will not normally consume sufficient fluid to meet their needs, and they require frequent reminders of the need to drink. Frequent drinking of small volumes is probably the most effective strategy. Care must be taken to ensure adequate hydration prior to training or competition by stressing fluid intake in the few hours prior to exercise.

MICRONUTRIENTS

Supplementation of the diet with vitamins or minerals is not generally warranted in athletes, irrespective of age: any additional demand for these nutrients imposed by training should be met if the energy intake is sufficient to meet the additional energy expenditure incurred in training and competition and if a varied diet is consumed. Not all athletes meet these conditions: some restrict energy intake to reduce body (fat) mass, and it is not unusual to find that the diet of athletes is somewhat monotonous. In spite of this, however, deficiency states are rare, although their effects on health as well as on athletic performance are potentially serious where they do exist.

Most reports of deficiencies among athletes, both adult and child, are in fact nothing more than reports of dietary surveys, where the recorded intake fails to meet the Recommended Daily Allowance (RDA). There are several reasons why this finding might occur, including under-reporting of intake by some athletes, and the fact that published food composition tables are often incomplete, with no data for the micronutrient content of many food items: for this latter reason, intake of some micronutrients—chromium is a good example—is grossly underestimated by most diet analysis programmes. Failure to meet the RDA is not, in any case, an indication of the existence of an inadequate intake. A deficiency is not established until and unless well-defined diagnostic criteria are met, and supplementation is not warranted unless a deficiency is known to exist.

Two micronutrients for which there is reason to suspect a significant possibility of deficiency are iron and calcium, but there are no good data on the prevalence of such deficiencies. The reasons for the possibility of deficiencies have been well rehearsed (Drinkwater, 1984: Deakin, 1994), and there is a need to be alert to the possibility that children—whether athletic or not—may fail to meet their requirements. If such a situation can be shown to exist, the first step must be an examination of the diet and recommendation of changes which can make good the shortfall through an increased intake of foods which will supply the required nutrients. Only in the last resort should supplementation be considered, and then it is likely to be no more than a short term solution.

REFERENCES

American College of Pediatrics (1982) Climatic heat stress and the exercising child. Pediatrics 69: 808–809

Bar-Or O (1989) Temperature regulation during exercise in children and adolescents. In: C Gisolfi, DR Lamb (eds) Perspectives in Exercise and Sports Medicine, Vol 2, Indianapolis, Benchmark Press. pp335–367

Bar-Or O, VB Unnithan (1994) Nutritional requirements of young soccer players. J Sports Sci 12: S39-S42

Bar-Or O, R Dotan, O Inbar, A Rotshtein, H Zonder (1980) Voluntary hypohydration in 10 to 12 year old boys. J Appl Physiol 48: 104–108

Bloomfield J, BA Blanksby, TR Ackland. (1990) Morphological and physiological growth of competitive swimmers and non-competitors through adolescence. Aust J Sci Med Sport 22: 4–12

Brewer J, PD Balsom, JA Davis, B Ekblom. (1992) The influence of birth date and physical development on the selection of male junior international soccer squad. J Sports Sci 10: 561–562

Buckler JMH, DA Brodie (1977) Growth and maturity characteristics of schoolboy gymnasts. Ann Hum Biol 4: 455–463

Burke LM and V Deakin. (1994) Clinical Sports Nutrition. McGraw Hill, Sydney

Costill DL (1988) Carbohydrates for exercise: dietary demands for optimal performance. Int J Sports Med 9: 1–18

Coyle EF (1992) Timing and method of increased carbohydrate intake to cope with heavy training, competition and recovery. J Sports Sci 9 (Special Issue): 29–52

Deakin V (1994) Iron deficiency in athletes: identification, prevention and dietary treatment. In: Burke LM and V Deakin. Clinical Sports Nutrition. McGraw Hill, Sydney. pp 174—199

Drinkwater BL, K Nilson, CH Chestnut (1984) Bone mineral content of amenorrheic and eumenorrheic athletes. N Engl J Med 311: 277–281

Eriksson, B (1972) Physical training, oxygen supply amd muscle metablism in 11–13 year old boys. Acta Physiol Scand (Suppl) 384: 1–48

Eriksson BO, PD Gollnick, B Saltin. (1973) Muscle metabolism and enzyme activities after training in boys 11–13 years old. Acta Physiol Scand 87: 485–497

Inbar O, O Bar-Or (1986) Anaerobic characteristics in male children and adolescents. Med Sci Sports Ex 18: 264–269

Lemon, P (1992) Effect of exercise on protein requirements. J Sports Sci 9 (Special Issue): 53–70

Lindholm C, K Hagenfeldt, BM Ringertz. (1994) Pubertal development in elite juvenile gymnasts. Effects of physical training. Acta Obs Gyn Scand 73: 269–273

Malina R (1994) Physical growth and biological maturation of young athletes. Ex Sports Sci Rev 22: 389–434

Maughan RJ (1994) Fluid and electrolyte loss and replacement in exercise. In: Oxford Textbook of Sports Medicine. Ed Harries, Williams, Stanish & Micheli. Oxford University press, New York. pp 82–93

McMiken DF (1975) Maximum aerobic power and physical dimensions of children. Ann Hum Biol 3: 141–147

Meyer F, O Bar-Or, D MacDougall, GJF Heigenhauser (1992) Sweat electrolyte loss during exercise in the heat: effects of gender and maturation. Med Sci Sports Ex 24: 776–781

Nickerson HJ, M Holubets, AD Tripp, WG Pierce (1985) Decreased iron stores in high school female runners. Am J Diseases Child 139: 1115–1119

O'Connor, H (1994) Special needs: children and adolescents in sport. In: Burke LM and V Deakin. Clinical Sports Nutrition. McGraw Hill, Sydney. pp 390–414

Rowland TW, SA Black, JF Kelleher (1987) Iron deficiency in adolescent endurance athletes. J Adolesc Health Care 8: 322–326

Theintz GE, H Howald, Y Alleman, PC Sizonenko. (1989) Growth and pubertal development of young female gymnasts and swimmers: a correlation with parental data. Int J Sports Med 10: 87–91

Warren MP, J Brooks-Gunn, LH Hamilton, LF Warren, WG Hamilton (1986) Scoliosis and fractures in young ballet dancers; relationship to delayed menarche and secondary amenorrhea. New Engl J Med 314: 1348–1353

8

FAT AND FAT DISTRIBUTION IN ELITE RHYTHMIC GYMNASTS

Marie-José Borst,[1] Willy Pieter,[2] and Nadeshda Yastrjembskaya[3]

[1] SYNTEX B.V.
Rijswijk, The Netherlands
[2] Lesgaft State Academy of Physical Culture, St. Petersburg, Russia and
Center for Research and Communication, Metro Manila, Philippines
[3] Kaliningrad State University
Kaliningrad, Russia

ABSTRACT

Contrary to artistic gymnastics, only a few studies exist on rhythmic gymnastics. Some of them have focused on body composition of these athletes, but none of them has included fat patterning. The purpose of this study was to assess relative body fat and fat patterning in American elite rhythmic gymnasts (n = 16) and to compare pre-menarcheal (mean age ± SEM: 13.30 ± 0.30 years) and post-menarcheal athletes (15.67 ± 0.62 years) on age, height, weight, body fat and fat distribution. In addition to age, height and weight of the athletes, the following skinfolds were taken: triceps, subscapular, supraspinale, abdominal, front thigh and medial calf. Analysis of variance was used to determine the differences between the pre- and post-menarcheal athletes on age, height, body weight and sum of six skinfolds. The post-menarcheal gymnasts were older and heavier than their pre-menarcheal colleagues. No other differences were found, although the post-menarcheal athletes also had a much higher sum of six skinfolds (55.00 versus 45.65 mm). A larger sample size may have shown the latter difference to be statistically significant as well. When compared to a combined sample of Olympic artistic gymnasts, the rhythmic gymnasts had a lower sum of six skinfolds (49.16 versus 51.05 mm) and a similar fat patterning. However, the Olympic gymnasts were also older (17.93 versus 14.19 years). The older rhythmic gymnasts clearly had more relative body fat than the Olympic artistic gymnasts, which may be related to the nature of the sport: there are no weight-bearing routines in rhythmic gymnastics.

INTRODUCTION

Scientific research on rhythmic sportive gymnastics (RSG) published in a Western European language is scarce and predominantly of recent origin. The larger part of this re-

Current Research in Sports Sciences, edited by Rogozkin and Maughan.
Plenum Press, New York, 1996

search is physiological in nature (e.g., Alexander, 1989; 1991a; 1991b; Alexander, Boreskie and Law, 1987; Badelon, Boulier, Fabre, Duvallet and Léglise, 1985; Imhof, 1986; Lindboe and Slettebø, 1984; Scheele, Rettenmeyer, Herzog, Steinbrück and Weicker, 1978), while others were more in the area of tests and measurement (Heinß, 1983; 1987; Yastrjembskaya and Pieter, 1995) or kinesiology (Dyhre-Poulsen, 1987).

Although body fat was within the realm of the physiological studies mentioned above, none has addressed fat distribution. For instance, Case, Fleck and Koehler (1980) found that the 1979 US modern rhythmic gymnastics team had a lower percentage total body fat than a group of artistic gymnasts as assessed by hydrostatic weighing: 11.24 % and 15.34%, respectively. Canadian elite rhythmic gymnasts fell in between these values with 13% (Alexander, 1989). As expected, young elite rhythmic gymnasts under 16 years of age have significantly lower biceps, hip, thigh and calf girths than those over 16 years, while they also have lower subscapular and abdominal skinfolds (Gionet, Babineau and Bryant, 1986). By the same token, elite rhythmic gymnasts were found to have a lower percentage body fat (12.15%) than their subelite counterparts (14.49%), although this difference failed to reach statistical significance (Alexander, 1991a).

Given the dearth of information on body fat and its distribution in rhythmic sportive gymnastics, the purpose of this study was to assess relative body fat and fat patterning in young elite rhythmic gymnasts and to compare premenarcheal and postmenarcheal athletes on age, height, weight, body fat and fat distribution.

METHODS

Subjects consisted of sixteen young high-performance athletes in rhythmic sportive gymnastics (RSG) as of August 1991, who were tested at the United States Olympic Training Center in Colorado Springs, Colorado, U.S.A. In addition to age, stature and body weight, the following skinfolds were taken with a Lange skinfold caliper: triceps, subscapular, suprailiac, umbilical, front thigh and medial calf. As suggested by Ross and Marfell-Jones (1991), no attempt was made to estimate the percentage of relative total body fat from these skinfold measures. Instead, sum of skinfolds was used to represent total body fat as was done with Olympic artistic gymnasts (Carter and Yuhasz, 1984). Stature was assessed with a wall-mounted stadiometer to the nearest 1 cm, while a calibrated electronic digital scale was employed to determine weight to the nearest 0.5 kg.

To get an impression of the relative fat distribution in the athletes, the following ratios were also calculated and included in the analysis: triceps/(triceps + subscapular skinfolds) and medial calf/(medial calf + subscapular + supraspinale skinfolds). These ratios were found to be rather sensitive indices and of high predictive value in assessing central or peripheral fat distribution in males and females (e.g., Kaplowitz, Mueller, Selwyn, Malina, Bailey and Mirwald, 1987; Shimotaka, Tobin, Muller, Elahi, Coon and Andres, 1989). In addition, body weight and skinfold patterning were scaled down to Phantom height to facilitate comparisons between groups while adjusting all measurements to a common stature (Ross and Marfell-Jones, 1991).

Analysis of variance was used to determine the differences between the premenarcheal and postmenarcheal rhythmic gymnasts on age, height, body weight, sum of six skinfolds, individual skinfolds, fat distribution, proportional skinfolds and body weight.

Table 1. Bio-demographical data and skinfolds of rhythmic sportive gymnasts

Variable	Total Group (mean ± SD; n=16)	Premenarcheal Group (mean ± SE; n=10)	Postmenarcheal Group (mean ± SE; n=6)
Age (years)	14.19 ± 1.64	13.30 ± 0.30	15.67 ± 0.62*
Experience (years)	5.22 ± 2.48	4.95 ± 0.75	5.67 ± 1.15
Competition exper. (yrs)	4.59 ± 2.20	4.45 ± 0.68	4.83 ± 1.01
Training freq. (ds/wk)	5.75 ± 0.45	5.67 ± 0.21	5.80 ± 0.13
Training sessions/day	1.16 ± 0.35	1.00 ± 0.00	1.42 ± 0.20
Training hours/session	3.88 ± 0.67	3.80 ± 0.08	4.00 ± 0.45
Stature (m)	1.59 ± 0.06	1.58 ± 0.02	1.61 ± 0.02
Weight (kg)	45.11 ± 7.05	41.50 ± 1.89	51.14 ± 1.57**
Triceps	9.4 ± 3.5	8.7 ± 0.9	10.7 ± 1.7
Subscapular	5.9 ± 1.2	5.3 ± 0.3	6.9 ± 0.3
Suprailiac	6.2 ± 1.5	6.0 ± 0.5	6.5 ± 0.7
Umbilical	6.6 ± 2.0	6.1 ± 0.5	7.4 ± 1.0
Front thigh	12.8 ± 4.5	11.6 ± 1.1	14.9 ± 2.3
Medial calf	8.3 ± 2.6	8.2 ± 0.8	8.6 ± 1.1
Σ 6 skinfolds	49.16 ± 12.77	45.65 ± 3.28	55.00 ± 6.19

*Significant difference between pre- and postmenarcheal gymnasts ($p < 0.002$).
**Significant difference between pre- and postmenarcheal gymnasts ($p < 0.003$).

RESULTS

Table 1 shows age, height, weight, skinfold thicknesses and sum of six skinfolds of premenarcheal and postmenarcheal rhythmic sportive gymnasts as well as the same variables for the group as a whole. As expected, the postmenarcheal rhythmic gymnasts were older ($p < 0.002$) and heavier ($p < 0.003$) than their premenarcheal colleagues. No other differences were found, although the postmenarcheal athletes also had a much higher sum of six skinfolds (55.00 versus 45.65 mm). A larger sample size may have shown the latter difference to be statistically significant as well.

No differences ($p > 0.05$) were found in terms of fat distribution between the premenarcheal and postmenarcheal rhythmic gymnasts (figure 1 shows the skinfold patterns of the two groups). When scaled to Phantom height, the postmenarcheal rhythmic gym-

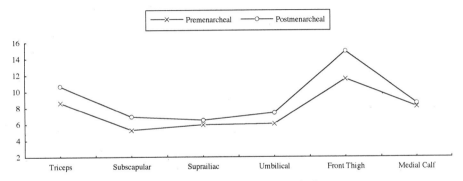

Figure 1. Skinfold patterning in American rhythmic gymnasts.

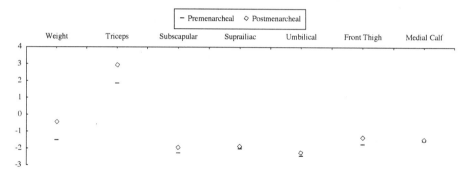

Figure 2. Proportional skinfold patterning in American rhythmic gymnasts.

nasts were still heavier ($p < 0.001$) than their younger colleagues, while they also had a larger proportional subscapular skinfold ($p < 0.005$) (see figure 2).

DISCUSSION

The premenarcheal rhythmic sportive gymnasts are close to the age at which their postmenarcheal colleagues had their menarche (age at menarche: 13.33 years) (Pieter, unpublished data). Interestingly, although postmenarcheal girls are found to be taller and heavier than their premenarcheal counterparts (Malina and Bouchard, 1991), the present rhythmic gymnasts only differed in absolute and proportional weight, with the premenarcheal gymnasts being lighter.

There were also no differences in absolute skinfold thicknesses, with the exception of the proportional subscapular skinfold. It may be that, although the premenarcheal rhythmic gymnasts were not different in height compared to their postmenarcheal colleagues, their body weight played a role in the absence of their menarche. It is hypothesized that weight and fatness may have to reach a certain threshold before menarche can occur (e.g., Frisch, 1987), although the available evidence at present is still equivocal (Malina, 1988; Plowman, Liu and Wells, 1991). For instance, Norwegian elite rhythmic gymnasts (mean age of 18.3 years with a range of 15 - 21 years) were of similar body weight and fat (as measured by the BMI) as their controls (mean age: 18.1 years; range: 16 - 20 years) (Lindboe and Slettebø, 1984). Yet, more rhythmic gymnasts had menstrual irregularities, including amenorrhea, compared to the control group. As well, peak weight velocity (PWV) in girls occurs somewhat later after peak height velocity (PHV) than in boys: about 0.3 and 0.9 years after PHV for girls as opposed to 0.2 and 0.4 years for boys (Malina and Bouchard, 1991). It may very well be that the present premenarcheal rhythmic gymnasts still have to experience their PWV and that their older counterparts are already at the end of theirs.

Canadian elite rhythmic gymnasts (14.9 ± 2.2 years) recorded a lower sum of six skinfolds (50.36 mm) (Alexander, 1991b) than their American postmenarcheal colleagues, but it was higher than the premenarcheal gymnasts. It is not clear what the maturity status of the Canadian gymnasts is, however. If the premenarcheal and postmenarcheal samples of the present study are combined, their sum of six skinfolds (49.16 mm) is not appreciably different from that of the Canadians. The Canadian gymnasts were taller by about 5 cm (1.64 m versus 1.59 m) compared to the combined American sample as well as heavier

(48.5 and 45.1 kg for the Canadian and American gymnasts, respectively). Given a similar proportion of premenarcheal and postmenarcheal athletes in the Canadian rhythmic gymnasts as in their American counterparts, it may be hypothesized that the Canadians were leaner, i.e., they had less fat for height than the American athletes.

It was not surprising to find no significant differences in fat patterning between the premenarcheal and postmenarcheal rhythmic gymnasts, since they are all subject to the same stringent requirements of high performance rhythmic sportive gymnastics. However, except for the medial calf, all skinfold values of the older gymnasts were higher, which is in accordance with normal growth and development (Malina and Bouchard, 1991). As far as relative fat patterning is concerned, it was anticipated that trunk fat would be higher than extremity fat prepubertally after which a similar increase in both trunk and extremity fat would occur (Malina and Bouchard, 1991). Although not statistically significant, the premenarcheal rhythmic gymnasts, recorded a higher ratio for trunk fat (0.611 versus 0.590 for the postmenarcheal gymnasts) as well as extremity fat (0.417 versus 0.384 for the postmenarcheal athletes).

Figure 3 shows the skinfold patterning of the American rhythmic gymnasts compared to their Canadian counterparts (Alexander, 1991b). Clearly, they are very similar in terms of their skinfold distribution, with the Canadians scoring a little higher on some skinfolds and the Americans on others. Typically, if the American gymnasts are divided into pre- and postmenarcheal categories, the former group appears to have lower skinfolds than their Canadian colleagues except for that of the medial calf. The postmenarcheal gymnasts showed higher skinfolds of the triceps and front thigh than their Canadian counterparts (see figure 4). As mentioned before, it is not known what the maturity status is of the Canadian rhythmic gymnasts. However, figure 4 does seem to suggest, that the older rhythmic gymnasts have more fat on almost all skinfold sites compared to their younger colleagues, which would be in line with normal growth and development (Malina and Bouchard, 1991).

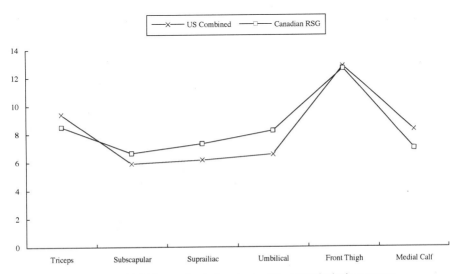

Figure 3. Skinfold patterning in American and Canadian rhythmic gymnasts.

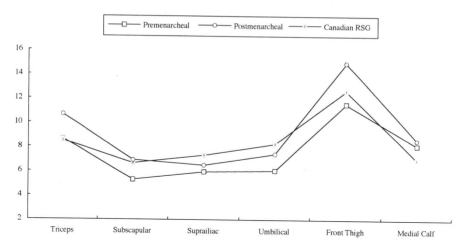

Figure 4. Skinfold patterning in premenarcheal, postmenarcheal and Canadian rhythmic gymnasts.

Even when scaled down to Phantom height, the postmenarcheal rhythmic gymnasts were still heavier than their younger counterparts, which seems to indicate they were subject to the normal effect of maturation (Malina and Bouchard, 1991). From figure 2 it can be seen that there was also a difference in triceps fat which did not reach statistical significance, however. This finding may be related to the small sample size. On the other hand, the difference in subscapular skinfold was statistically significant indeed. The exact reason for this is not clear. It may be that an increase in the subscapular skinfold as a result of maturity is less dependent on sample size than the triceps skinfold. Baumgartner and Roche (1988) found that the subscapular/triceps ratio in girls increased with age from 4 - 15 years, indicating that the subscapular skinfold increases to a larger extent than that of the triceps.

Compared to (West) German elite rhythmic gymnasts, the American postmenarcheal gymnasts were younger (the German gymnasts were 18.6 years of age) and shorter (Germans: 165.6 cm), but similar in body weight (Germans: 51.8 kg) (Rasim, 1982). In terms of individual skinfold thicknesses, the Germans recorded lower values for the triceps (Germans: 5.58 mm), the subscapular (Germans: 4.65 mm), the suprailiac (Germans: 2.98 mm) and the umbilical skinfold (Germans: 5.13 mm) (Rasim, 1982). The difference in skinfold thicknesses may not be related to training, since the Americans worked out more than their German counterparts (24 hours per week versus 13.3 hours for the Germans). The intensity of training may have played a role as opposed to the quantity of training as expressed by hours of training per week. Selection, of course, may yet be another reason for the difference.

The present sample of rhythmic sportive gymnasts as a whole compares favourably with elite artistic gymnasts from other countries. For instance, (West) German artistic gymnasts (14.3 years) were slightly shorter (156.3 cm) and of similar weight (46.1 kg) with matching triceps (8.2 mm) and subscapular (5.6 mm) skinfolds (Malina, Meleski and Shoup, 1982). On the other hand, 1978 French Canadian artistic gymnasts (15.3 years) were shorter (147.8 cm), lighter (39.7 kg) and had lower triceps (6.2 mm) as well as subscapular (5.6 mm) skinfold thicknesses (Malina et al., 1982). When compared to a combined sample of Olympic artistic gymnasts, the rhythmic gymnasts had a lower sum of six skinfolds (49.16 versus 51.05 mm for the Olympic artistic gymnasts) and a similar skinfold patterning (Carter and Yuhasz, 1984). However, the Olympic gymnasts were also older (17.93 years). The postmenarcheal rhythmic gymnasts, on the other hand, clearly had more relative body fat than the Olympic artistic gymnasts which may be related to the nature of the sport: there are no weight-bearing routines in rhythmic gymnastics.

ACKNOWLEDGMENTS

The authors are grateful to Jay T. Kearney, Ph.D., Head, Sports Science Division, United States Olympic Committee and Steven Fleck, Ph.D., Head, Sports Physiology Department. This project was partially funded by a research grant from the United States Olympic Committee. Finally, thanks is expressed to the research assistants at the United States Olympic Training Center in Colorado Springs where all the data were collected and to the second author's (WP) former students at the University of Oregon, Eugene, who also assisted in the data collection.

REFERENCES

Alexander, M. (1991a), A comparison of physiological characteristics of elite and subelite rhythmic gymnasts, Journal of Human Movement Studies, 20, 2: 49 - 69.

Alexander, M. (1991b), Physiological characteristics of top ranked rhythmic sportive gymnasts over three years, Journal of Human Movement Studies, 21, 3: 99 - 127.

Alexander, M. (1989), The physiological characteristics of elite rhythmic sportive gymnasts, Journal of Human Movement Studies, 17, 2: 49 - 69.

Alexander, M., Boreskie, S. R. and Law, S. (1987), Heart rate response and time motion analysis of rhythmic sportive gymnastics, Journal of Human Movement Studies, 13, 9: 473 - 489.

Badelon, B., Boulier, A., Fabre, J., Duvallet, A. and Léglise, M. (1985), Bilan rachimétrique des 250 meilleures G.R.S. (gymnastique rythmique et sportive) au cours du XIe Championnat du Monde à Strasbourg. Étude critique, Médecine du Sport, 59, 2: 41 - 43.

Baumgartner, R. N. and Roche, A. F. (1988), Tracking of fat pattern indices in childhood: The Melbourne Growth Study, Human Biology, 60, 4: 549 - 567.

Carter, J. E. L. and Yuhasz, M. S. (1984), Skinfolds and body composition of Olympic athletes, In: J. E. L. Carter (ed.), Physical Structure of Olympic Athletes. Part II, Basel: S. Karger, pp. 144 - 182.

Case, S., Fleck, S. and Koehler, P. (1980), Physiological and performance characteristics of the 1979 U.S. MRG team, International Gymnast, 22, 4: TS 10 - 11 (Technical Supplement vol 1, no. 2).

Dyhre-Poulsen, P. (1987), An analysis of splits leaps and gymnastic skill by physiological recordings, European Journal of Applied Physiology, 56, 4: 390 - 397.

Frisch, R. E. (1987), Body fat, menarche, fitness and fertility, Human Reproduction, 2, 6: 521 - 533.

Gionet, N., Babineau, C. and Bryant, D. (1986), Anthropometric and flexibility evaluation on young elite rhythmic sportive gymnasts, Canadian Journal of Sport Sciences, 11, 3: 15P.

Heinß, M. (1987), Die Erfassung von Wettkampfübungen mit Hilfe graphischer Zeichen - eine Möglichkeit zur Erhöhung der Objektivität der ästhetischen Bewertung von Leistungen in der Rhythmischen Sportgymnastik, Wissenschaftliche Zeitschrift der Deutschen Hochschule für Körperkultur Leipzig, 28, 1: 46 - 55.

Heinß, M. (1983), Zeichentheoretische Ansätze einer Methodenentwicklung zur weitern Objektivierung der Diagnose von Wettkampfübungen der Rhythmischen Sportgymnastik, Wissenschaftliche Zeitschrift der Deutschen Hochschule für Körperkultur Leipzig, 24, 2: 75 - 86.

Imhof, U. (1986), Telemetrische Herzfrequenzmessungen in der Rhythmischen Sportgymnastik, Schweizerische Zeitschrift für Sportmedizin, 34, 2: 73 - 76.

Kaplowitz, H. J., Mueller, W. H., Selwyn, B. J., Malina, R. M., Bailey, D. A. and Mirwald, R. L. (1987), Sensitivities, specificities, and positive predictive values of simple indices of body fat distribution, Human Biology, 59, 5: 809 - 825.

Lindboe, C. F. and Slettebø, M. (1984), Are young female gymnasts malnourished? European Journal of Applied Physiology, 52, 4: 457 - 462.

Malina, R. M. (1988), Growth and maturation of young athletes: biological and social considerations, in F. L. Smoll, R. A. Magill and M. J. Ash (eds.), Children in Sport, Champaign, IL: Human Kinetics Books, pp. 83 - 101.

Malina, R. M. and Bouchard, C. (1991), Growth, Maturation, and Physical Activity, Champaign, IL: Human Kinetics Books.

Malina, R. M., Meleski, B. W. and Shoup, R. F. (1982), Anthropometric, body composition, and maturity characteristics of selected school-age athletes, Pediatric Clinics of North America, 29, 6: 1305 - 1323.

Plowman, S. A., Liu, N. Y. and Wells, C. L. (1991), Body composition and sexual maturation in premenarcheal athletes and nonathletes, Medicine and Science in Sports and Exercise, 23, 1: 23 - 29.

Rasim, M. (1982), Die fettfreie Körpermasse bei deutschen und japanischen Kunstturnern und Kunstturnerinnen sowie deutschen Sportlerinnen in der Rhythmischen Sportgymnastik der nationalen Spitzenklasse, Leistungssport, 12, 1: 67 - 80.

Ross, W. and Marfell-Jones, M. (1991), Kinanthropometry, In J. D. MacDougall, H. A. Wenger and H. J. Green (eds.), Physiological testing of the High-Performance Athlete, Champaign, IL: Human Kinetics Books, pp. 223 - 308.

Scheele, K., Rettenmeyer, A., Herzog, W., Steinbrück, K. and Weicker, H. (1978), Die organische Leistungsfähigkeit in den kompositorischen Sportarten, Österreichisches Journal für Sportmedizin, 8, 3: 5 - 8.

Shimotaka, H., Tobin, J. D., Muller, D. C., Elahi, D., Coon, P. J. and Andres, R. (1989) Studies in the distribution of body fat: I. Effects of age, sex, and obesity, Journal of Gerontology, 44, M66 - 73.

Yastrjembskaya, N. M. and Pieter, W. (1995), Element complexity: basis for the construction of a method of evaluation in rhythmic gymnastics, submitted for publication.

9

BODY COMPOSITION AND NUTRITIONAL INTAKE IN BELGIAN VOLLEYBALL PLAYERS

Marina Goris, Christophe Delecluse, Rudi Diels, Ruth Loos, and Sigi Celis

Faculty of Physical Education and Physiotherapie
Department of Kinesiology
Catholic University Leuven
Tervuursevest 101, 3001 Leuven, Belgium

INTRODUCTION

Both theoretical considerations and research suggest that the composition of the body may significantly affect performance in sports (Williams 1989). Although research has not revealed a specific percentage of body fat or lean body mass ideal for any given sport, it has provided enough data to make some generalities. In general, research supports the finding that excess body fat may impair performance in sports, where the body needs to be moved rapidly or efficiently, such as volleyball.

An adequate diet, in terms of quantity and quality, before, during and after training and competition will maximize performance. Total energy intake must be raised to meet the increased energy expended during training and maintenance of energy balance can be assessed by monitoring body weight, body composition and food intake.

Therefore the aim of this study is to investigate the body composition and nutritional habits in volleyball players and to compare elite players with players of a lower level for these variables.

METHODS

Subjects

We studied 18 male Belgian elite volleyball players and 22 players of a lower competitive level. All subjects volunteered to participate in this study. The elite players were all professionals and were members of the national volleyball team. They were training approximately three to four hours a day. The players of the lower level trained only three times a week.

Current Research in Sports Sciences, edited by Rogozkin and Maughan.
Plenum Press, New York, 1996

Body Composition

Body composition was assessed by the hydrostatic weighing method with residual volume being measured by the helium dilution technique. Density was converted to percent fat by using Siri's equation (1956): %fat=((4.95 - 4.50)/density)*100.

Skinfolds

Twelve skinfolds were measured with a HERO-caliper according to standard procedures (Brown and Jones 1977).

Nutritional Habits

For seven days of a typical week the players recorded all food, drink and nutritional supplements ingested. They were instructed to eat as usual but to record as accurately as possible the quantity and type of food consumed. The questionnaire was developed and validated by Nutricia. Each diet was coded and analysed by the Becel nutritional computer program. Where possible, each nutrient was compared to the recommended daily allowance (RDA) established by the National Research Council.

RESULTS AND DISCUSSION

General Characteristics of the Subjects

Table 1 details the general characteristics of the players. There is no significant difference for age and BMI between the two groups. The elite players are significantly (p<0.01) taller and heavier than the players of the lower level.

Several researchers reported data of elite volleyball players (table 2). The mean result for height and weight for our elite players are in good agreement with the values of elite players over the last ten years. As can be seen in table 2; the mean height and weight in elite players increased over the years.

Body Composition

The elite players had a mean value of 9.3% body fat, this value is significant (p<0.01) lower than the value in the players of the lower level who had 12.8% (table 3).

The elite players have a significant higher fat free mass than the players of the lower level. It can be concluded that the heigher body weight of elite players is due to a higher

Table 1. Descriptive characteristics of Belgian male volleyball players

Variables	Elite players N=18		Lower level N=22		
	$\bar{X}$	SD	$\bar{X}$	SD	$\bar{X}$ Diff
Age (yrs)	25.6	3.4	27.2	5.9	NS
Height (cm)	195.9	4.9	184.7	7.2	**
Weight (kg)	89.8	5.2	77.8	8.9	**
BMI (W/H^2)	23.4	1.4	22.8	2.2	NS

**= p<0.01.

Table 2. Age, height and weight in elite volleyball players

References	Subjects	N	Age (yrs)	Height (cm)	Weight (kg)
Hirata (1964)	OG '64	116	26.2	183.8	79.0
Medved (1966)				181.5	71.0
Carter (1984)	OG '72	144	24.9	188.5	83.2
	OG '76	130	24.9	189.5	85.5
Hirata (1979)	OG '76			190.0	
Gerard en Lardinoit (1989)	OG '80			192.0	
Puhl et al. (1982)	USA '82	8	26.1	192.7	85.5
Volleyball magazine (1984)	EC '83			196.0	
	EC '87			196.9	
Gérard & Lardinoit (1988)	OG '88	141	25.2	193.8	86.9
Ejem (1991)	OG '88	144	25.2	193.8	86.9
	WC '90	188	24.9	194.2	86.9

fat free mass. Fleck (1983) reported in a study on 12 elite USA volleyball players a mean value of 10.3% body fat, also assessed by the hydrostatic technique.

The elite players have a significantly (p<0.01) higher residual lung volume, while they are also taller. It can be seen as normal.

Skinfolds

There was no significant difference between the two groups for the sum of twelve skinfolds (table 3). This means that the elite players have less internal fat than the players of the lower level.

Nutritional Data

The average calorie intake in both groups is nearly the same 2980 Kcal/day (table 4).

There is no significant difference in the proportion of proteins, fats and carbohydrates. But in both groups the intake of fats is too high (32.5% for the elite and 35.3% for the lower level players).

The mean alcohol intake was significantly lower in the elite players than in the other group, respectively 0.4% and 4.1% of the total energy intake. When the body weight is taken into account, the elite players have a mean daily calorie intake of 33.2 Kcal/kg body

Table 3. Body composition in Belgian male volleyball players

Variables	Elite players (N=22) X	SD	Lower level (N=22) X	SD	X Diff
% fat	9.3	2.3	12.8	3.4	**
Fat (kg)	8.4	2.4	10.1	3.5	NS
Fat free mass (kg)	81.4	4.1	67.7	6.5	**
RV (ml)	1377	172	1166	131	**
Sum 12 skinfolds (mm)	112	41	118	39	NS

** = p < 0.01

Table 4. Seven-day nutritional record

| | Elite players (N=18) | | Lower level (N=22) | | |
	$\overline{X}$	SD	$\overline{X}$	SD	$\overline{X}$ Diff
Kcal/day	2983	844	2970	726	NS
Mj/day	12.5	3.5	12.4	3.0	NS
Protein (%)	16.9	5.2	14.7	2.5	NS
Fat (%)	32.5	6.1	35.3	3.7	NS
Carbohydrate (%)	50.2	8.0	45.9	5.5	NS
Alcohol (%)	0.4	0.7	4.1	1.1	**

**=p<0.01.

weight and the other group 38.2 Kcal/kg. This is for both groups under the recommended value of 45.0 Kcal/kg by Inzinger (1985) for athletes involved in team games with intensive training programmes. Maybe the athletes underreported their food intake or the recommended value is too high. Konopka (1984) and Hemerijck & Vanneste (1987) advise for athletes involved in this kind of exercise a proportional energy intake of 18% proteins, 30% fats and 52% carbohydrates. The values of the elite players in this study are in good agreement with these recommendations.

The elite players have a mean protein intake of 1.36 g/kg body weight/day, and the other players 1.37 g/kg; this is within the range recommended for atheletes, from 0.8 g/kg to 2.0 g/kg.

The elite players have a mean fat intake of 109 g/day, the other group 177 g/day; the carbohydrate intake is respectively 374 g/day and 340 g/day.

The average water intake is 2235 ml in the elite players and 2114 ml in the other group. The daily recommendation is 2500 ml/day or even more for top athletes in team games. Although a player loses significant amounts of water through perspiration it is overlooked as an important nutrient.

Vitamins and Minerals

The players of the lower level have a significantly lower intake for vitamin B1 and B6 than the elite players (table 5). In both groups the intake for vitamin C is far higher than the daily recommendations. In general the intake of the vitamin B is below the recommendations for athletes in both groups.

Table 5. Vitamin intake (mg/day) in male volleyball players

| | Elite players (N=18) | | Lower level (N=22) | | |
	$\overline{X}$	SD	$\overline{X}$	SD	$\overline{X}$ diff
Retinol (vit A)	1.1	0.7	0.9	0.3	NS
Thiamin (vit B1)	1.5	0.4	1.2	0.3	**
Riboflavin (vit B2)	1.8	0.7	1.5	0.5	NS
Pyridoxine (vit B6)	1.9	0.6	1.6	0.4	**
Niacin (vit PP)	16.8	5.9	14.8	5.0	NS
Ascorbic acid (vit C)	152	100	100	44	NS

**=p<0.01

Table 6. Mineral intake (mg/day) in male volleyball players

Variables	Elite players (N=18)		Lower level (N=22)		$\bar{X}$ diff
	$\bar{X}$	SD	$\bar{X}$	SD	
Potassium	4209	1217	3696	941	NS
Calcium	1282	644	1028	381	NS
Phosphorus	2069	674	1827	494	NS
Iron	18.5	4.7	15.1	3.7	**

** = $p < 0.01$

The average mineral intake is in accordance with the recommended allowance (table 6). The elite players have a significant higher iron intake than the players of the lower level.

CONCLUSIONS

This study showed that:

• elite volleyball players are taller, have a higher body weight, less body fat and a higher fat free mass than players of a lower level.
• the average energy intake of elite volleyball players is lower than the estimated energy requirements.

In general the nutritional intake is close to the recommended values but still some players can improve their nutritional pattern.

REFERENCES

Brown W; Jones PRM, 1977, The distribution of body fat in relation to habitual activity, Annals of human biology 4: 537–550.

Carter JEL, 1984a, Physical structure of Olympic athletes, Part II: Kinanthropometry of Olympic athletes, Basel: Karger, 245 p.

Carter JEL (ed), 1984b, Age and body size of Olympic athletes, Medicine sport science 18: 53–79.

Ejem M, 1991, Principal somatic parameters of volleyball players, International volley tech (1): 23–38.

Fleck SJ, 1983, Body composition of elite American athletes, The American journal of sports medicine 11(6): 398–403.

Gérard P; Lardinoit T, 1989, Seoel 1988, leeftijd, geslacht, gewicht en algemeen voorkomen van de deelnemers, Volleybalmagazine 17(6): 8–9.

Gérard P; Lardinoit T, 1989, Seoel 1988, leeftijd, gewicht en algemeen voorkomen van de vrouwelijke speelsters, Volleybalmagazine 17(7): 14–15.

Goris M, 1984, Toepassing van de fysiologische antropometrie in de studie van de lichaamssamenstelling en van trainingseffecten bij jonge volwassenen, 381 p. (leuven: K.U.Leuven; doctoraatsproefschrift lichamelijke opleiding).

Hemerijck N; Vanneste N, 1987, De voeding bij de sportbeoefenaar, Poperinge: RSK, 183 p.

Hirata K, 1966, Physique and age of Tokio Olympic Champions, The journal of sports medicine and physical fitness 6(4): 207–223.

Hirata K; Horvath SM (eds), 1979, Selection of Olympic champions (volume 2), Santa Barbara: University of California Santa Barbara, 420 p.

Inzinger M, 1985, Nutrition training, (Telemundi; London; 24 juli 1985).

Konopka P, 1984, Sport en voeding: prestatieverbetering door een uitgekiend voedingspatroon tijdens de training, Rijswijk: Elmar, 236 p.

Medved R, 1966, Body height and predisposition for certain sports, The journal of sportsmedicine and physical fitness 6: 89–91.

Puhl J; Case S; Fleck S; Van Handel P, 1982, Physical and physiological characteristics of elite volleyball players, Research quarterly for exercise and sport 53(3): 257–262.

Siri WE, 1956, The gross composition of the body, in Lawrence JH; Tobias CA (eds), Advances in biological and medical science, New York: Academic Press, 239–280.

Volleybalmagazine, 1984, Lengte, Volleybalmagazine 12(1): 22.

Williams MH, 1989, Beyond training: how athletes enhance performance legally and illegally, Champaign: Leisure Press, 215p.

10

A COMPUTER SYSTEM FOR ANALYSIS AND CORRECTION OF ANTIOXIDANT INTAKE

A. Ivanov

Research Institute of Physical Culture
Dynamo Ave. 2, 197042, St. Petersburg, Russia

INTRODUCTION

Free radicals have been proposed to be involved in at least 50 diseases. Nutrition/epidemiology studies indicate that antioxidants may be protective against cardiovascular diseases and some types of cancer (Stainberg, 1993). Considerable intake of B-carotene, vitamins A, E, C and polyunsaturated fatty acids(PUFA) reduce cancer death by 13% and overall mortality by 10%. It seems likely that B-1, B-2 and B-6 vitamins may have played a role in reducing the level of these and some other diseases too (Sperduto et al., 1993). Therefore, the problem of an adequate antioxidant supply is important. At the same time the determination of the optimal individual level of antioxidant intake is difficult to ascertain. The nutrients required by an individual is related to a number of factors such as health habits, (smoking and other intoxications), profession, mental and physical conditions. The individuals nutrient needs may be significantly different from the Recommended Dietary allowance (RDA) because the assessment of personal information including diet is necessary. Computer systems are widely used in the epidemiological/nutrition research (Block and Hartman, 1989; Feskanich et al., 1989). The aim of this project was to develop the Antioxidant Computer Program (ACP), computerized method for analysis and correction of antioxidant intake.

MATERIALS AND METHODS

The ACP is the BASIC computer program for IBM compatible computers. It has been developed as a computerized interview tool analysis and correction of antioxidant intake. There are the Russian and English versions of the program.

Antioxidants are an important part of the diet. For the correction of food intake the exact value of individual food and dietary supplement intake and body requirements for energy and nutrients are necessary. Because the computerized interview must give complete, valid and reproducible food data requirements, the program questions (input data)

Current Research in Sports Sciences, edited by Rogozkin and Maughan.
Plenum Press, New York, 1996

61

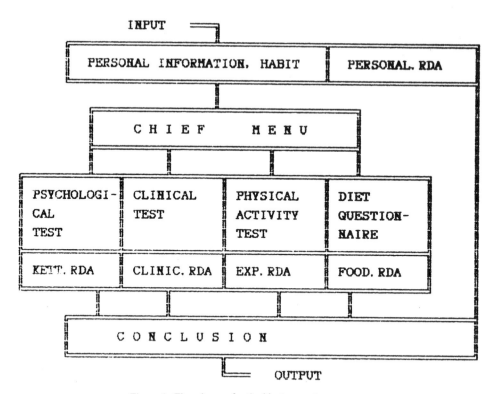

Figure 1. The scheme of antioxidant computer program.

must be understood by non nutritionists and the conclusion (output data) must be useful for practical correction of the diet. Therefore, we have made the ACP with independent blocks for assessment of diet requirements, diet regime and food intake (fig.1).

Each block is a single computer program containing a working program ("exe" and "bat" files) and a data file. Each block maybe used separately but all blocks are related to the Chief menu program. The ACP begins with the personal and habit information and concludes with the final output. Program blocks input results into data-files ("personal.rda", "kett.rda", "clinic.rda","exp.rda" and "food.rda"). The conclusion block processes the data-file information and outputs recommendations onto the screen, text-file or printer. The processing of information from one or more data-files is possible. Full questioning for one block takes about 40 minutes.

Personal Information, Habit

The personal and habit information block includes 12 questions on name, date, sex, age and anthropometric measures (body weight and height). This information is necessary for other blocks of the program and for determination of the recommended level of energy and nutrient intake (RDA). The habit information includes questions on intoxication increasing the entrance of free radicals into the body: smoke, air condition in the region where respondent lives, professional information. The program calculates the RDA and input data in the "personal.rda" file. Also, the program calculates the individual "free radical load" level and assigns it to be low, average or high. We recommend to increase

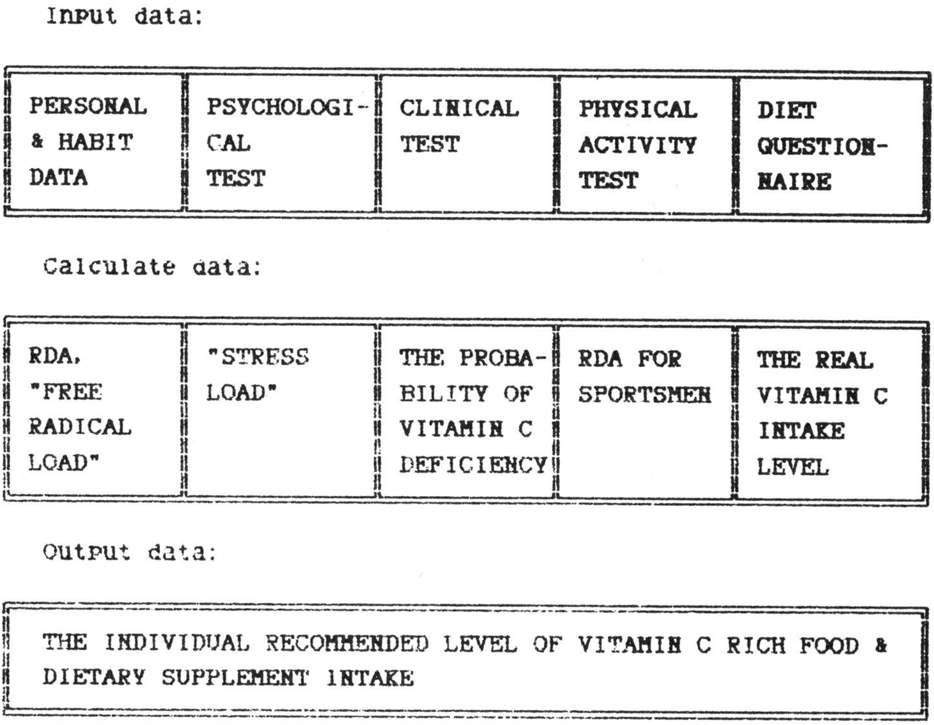

Input data:

PERSONAL & HABIT DATA	PSYCHOLOGI-CAL TEST	CLINICAL TEST	PHYSICAL ACTIVITY TEST	DIET QUESTION-NAIRE

Calculate data:

RDA, "FREE RADICAL LOAD"	"STRESS LOAD"	THE PROBA-BILITY OF VITAMIN C DEFICIENCY	RDA FOR SPORTSMEN	THE REAL VITAMIN C INTAKE LEVEL

Output data:

THE INDIVIDUAL RECOMMENDED LEVEL OF VITAMIN C RICH FOOD & DIETARY SUPPLEMENT INTAKE

Figure 2. The scheme of vitamin C intake correction.

vitamin C in the diet to 200 mg/day when the "free radical load" level is high. The general conclusion is based not only on personal and habit information but also includes data from other blocks (fig.2).

Psychological Test

We have included 105-questions of a psychological nature for the determination of character formation. The first aim of the personality inventory is to value the respondents ability to give valid information. The second aim is to value the "stress" parameter. We have related the "stress" parameter with such characteristics as emotional instability, anxiety, tension etc. We have valued the "stress" parameter in grades from 1 to 10. A high level of the "stress" parameter is from 8 to 10 and leads to the proposal "antioxidant diet" of the respondent.

Clinical Test

About 90 clinical symptoms of nutrient deficiency have been related with biochemical and physiological tests in the previous study (Rogozkin and Ivanov, 1992). Results allowed us to form the knowledge base for the assessment of antioxidant requirements. The knowledge base includes the information on sex, age and body weight and 17 questions on clinical symptoms of B-carotene, A, E, C, B-1, B-2, B-6 vitamins, PUFA and iron deficiency. We also include the determination of non antioxidants (eg iron) because the clinical symptoms are not specific for separate nutrients. The program calculates the probability of deficiency for all nutrients and inputs significant values to the data-file.

Physical Activity

The physical activity block contains 9 questions on life style, profession, sport and the normal day routine. The personal information and life style data allow calculation of energy expenditure. After this a respondent answers the daily routine questions. He must choose a corresponding level of physical activity for a period of every 15 minutes of his usual day. The interview is repeated twice, once for working days and once for holidays. The normal daily routine questionnaire allows calculation of the real energy expenditure.

Diet Questionnaire

We have used the 1-month food frequency method with the value of portion sizes for assessment of usual diet. The questionnaire was developed in the previous study with different age groups(Ivanov, 1993; Ivanov et al., 1993; Steele et al., 1994). The questionnaire includes 69 groups of food and dishes giving more than 90% of the nutrients for the average diet in St. Petersburg. The computer program proposes 5 versions of portion sizes for each food group. The interview may be repeated twice for working days and holidays if necessary. The calculation of average energy intake of 19 nutrients is possible. Data on antioxidant intake excluding selenium is possible but unfortunately we have no full and valid data about selenium content in ready-to-eat meals in the Russian food composition tables (Skurihin and Volgarev, 1984).

RESULTS

The computer system is used for correction of peoples food intake when they begin to undertake physical exercise. The respondents were women who were either obesity or were normal weight and were starting to do medical gymnastics or take part in sport. The results of the computer test were analysed for 201 women aged 20–49 years old. It was shown that energy intake was 34.2% higher than the RDA data ($p<0.01$). Real energy expenditure from the physical activity block was 35.5% higher than the RDA data ($p<0.01$). There was no significant difference between the energy intake and the energy expenditure, but the average standard deviation for energy expenditure value was significant (maximum to 513kcal). It is likely that the physical activity block reflects the normal value of energy expenditure more than the diet questionnaire block reflects the normal value of food intake. Analysis of the clinical symptom block shows that 33.3% of respondents have significant likelihood of vitamin C deficiency. At the same time, 22.0% of respondents have an intake of less than 60 mg per day of vitamin C from the diet questionnaire. 13.2% of participants have other characteristics of antioxidant deficiency and only 0.3% of them have a sufficient intake of these nutrients.

CONCLUSION

The computer program processes results from all the data files and calculates the recommended level of antioxidant intake. The program then gives a conclusion including the recommended intake level of common food groups (17 groups) and recommendations on dietary supplement intake. The special "antioxidant diet" can be included, if it is necessary. It contains the high quantity of antioxidant-rich food, plant protein rich products and "antistress" dietary supplements.

REFERENCES

Block, G., Hartman, A., 1989, Issues in Reproducibility and Validity of Dietary Studies, Am. J. Clin. Nutr. 50: 1133–1138.

Feskanich, D., Sielaff, B. H., Buzzard, I. M., 1989, Computerized Collection and Analysis of Dietary Intake Information, Computer Methods and Programs in Biomedicine, 30: 47–57.

Ivanov, A., 1993, Dietary Intake in Women of St. Petersburg, International J. of Sports Medicine, 5: 295.

Ivanov, A., Pshendin, A., Rogozkin, V., 1993, The Assessment of Food Intake for School Children in St. Petersburg and Its Region, Abstracts of the European Meeting on Current Research into Eating Practices, Potsdam: 42.

Rogozkin, V., Ivanov, A., 1992, Relation Between the Intake of Vitamins as Calculated from the Cross - Check Dietary History Methods and the Relative Amounts of These Vitamins in Blood and Urine, Abstracts of the 4th Meeting on Nutritional Epidemiology, Berlin: 16.

Skurihin, J. M., Volgarev, M. N., 1984, Book 3: Composition Tables of Basic Nutrients and Energy Value of Ready-to-Eat Meals and Culinary Products.

Sperduto, R. D. et al., 1993, The Linxian Cataract Studies. Two Nutrition Intervention Trials, Arch Ophthalmol. 111: 1246–1253.

Stainberg, D., 1993, Antioxidant Vitamins and Coronary Heart Disease, N. Engl. J. Med. 328: 1487–1489.

Steele, M., Utenko, V. et al., 1994, Dietary Nutrition Intake Comparisons for Rural and Urban Russian Boys, Ages 6, 9 and 15 Years, Living in St.Petersburg and Surrounding Areas, Am. J. Human Biology 6: 153–159.

11

MENTAL TRAINING FOR SPORT AND LIFE

L.-E. Uneståhl

Scandinavian International University
Sommarrovagen, 6, S-701.43, Box 155, Örebro, Sweden.

INTRODUCTION

1. 65% of the top athletes in Sweden now use forms of mental training and mental preparation and a similar investigation in 1980 showed 30% used this. Large variations exist between various national teams and individual sports use more mental training than team sports. The Swedish Model is characterized best in sports like tennis, table tennis, golf etc.
2. Mental Training programs have so far been used by over two million Swedes, which translates to 25% of the population. This figure covers various frequencies in usage from single sessions to many years of regular training.
3. All school children are supposed to receive basic mental training. Sweden was the first country (and is still the only one), who has included basic mental training into the curriculum, statewide. For the future this will mean that every Swede will have experience of mental training.
4. Mental training programs (muscular relaxation, self-hypnosis, self-image training, motivation, sleep) can be acquired by everyone from every pharmacy in Sweden.
5. The most common areas, where people come in contact with mental training programs are: Schools, Sport, Military, Personal growth, Personal development, Management training, Health care, Behaviour change (smoking, weight, sleep) and Psychosomatic problems.
6. Sweden was the first country that founded a National Society for Mental Training. It consists of a main body and six special sections: Sport, Business and Public administration, Education, Health care, Art and Stage performance, Research.

RESEARCH BACKGROUND AND METHODOLOGY OF MENTAL TRAINING

The systematic training of mental skills in sports and life are based on two areas of research:

Current Research in Sports Sciences, edited by Rogozkin and Maughan.
Plenum Press, New York, 1996

67

    I. Investigation of a state of awareness and consciousness, appropriate for optimal change and growth.

   II. Identification of the right content of mental training. What skills are important for Peak-performance and Wellness in Sport and Life?

Following is a short summary of some research findings from the sixties. The research is described in 46 research reports from Uppsala University and summarized in the book: Hypnosis and Posthypnotic suggestions (Uneståhl, 1973, 1992).

1. Regular, systematic and long-term self-hypnotic training was superior to hetero-hypnosis in a variety of measured dimensions.
2. Audio taped hypnotic inductions were as effective as inductions given by a present hypnotizer, measured on a standardized scale of hypnotic susceptibility (the Stanford scales).
3. Long-term imagery training gave a significant increase in imagery skills, measured by standardized scales for imagery vividness and control.
4. Long-term training in relaxation and imagery gave a significant increase in hypnotic skills, measured by the Stanford scales of hypnotic susceptibility.
5. Hypnotic alterations of bioelectric, cardiovascular, respiratory, vasomotor, gastrointestinal, endocrine and metabolic functions were larger and more precise in comparison with non-hypnotic alterations where the same techniques were used.
6. Relaxation is a common and sometimes the basic element in hypnotic induction procedures. It is also a common feature of the hypnotic experience. In spite of this, however, relaxation is neither a necessary element in induction procedures nor is it a necessary dimension of the hypnotic state.
7. Hypnotic susceptibility scores had a significant positive correlation with ideo-motor skills, but a zero-correlation with secondary suggestibility (gullibility).
8. Neurodynamic markers and biochemical correlates of the processes of mental relaxation (Bundzen, Uneståhl, 1994).

Mental Skills for Peak Performance

Terms like success and progress are more easily defined in sports, compared with life in general. Thus, sports were the first area of investigation to find the relevant dimensions behind good performance. The national teams in shooting, track and field athletics, swimming, judo, skiing, handball and soccer were investigated and compared with athletes and players of a lower caliber. (Uneståhl 1973 - 1992). The 4 factors showing significant differences were:

1. Self-image (Self-confidence, self-evaluation, self-esteem);
2. Goal-images (Knowing where to go and to feel committed to the goal(s));
3. Attitude (Reality-testing and reality-interpretation);
4. Control (To identify, produce and control mental skills like relaxation, feelings etc.);

IPF - The Ideal Performing Feeling

Each one of these four factors were examined in order to find the optimal contribution to IPS. The athletes' descriptions of the ideal feeling had for instance remarkable similarities with a hypnotic state.

PRINCIPLES OF INDUCTION

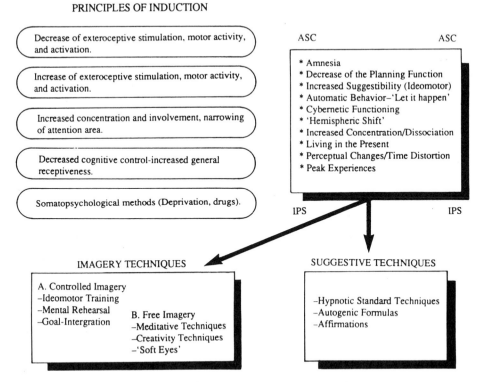

Figure 1. Common dimensions in ASC and IPS.

Mental Skills for Wellness

Similar studies during the 1980's of people with various length of yearly absence from work due to illness and patients (for instance cancer patients) with different lengths of illness showed that the same four areas were important for Wellness and for IHS - the ideal healing state (Unestähl 1980–89). In the same way as in the investigations about IPS, the four dimensions have been isolated and analyzed in order to find each factor's contribution to the total effect.

Mental Skills Training

Based on these studies of peak-performance the Inner Mental Training (IMT)-programs were developed, evaluated and modified during the 1970's. Field studies and laboratory experiments were carried out on the whole training system as well as on single parts of the training. When exploratory studies pointed to similarities between IPS, ILS (Ideal Learning State) and IHS, the IMT-programs began to be used in schools, business and in clinical settings in the same way as they had been used in sports.

The training programs were always self-instructional and did not require expertise in order to be used. They were "experiential" in nature and based on a combination of learning common principles and individual applications due to the needs of the individual and the demands of the situation.

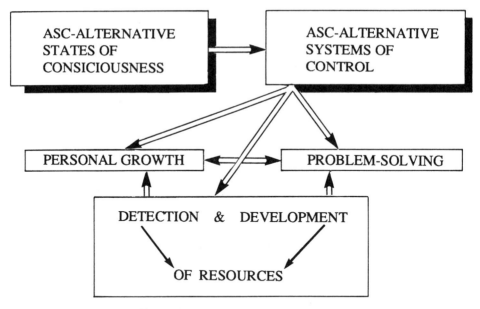

Figure 2. ASC in growth and problem solving.

Definitions

Inner Mental Training (IMT) is defined as a systematic, long-term and development training of mental skills and attitudes aiming to Peak-performance and Wellness. Inner Mental Preparation (IMP) is defined as pre-trained mental skills and procedures, which are intended to become effective at certain pre-decided occasions.

Phases

IMT is divided into three phases, which are trained in the following order:

A. *Mental Conditioning.* Learning muscular and mental relaxation and a special state of mind, an alternative state of consciousness (ASC) which gives the base for control and positive change.

B. *Mental Technique Training.* Learning alternative systems of self-control and self-directing techniques like suggestions, autogenic formula or imagery techniques, all of which are more effective if combined with ASC.

C. *Mental Strength Training.* The mental skills in A and B are combined and applied to areas like motivation, emotional reactions and mood states, attitudes, concentration etc.

The Philosophy of Mental Skill Training

1. The principles of Peak performance and Wellness can be explored, controlled and developed.
2. Mental skills should be looked at and treated in the same way as physical skills.
3. Human growth and development is directed by trainable factors in the human mind.

4. The sports related growth model (SGM) starts with visions, missions and performance goals.

5. SGM neglects the history, accepts and enjoys the present and is oriented to and energized by the future.

6. SGM sees human beings more as a cause to the future than as a result of the past.

7. IPS, IHS and the ideal reality state (IRS) are most easily reached, maintained and developed through a hypnotic-like state, where the reality-testing is weakened.

From Shrink and Negative Sport to Positive Sport

The traditional applied Sport Psychologist is often looked upon as a shrink, the expert in problem shrinking. This is well in accord with the old role of a coach, that of a fault seeker and problem solver. So far, the main reason in many countries for asking a Sport Psychologist to work with athletes or teams have been for fault seeking and problem corrections. This is in agreement with the common models in society where problem solving approaches are much more common then preventive or growth models.

There are numerous studies, articles and books written about negative feelings such as anxiety, aggression and depression. In comparison astonishingly little is written about positive feelings. Scientists have almost been ashamed to look into areas like happiness, joy and exhilaration. The same is true of news media. The constant selection of negative news may bring about dangerous illusions of reality together with paralysing feelings of helplessness.

Unfortunately, this has also been true of Sports Psychology. Compare, for instance, the number of studies about anxiety in Sport with those about peak-experiences. The future Sports Psychologist, however, (like the future coach) will hopefully be looked upon as a stretch, an expert in expanding human capacities and capabilities. This human potential approach is in full agreement with the main purpose of mental training. Mental Training is not primarily a clinical method designed for athletes/persons with problems. Mental training is offered as a general method for everyone who wants to improve and grow as an athlete or as a human being. This educational approach, however, will also have important contributions to clinical settings in order to prevent problem-fixation and institutionalization. This also brings sports psychology and mental training into the frontline for a change towards a more positive society with emphasis on opportunities and possibilities instead of barriers, problems and obstacles.

The Use of Alternative States of Consciousness (ASC-1) in Learning Mental Skills

Mental relaxation or self-hypnosis is an essential and important part of the basic mental training. Alternative systems of control (ASC-2) as well as Alternative systems of Change (ASC-3) and growth are both dependent on self-hypnosis in order to become effective. Thus, basic mental training is a necessary base before learning mental skills like concentration, mental toughness, motivation etc.

Learning Self-Hypnosis

Different ways of learning self-hypnosis can be distributed along two axes, degree of outside guidance and degree of obtained self-control.

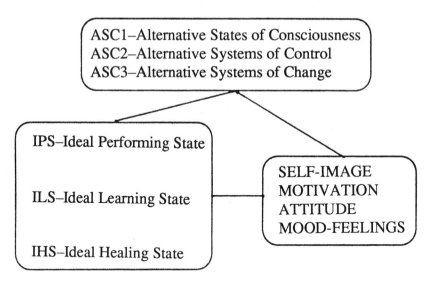

Figure 3. Principles of ideal states.

Mental Skills Training in Various Groups

Early Life-Skills Training. The research project Relaxation in Swedish schools led in 1980, after 6 years of research, to the inclusion of basic mental training in the Swedish school curriculum. During the 1980's a mental training program, called the seven C's, has been evaluated in various age-groups. The seven C's contains seven child-like capacities: Confidence, Concentration, Control, Calmness, Commitment, Creativity and Cheerfulness.

During 1988 a program for a complete life-skills training in schools was worked out and is now under evaluation. It starts during the first year of elementary schools and finish during the last year of high school.

The contents are as follows:

Year 1.   Basic Mental Training (Muscular and Mental Relaxation, Triggers)
Year 2–3. Mental Training (The seven C's)
Year 4–6. Training of basic Health habits (Australian Health Development Foundation)
Year 7–8. Communication skills, Relation training, Teamwork
Year 9–11. Decision making training (Leon Mann, Australia)
Year 12.   Leadership- and Parentship-training

Military. Construction and evaluation of a mental preparation training program applied to the Swedish male population during the military service started during 1988. The purpose is to continue and develop the earlier school training.

Business. A large number of Swedish companies (Volvo, SAS, SKF, IKEA, Unilever, Telecomp., etc.) have gone through a mental training program specially designed to their specific needs. Typically, the program started with the individual development (relaxed efficiency and increased work/life satisfaction) as a base for better team spirit and followed by organizational growth. The training, which aims for lasting and long-term effects, was intended for everyone in the company. In order to reach everyone on all levels a large number of supervisors or mental trainers were educated.

University. In many countries and in many disciplines there exist unfortunate gaps between science and application. Universities are often critisized for their isolation from relevant questions and needs in society. Discussions about education are often more directed to administrative issues than to overall goals.

Swedish Universities statues state personal development as the overall goal for all university education. They also state that the two major avenues towards this goal are knowledge and skills. So far, however, very little of skill-development has been included in the courses. This was one of the reasons a 2-year part-time university course, called "Personal development through mental training" was started at Örebro University in the fall of 1987. Today more than 1500 students are taking this course, which are provided to people all over Sweden with the help of modern technology (video, data etc.).

This home university idea is based on certain educational principles, related to mental training. Examples are the following:

1. *The law of state bound learning.* The law says that the learning process is connected and often conditioned to the external environment, to the internal mood state and to the characteristics of the learning situation.
2. *The Zeigarnik effect.* It says that as all learning is goal-related, the type of goals (learning for exam or for life) will decide the long-term recall as well as the integration and application of the knowledge.

Both of these laws, of course, speak for a home - or work-university concept, especially in mental skills learning.

This distant learning course consists of 4 parts:

1. Literature selected from several disciplines
2. Videos with lessons, interviews, demonstrations
3. Training programs for 5 days a week training
4. Continuous measurements of personal growth as an individual feedback

The following training programs for learning mental skills are practised during the course:

1. Muscular relaxation I - II
2. Mental relaxation
3. Self-hypnosis
4. Imagery skill training
5. Visualization
6. Mental rehearsal
7. Affirmations
8. Self-image training
9. Goal-settings
10. Goal-integration
11. Concentration training
12. Mental toughness
13. Desensitization

A

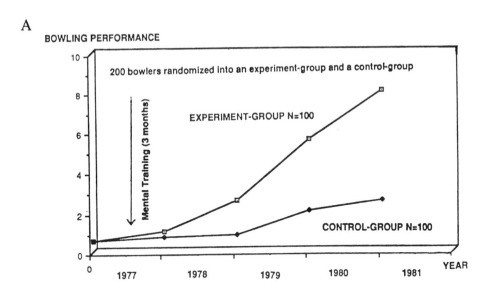

B

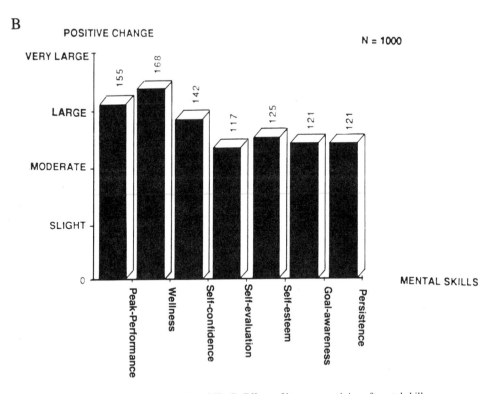

Figure 4. A. Effects on bowling skills. B. Effects of long-term training of mental skills.

14. Problem solving
15. Creativity training
16. Ideomotor training
17. Mind-Body training
18. Behaviour change
19. Habit formation
20. Attitude training
21. Communication skills
22. Team training
23. Humour training
24. Inner joy
25. Life quality

EVALUATIONS

N = 1000. Measurements before and after 1 - 2 years of systematic training of mental skills. 50 training areas. Minimum training time per program is 1 week. Some results are shown in Figure 4.

CONCLUSION

In a time where the benefits of sport are questioned by an increasing number of people it should be important to point to the role of sport and Mental Training (MT) as a device for personal growth or as a model for life. Here are some reasons why sport and MT ought to be a model for society.

1. Athletes, especially in individual sport, is like MT directed by a developmental model.
2. Thus, sport and MT emphasizes growth related interventions while society is mostly based on problem based interventions.
3. Society reads trends and makes history-based guesses about the future while sport and MT emphasizes creative solutions without historic bases.
4. The term "Future now" implies that future and present self-images can and must co-exist in a parallel and change-driving way.
5. As growth, progress and success is more easy defined in Sport it also serves as a model area for Life, where goals also need to be specific and measurable.
6. The SGM (sport related growth model) is complemented with a training model, where everyone can train those physical and mental skills, which are important for Peak-Performance and Wellness.
7. The search for alternative systems of self-control ought to be guided by the principle that the term "impossible" could not be used about the future.

REFERENCES

Back, E. & Uneståhl, L.-E., 1989, Immunologiska förändringar genom posthypnotiska suggestioner (Immunological changes through posthypnotic suggestions), Örebro University.

Breife, S. & Uneståhl, L.-E., 1982, Effekter av 3 månaders Mental Träning i bowling (Effects on bowling skills under and after 3-months of Mental Training), Gothenburg and Örebro University.

Bundzen, P. & Uneståhl, L.-E., 1994, Thinking with images in alternative state of consciousness - neurodynamic markers, Inter. symposium "Physiological and biochemical basis of brain activity", St.Petersburg :15.

Ljungdahl, L. & Uneståhl, L.-E., 1989, The use of systematic humor training in chronic pain patients, Örebro University, JAMA.

Setterlind, S. & Uneståhl, L.-E., 1983, Introduction of relaxation training in swedish schools, Gothenburg and Örebro University.

Schmid, A. & Peper, E., 1983, Do your things when it counts, In: Uneståhl, L.-E. (Ed.).: The Mental Aspects og gymnastibs, Veje Publ.

Uneståhl, L.-E., 1973, Hypnosis and posthypnotic suggestions, Veje Publ.

Uneståhl, L.-E., 1975, Hypnosis in the seventies, Veje Publ.

Uneståhl, L.-E., 1986, Sport Psychology in Theory and Practice, Veje Publ.

Uneståhl, L.-E., 1986, Contemporary Sport Psychology, Veje Publ.

Uneståhl, L.-E., 1989, Effects of systematic and long-term training of Mental Skills, Örebro University.

Uneståhl, L.-E., 1992, Mental Skills for Sport and Life, In: "Sport and Health", St.Petersburg :29–36.

12

PROFESSIONAL SPORT AS SUB-CULTURE

N. I. Ponomarev

Academy of Physical Education
Decabristov St. 35, 190121 St. Petersburg, Russia

INTRODUCTION

Aim of research - description of modern professional sport as a sub-culture phenomenon. Sub-culture can be thought of as the culture of social community of a group with common specific features. The conception of sub-culture helps us to learn the process of culture differentiation and integration including physical training and different types of athletic activity.

Research methods include: theoretical analysis and the use of knowledge of modern phenomena for the interpretation and reconstruction of the events and activity of the past.

RESULTS AND DISCUSSION

1. Professional Sport Has Its Own Features

In sport, the criteria of professionalism is the presence of working relations between the sportsman and the employer, fixed in the form of a contract that regulates the relations between the employer and professional sportsman. This is the main and defining feature. The results of a professional sportsmans activity can be regarded as an article of trade and the professional sportsman is the only person for whom sport is his only type of labour activity ie his profession. Professional sport is itself the product of modern society and its culture and its type of business is an important part of industry. Special legislative documents approved in some Western countries can be the evidence of this fact. For example the Department of Trade in the USA refers to professional sport as an industry in their classification of business types. Professional sports and their clubs are commercial organizations and professional sport itself is a branch of business which follows economic laws. In the USA professional boxing and football are commercial clubs and leagues exists. The manpower market ie sale of individual sportsmen and teams are legalized and provisions intended to regulate the rights of professional sportsmen are implemented. In professional sport there operates a "system of limits" which is the way to freeze wages of sportsmen and prolong the period of contract validity for an unlimited period time. The "system of

Current Research in Sports Sciences, edited by Rogozkin and Maughan.
Plenum Press, New York, 1996

77

limits" represents the set of provisions and codes of professional leagues intended to limit the sportsmen activities and opportunities in the period of hire and presénce in the clubs. In some countries the professionals form a trade union. The professional sportsman creates his goods in the form of record and results i.e. place gained in competition or as a service i.e. an athletic show. As written by J. Loy(1972) the famous American sociologist "may be sport is the ideal testing sphere for the test of many social theories". The problems of professional sport of the USA are widely expressed in modern American sociological literature and also in publications of Russian authors (Serebryakov, 1977; Guskov,1983). The economic policy of professional clubs and leagues in its dependency from the antitrust laws of the USA take a special place in literature about professional sport. From dissertations concerning the economic aspects of professional sport the work of G. Ross "Essays on the Economics of the Professional Team Sport Industry" written in New-York University in 1974 should be noted. In this work the analysis of economic questions, the structure and the machinery of manpower in professional sport is analysed. The results obtained by G. Ross show that the "system of limits" and the admission of sportsmen to clubs does not change the distribution of players in comparison with that existing in the absence of such rules. In practice such limits allow the club owners to get the major part of the profit created by the players, and at the same time reduces the cost of the sportsmen themselves. The conclusions of G. Ross's dissertation proves that the profits of different "internal" relations that exist between the club and the company to which it belongs and also in such cases when the club owners put on a mask on the profit, by writing off considerable amount for their own salary.

2. Professional Sport Is Complicated and Not Simple

The positive features of professional sport are: the search of new methods of training, selection of athletes, organization and sale of industry. The non-training effective means widely used are those such as: personal hygiene, rational daily routine, special diet, traumas prevention and additional measures for efficiency of recovery i.e. massage, hydrolic treatment, sauna and heat chambers, treating with ultra-violet light, toned up air, means of physiotherapy and psychology, special complexes for fast adaptation of sportsmen to extreme conditions i.e. hot climate, high altitude, fast change of climate and time zones and rehabilitation measures after injures and disease. The Russian authors Yu N Vavilov and A P Laptev (1990) wrote that the training of a professional is to be implemented under the leadership of an experienced coach who is engaged only in working with sportsmen. All the other functions i.e. day to day organization, financial concerns etc are to be implemented by a manager. The coach continually raises his professional skill by independent work, course studies, consultations with experts etc. The professional coach knows his sportsmen very well and seeks to attain high sporting results. In terms of training, the main stages are fixed without details but training is implemented creatively with professional improvisation. In this case the coach is emancipated and assumes the responsibility. The professional coach is given confidence and his activity is evaluated only with final specific sport results during the period of his work. As a rule professional sportsmen are trained in a decentralized way. It gives the chance of individualizing the training process greatly, to provide normal conditions of life in a family, to reduce the psychological tension, to create the best conditions for training and rehabilitation to improve training for competitions.

3. At the Same Time Professional Sport Is Connected with Several Negative Features

The main negative feature is the slogan "Victory at Any Price is the Only Aim" is wide spread. Professional sport is an area where motivation and the urge to win coincide and this is characteristic of any professional activity. The sporting result is regarded mainly as the means of attaining material welfare and maximal profit. It is not by chance that the French scientist I. Dumazedier (1973) ascertains that the gap between the sporting idea formulated by P. Kuberten and the actual sporting activity in modern society increases. The practice of professional sport testifies that it influences the world sport movement, the Olympic movement, as the phenomen common to all mankind. This influence has the following branches: The intensified commercialization of activity as an attribute of professional sport leads to loads that sometimes exceed real biological human resources, to unhealthy excitement of fans growing to vandalism and nationalistic extremism. In such circumstances the sanity, education and cultural functions of sport are reduced. An extremely negative procedure in the activity of professional sportsmen is the use of stimulating remedies such as anabolic steroids, corticosteroids, autohemotransfusion, ß-blockers, injections of vitamins etc. These not only ruin the health of sportsmen but also create inequality in competition and break the basis of competitive relations between sportsmen. Another factor of anxiety is that children and youths are engaged in professional sport and their training leads to physical and spiritual development loss. Sometimes in professional sport the social and biological requirements of motor activity do not coincide. This makes sport management a cultural phenomenon. The health status of persons that finishing active sports prematurely arouse concern. The emotional and prestigious nature of professional sport, the reaction of spectators and of mass media assumes the occurrence of disputed situations. The display of such situations depends on the level of the sportsman's moral consciousness, the causes of his activity and cultural surroundings. Behaviour beyond the rules is the show of so called misadaptive aggression that characterizes the immature. Distribution of money as payment can be one reason for conflicts and aggressive behaviour. Commercialization, as a main feature of professional sport, substitutes human development for the production of services and profit and in general in breaks the fundamental values and functions of the sporting movement. In such a case, sporting victories do not reflect efforts directed for human perfection but for efficiency of the training system for champions and record-breakers. However professional sport has the possibility of developing human aspects connected with the Olympic movement. This can be in the form of purposeful activity. Concerning professional sport itself, it is a social defence for sportsmen providing pensions, social insurance etc. Humanization of relations with professional sportsmen strengthens the idea of "sport for everyone". In the light of the above mentioned, it is correct to put a question about the creation of Humanization Committees of professional sports attached to International Olympic Committee, National Olympic Committees and International Sport Federations.

4. During the Last Few Years, Professional Sport Has Begun to Develop and Was Legislatively Approved in Russia

One should remember that it is difficult to find another branch of human activity that can arouse so many emotions as sport. It is an economical, social and cultural phenomenon and at the same time it is the method of creative work and human improvement.

One of the most important aspects of its development is the solution of the professionalism problem. It is closely connected with the future of sport as asocial institution of society. At the same time it is so complicated that it effects almost all sides of sport i.e. financial, ideological, social, educational and cultural. If properly approached, the professional sport will represent hardly any social, ideological or financial danger for the subsequent development of the country. On the contrary it has several advantages that should be carefully analysed and evaluated. As the marketable relations are being developed in Russia, the commercialization of the highest achievements in sport will grow continuously. Public, amateur and professional sport is required in the country. The question of commercialization and the criteria for its admissibility becomes acute and topical. Here, social justice should become the starting point. Being independent organizations the sport clubs have social agreements with the leaders of the local town about the conditions of its appearance in public, including the provision with gyms and other sport sites and the distribution of profit received from competitions etc. Such sports clubs can be regarded as an important source of financial receipt in the development process of a town or even region. The money received in such way can be used the development of the social atmosphere. The modern economic and political situation made it possible for professionalism to appear in sport of the country, i.e. the work of sportsmen in foreign clubs according to the contract, the use of commercial advertising and sponsors. The main problem is the creation of a social management system in the range of physical culture and sport legislation. Professional sport should not violate moral and ethical norms and laws. It is necessary to make arrangements in order to neutralize the negative aspects of professional sport especially in use of drugs. Attention should be paid to health of professional sportsmen.

REFERENCES

Dumazedier, I., 1973, Sport and Activities, Int.Rev. Sport Soc. 9:37–42.
Guskov, S. I., 1983, Review of Social Research in USA Theor. Pract.Phys.Cult. 1:45–47.
Loy, J. W., 1972, A Case for the Sociology of Sport, J.Health Phys. Educ. Recr., 43:17–24.
Ponomarev, N. I., 1978, About System Analysis of Sport Int.Rev.Sport Soc. 1:34–39.
Ponomarev, N. I., 1980, Sport as a Show. Int. Rev. Sport Soc. 3:44–48.
Ponomarev, N. I., 1981, Sport as Society, M.Progress, :180.
Ross, C., 1974, Essays on the Economic of the Professional Team Sport Industry, The City Center of New York.
Serebryakov, A. B., 1977, Problems of Professional Sport in Subjects of Dissertation in USA Theor. Prac.Phys.Cult. 7:58–60.
Vavilov, Y. M., Laptev, A. P., 1990, Peculiarities of Professional Athletes' Performance. Theor.Prac.Phys.Cult. 7:34–36.

13

PSYCHO-PEDAGOGICAL ANALYSIS OF INDIVIDUAL STYLE OF OPERATIONAL PROBLEM SOLVING

A. V. Rodionov

Russian Research Institute of Physical Culture
Kazakova Str. 18, 103064, Moscow, Russia

INTRODUCTION

This research refers to the category of "psycho-pedagogical" studies because it deals with the problems of so called "psycho-tactical" training of athletes. Tactical training is certainly a pedagogical problem, but tactics based on operative decision making (DM), and decision making regularities is mainly a psychological problem. Here, as in no other complex problem, pedagogical and psychological phenomena blend. The process of operative problem solving chiefly depends on the structure of individual technico-tactical skills, formed on the basis of special physical and psychological qualities, personality traits and psychophysical features. These qualities are considered to be the core of individual style of activity (ISA) formation (Vjatkin,1993). Individual style of activity is the result of a person's interaction with an object and social environment and it is an inclination to original operative problem solving. An individual style clearly manifests itself in sports, games and combats:

- they proceed in non-stationary conditions,
- decisions to evaluate the situation are made under time limits,
- activity proceeds with the necessity of solving tasks in the face of changing situations which accompany the fight (Rodionov, 1973)

Many decision-making components are displayed in these kinds of sport: 1) analysis of the opponent's actions; 2) assessment of probable actions; 3) reflex reactions; 4) operative image construction and transition from the cognitive level to actions. In this paper, we present results of research in basketball and volleyball.

METHODS

Basketball players were of the age range of 12–14 and 15–16 years, and the volleyball players were 17–18 years old. In the first instance, we observed the beginning of indi-

Current Research in Sports Sciences, edited by Rogozkin and Maughan.
Plenum Press, New York, 1996

81

vidual style, in the second, the most important part was its ending, and in the third, of the style in all phases of activity were observed. Research methods in all three series were the same: personality traits and psychodynamic features; diagnosis of perceptive, psycho-motor and cognitive qualities of athletes; recording of tactical operations performed during competition.

RESULTS AND DISCUSSION

The results of three independent experiments are presented in this work. Three groups of subjects differing in age, level of sport participation and sport specialization. The first experiment, with 12–14 years old basketball players was carried out together with O I Lukonina. It was hypothesised that the age of 12–14 years is advisable for the beginning of a basketball player's specialization according to her physical characteristics and she is about to form an individual style of playing. Measurements of young basketball players of 12–14 years (n=17) and junior players of 15–16 years (n=18) were compared. The latter served as an "initial model." Assessment of personality factors by an adapted 14-factors Kettel test revealed that 12–14 year old players were characterized by such personality traits as intellect, egocentrism, anxiety, frustration. It is necessary to point out that a temperament such as sensitivity got the highest mean score (82.4%). A high level of self-discipline was revealed in athletes of both groups and is apparently connected with the importance of this trait for team sports activity even at the age of 12–14. The effect of specificity of sport activity and social experience may explain the fact that the girls of 15–16 years are less different according to this trait. No significant difference was found between the two groups in all variables according to comparative analysis of data (t-test). Only in the variable of dominance was a significant difference was observed, which indicates a greater importance of this factor to the adolescent basketball players. It is possible that the structure of the group or team changes with the age, but the separate personality traits are formed earlier (before the age of 12). The Lusher subtest "The Main Colours Table" was suggested to reveal individual personality patterns of young basketball players. The technique of testing and the procedure of the data analysis are described in the work of L.N.Sobchik(1990). As a result of testing the information about each athlete's predisposition to a definite colour set was obtained. Interpretation was carried out according to structural meaning of the colour patterns and pairs depending on their functional distribution, connected with the preference and avoidance. Preference of the red colour was traced in basketball players. This points to the expression of vigour, nervous and harmonious activity, strive to success.

Comparative analysis of psycho-vegetative activity of middle and senior aged basketball players revealed a predominance of ergotrop syndrome in frames of vegetative balance in both groups and the higher level of mental working capacity in the group of senior players (x=9.3) in comparison with the middle group of basketball players (x=10.1). Though no significant difference according to Mann-Whitney criterion was found (U=89.5; p>0.05). Psycho-vegetative activity coefficient and mental working capacity variables correlate at the level of r =-0.54. They characterize different psychological mechanisms. Psycho-vegetative activity coefficient correlates with the anxiety test according to Lusher test (r=0.54; p<0.05). This activity is often a specific manifestation of anxiety. The latter is connected with the ability of self-regulation (r=-0.65; p<0.05) and those athletes who experience high anxiety can not manage their states both because of inadequate evaluation of the situation and their actual state. The following types of individual styles of playing activity, conditioned by neurodynamic properties and displayed in the structure of the means of operative task solving during the game, were distinguished: a) non-insurant, improvisatory type, characterized by high lability of nervous processes, pronounced psychophysiological reactions on training and competitory loads; b) insurant algorithmic type, characterized by less lability and by a balanced and mobile nervous system, by adequate psychophysiological reaction to loads; c) non-insurant algorithmic type, characterized by mobility of nervous processes, taking the middle position between the upper two groups in psychophysiological reactions. In playing activity the non-insurant improvising type are distinguished by high searching activity for playing actions, by quick adequate shifts from one action to another according to the changes in playing situation, by improvisation in positional attack, by disposition to original task-solving in attack and defence. Those with insurant algorithmic style base their interaction on the system of expectation of the opponent's actions in defence, they do not shift effectively enough under changing playing situations. Non-insurant algorithmic style of playing activity in young basketball players distinguishes itself by its disposition to base decision making on a subjective probability model by an inclination to habitual actions without creativity. Thus, the specificity of the playing activity in basketball optimizes conditions for mental and personality development. These traits and properties influence the style of the playing pattern of young basketball players. Individualization in the training of young basketball players may uncover the potential possibilities of the athlete on the basis of pedagogical intervention for the sake of necessary traits and quality development, adequate management and technico-tactical skills enhancement. All this can provide effective competitive activity. Exercises which promote formation of an adequate style of competitive activity are differentiated according to the following factors: 1) assistance - opposition: 2) preconcerted - unprepared actions; 3) standard - improvised playing task decisions; 4) limited and unlimited situations. A complex of exercises enhancing the development of mental qualities and formation of tactical skills were developed for every group of individual style of playing activity.

The second part of this work was carried out together with M.J.Kim. 46 volleyball players (17–18 years old) served as subjects in this experiment. Individual personality traits were assessed by Kettell 16 PF. On the basis of analysis of the integral variable of playing activity, two polar groups of athletes were distinguished: 1. "effective" players (26 persons) and 2. "ineffective" players (15 persons). The significant difference between the two groups them was found by coefficient of variation ($p<0.05$). Five athletes midway between the groups were excluded. Effective players are distinguished by a strong dominance and self-control (E, Q factors). Their nervous system was more mobile and they reacted to signals more precisely. As far as playing parameters were concerned, the greater difference was found in the quality of ball-taking and attack. This division of the volleyball players into effective and ineffective did not give enough grounds for an individual based approach to tactical training. It was necessary to choose the methodical approach which could differentiate the players according to the types of ability manifested. One of the techniques of automatic classification and image identification was used. The athletes were divided into three taxones, with the greater difference between the first and the third taxones (Table 1)

The group of athletes from the first taxon were called "cognitive" because they scored high in B (16 PF), S, OT. These variables are connected with a high level of cognitive activity. The second group was called "reactive," because the athletes scored high senso-motor reactions and in factors C, E (16PF) CA, CR, AR. The third group was called "prognostic," as the athletes had high dimensions in probability prognosis. Additionally,

Table 1. Individual differences of volleyball players, distributed in three taxones

n		T-1(n=12) "cognitive"	T-2(n=19) "reactive"	T-3(n=10) "prognostic"	P1-2	P1-3	P2-3
			Various taxones				
1	B	8.41	7.03	7.54	<0.05	—	—
2	C	6.10	7.63	5.84	<0.05	—	<0.01
3	E	7.02	7.95	5.69	—	<0.05	<0.01
4	CA	1.49	1.67	1.34	—	—	<0.05
5	CR	228.47	216.34	249.19	—	—	<0.05
6	CRA	252.41	268.39	277.14	—	<0.05	—
7	S	2.21	1.68	1.84	<0.05	—	—
8	AR	53.19	41.65	59.20	<0.01	—	<0.01
9	OT	0.18	0.47	0.32	<0.05	—	—
10	PR	217.40	238.72	197.61	<0.05	—	<0.01
11	PRE	202.31	246.35	191.11	<0.01	—	<0.01
	CE	1.88	1.61	1.49	—	<0.05	—

CA: coefficient of approximation
CR: choice reaction
CRA: choice reaction in alteration
S: speed of information processing
AP: anticipation reaction
OT: operative thinking
PR: probability reaction
PRE: probability reaction (easy variant)
CE: coefficient of efficiency of playing activity

they were poor in dominance and had a middle position in sensomotor and perceptive qualities. Individual playing activity in volleyball is a complex phenomenon with interdependent connections between the elements. That is why it is impossible to recommend a universal means of management. A number of exercises are necessary for the enhancement of different qualities and skills. The means of training can be conventionally divided into three groups: 1. model playing situations with task-solving, when the process of decision optimizes manifestation of definite personality traits; 2. operative tasks, in which athletes must display definite qualities in their solution; 3. individual playing activity skill formation. According to these groups of training there are those which: (1) optimize the manifestation of personality traits, (2) develop special abilities, (3) form playing activity skills.

CONCLUSION

Application of a differentiated approach in combination with competitive activity optimises the manifestation of some personality and psychodynamic traits, which, on one hand, promote the development of special abilities, and, on the other hand, form the basis for improvement of the individual style of the athlete's activity. These two investigations emphasized that manifestations of individual style of activity are diverse and they are not conditioned by a single factor. The variety of factors illustrate for example where intellectual qualities may differ in their characteristics, allows us to combine some of them under the term of "operative intellect," and others "reflexive intellect." Often researchers examine only the neurodynamic "core" and other manifestations of the style, missing its middle part. Mechanisms of the individual style of activity formation conceal: 1) a system of personality traits which are combined with the "core" through temperament; 2) in structural-dynamic correlation of mental qualities in perception, psychomotorics and intellect, which are the manifestation of special abilities of the athlete; 3) in individualized technico-tactical skills formed on this basis.

REFERENCES

Rodionov, A. V., 1973, Psychodiagnostic of sport ability, M., FIS.
Sobchik, L. N., 1990, Methods of psychological diagnostic, M.:87.
Vjatkin, B. A., 1993, Sport and developing individuality of man, Theor. Prac.Phys.Cult. 2:1–5.

14

PREDICTION OF THE EFFICIENCY OF SKILLED ATHLETES

V. I. Balandin

Research Institute of Physical Culture
Dynamo Ave. 2, 197042, St. Petersburg, Russia

INTRODUCTION

Prediction of sporting achievements is considered to be one of the most important elements in the of management of the system of preparation of elite athletes, as a function of the scientific management of sport, as the most important part of the problem of working out a strategic conception of development, organisation and management of the sporting movement of the country (Pereverzin, 1972; Semenov, 1983; Kuznetsov, 1984 et al.). According to the existing terminology, prediction is a scientifically-based judgement on the possible future status of the subject, as well as on the probable ways and date of his achievement on the base of a study of the past and present time. In other words, predictions are based on a well-grounded judgement on the future state, development and final results of an event, a phenomenon, a process at a certain time and on certain conditions in future. Following on from the above mentioned prediction is cognitive activity of man, directed to the prediction of development of an object on the basis of analyses of the tendency of its development. Prediction is a mental anticipation of desirable results in sporting activity. It is an element of management and is an informative process. It is based on the sources of information that include information on the future. At present there are more than a hundred methods of prediction, which could be brought together into several groups. With regard to sport, the most frequently used are methods of those of extrapolation and analogy (Balandin et al., 1986). The most important task is prediction of an athlete's or team's readiness for competition. Assessment of the functional state of athletes and prediction of their functional readiness are of importance. Functional readiness is a dynamic relation of functional spheres of the athlete which secure the efficiency of his competitive activity.

MATERIALS AND METHODS

In order to assess the functional readiness of an athlete for an important competition his overall state in the final stage of preparation was evaluated. A test battery used to assess the

Current Research in Sports Sciences, edited by Rogozkin and Maughan.
Plenum Press, New York, 1996

87

vegetative (energetical) sphere was estimated with heart rate and blood pressure measurements and their derivatives, by the functional Rufier's test (30 sitdowns in 45 seconds), the Genche test (expiration delay), step-test, and index of myocardial strain (ECG) measurements. The objective psycho-physiological sphere was assessed with Romberg's test (vestibular stability). Reflexometry included the following tests: reaction on a moving object (RMO), speed of sensor-motor reactions, muscular feeling, mobility of neuro-muscular apparatus (with tepping test), speed of information workup, as well as operative memory and quality of attention. The subjective psycho-physiological status was evaluated with questionnaire methods SAM (self-feeling, activity, mood), and higher nervous activity, psycho-nervous adaptation and anxiety reaction by Spilberger. At the final stage of preparation for important competitions the athletes were studied 2 or 3 times: at the beginning of the stage and before the new micro-cycles of training. Each exponent was evaluated with 9-point scale. This gave a qualitative estimation of all three functional spheres. The method of formation of the scale is described in text-book by V.Balandin (1991).

RESULTS AND DISCUSSION

The general estimation of functional spheres is represented as a difference in the marks of the first and the following stages (+ or -) (see Table 1.).

Table 1. Scheme of assessing athlete's functional readiness for competitions

Functional state	N	Functional spheres	Dynamic of marks	Diagnostic syndrome
Good	1	Energetical	+	State of good sport performance
		OPPS	+	
		SPPS	+	
	2	Energetical	−	Incomplete recovery of physical load
		OPPS	+	
		SPPS	+	
	3	Energetical	+	Incomplete recovery of mental load
		OPPS	+	
		SPPS	−	
Satisfactory	4	Energetical	−	Initial stage of physical exhaustion
		OPPS	+	
		SPPS	+	
	5	Energetical	+	Initial stage of mental exhaustion
		OPPS	−	
		SPPS	+	
Poor	6	Energetical	−	Physical overstrain
		OPPS	+	
		SPPS	−	
	7	Energetical	+	Mental overstrain
		OPPS	−	
		SPPS	−	
	8	Energetical	−	Initial stage of general overstrain
		OPPS	−	
		SPPS	+	
	9	Energetical	−	General overstrain (overtraining)
		OPPS	−	
		SPPS	−	

Note: OPPS: Objective psycho-physiological state; SPPS: Subjective psycho-physiological state.

Good functional status is evaluated by either a positive balance of all three functional spheres or by an insignificant fall of objective or subjective psycho-physiological state. This fall can testify of an incomplete recovery after physical and psycho-nervous loads. This state results from insufficient rest and with no significant alterations in the training regimen. A satisfactory functional state depends on the prevention of physical or mental exhaustion in its initial stage. If there are signs of physical, mental or general overstrain it means poor functional state and suggests that the athlete should be removed from the training and it sometimes requires medical treatment. As a rule, after overstrain an athlete rarely returns to elite sport and usually can't achieve high performance (Gorbunov, 1994).

For each athlete, accordingly to his functional state, a set of recommendations is made. Those recommendations give suggestions for alterations in the quantity and intensity of training loads and mental strain, as well as use of pedagogical, psychological and medical means of recovery (Table 2).

Carrying out of these recommendations permits coaches and athletes to avoid overtraining and to approach competitions in a good functional state. Comparisons between the state of functional readiness and successfulness in competitions allows a prediction of the efficiency of competitive activity. The most acceptable for these purposes are methods of extrapolation and analogy. The method of analogy is more feasible if there is a possibility of comparing an athlete's actual state with one that secured him success in competition. The method of extrapolation is used if there are a set of data (no less than three). In that case it is possible to extrapolate the curve to the period of competition. Verification of the prediction of functional readiness has proved the success of these methods in participants in cycle competitions and fighting sports. In 1990, the national fencing team (30 athletes), in which functional readiness was predicted using this method, won the World Championship in Budapest and, after a 5 year interval, won National Cup. Thus the monitoring of athlete's functional state (as shown in tables 1 and 2) as well as pedagogical observations allows us to establish 9 levels of his functional readiness and therefore avoids eventual pathological states, including overstrain (overtraining).

Table 2. Recommendations for correction of an athletes's functional state

Diagnostic syndrome	Recommendations
1. State of good sport performance	Physical loads without limitation, possibility for its increase.
2. Incomplete recovery of physical load	Common regimen. Abstention from increase of physical loads. Sufficient rest, use of recovery means.
3. Incomplete recovery of mental load	Psycho-regulatory therapy. Switching over to other kinds of activity. Increase of emotional kinds of physical loads.
4. Initial stage of physical exhaustion	Variations of quantity and intensity of loads, release of extra training sessions, use of supplementary means of recovery.
5. Initial stage of mental exhaustion	Psycho-regulation therapy and correction, use of pharmacological and other means of recovery.
6. Physical overstrain	Reduction of quantity and intensity of physical loads, release of training session during 1-3 days, use of pedagogical and pharmacological means of recovery
7. Mental overstrain	Switch over to other kinds of activity. Free regimen of training, psychotherapy, use of medical means of recovery, release of training during 1-3 days
8. Initial stage of general overstrain	Release of training and competitions, free regimen, use of complex means of recovery.
9. General overstrain (overtraining)	Hospitalization and medical treatment

CONCLUSION

This method was successfully applied to a number of elite athletes in various kinds of sport. The method of diagnosis could be considerably simplified by using only one exponent for each of the three functional spheres, but doing so would reduce the reliability of the method. Therefore, it is advisable to evaluate each sphere with two or three exponents of the most important kind for each type of sport. Our 9 points scales make easier the process of general evaluation of the three functional spheres and the decision of an athlete's functional readiness for competition.

REFERENCES

Balandin, V. I., Bludov, Ju. M., 1986, Prognozirovanie v sporte, M., FiS :193.

Balandin, V. I., 1991, Phunkcionalnaja gotovnost sportsmenov i metody ejo diagnostiki, L., komitet po FCiS :26.

Gorbunov, G. D., 1994, Psychopedagogika fisicheskogo vospitania i sporta, Autoreferat dissertacii doctora pedagogicheskich nauk, Saint-Petersburg.

Pereverzin, I. I., 1972, Prognozirovanie i planirovanie fizicheskoj cultury, M., FiS.

Semenov, G. P., 1983, Prognozirovanie sportivnich dostigenij i osnovnich pokazatelej razvitija vidov sporta v SSSR, Teorija i praktika phisicheskoj cultury 11:59-61.

Kuznetsov, V. V., 1984, Obshzie zakonomernosti i perspektivi razvitija teorii sistemi sportivnoj podgotovki, Metodologicheskije problemi sovershenstvovanija sistemi sportivnoj podgotovki kvalificirovannich sportsmenov, Moscow : 6-21.

15

MENTAL RELAXATION–NEURO-DYNAMIC MARKERS AND PSYCHOPHYSIOLOGICAL MECHANISMS

P. Bundzen, P. Leisner, A. Malinin, and L.-E. Uneståhl

Research Institute of Physical Culture
Dynamo Ave. 2, 197042, St. Petersburg, Russia
Scandinavian International University
Sommarrovagen, 6, S-701.43, Box 155, Orebro, Sweden

INTRODUCTION

In the last few years there have been more active attempts at studying the neuro-physiological mechanisms of consciousness and, in particular, alternative states of consciousness that are formed under the conditions of hypnosis, self-hypnosis and mental relaxation (Grawford et al., 1989; Lebedeva, Dobronravova, 1990; Sabourin et al., 1990; Bundzen, Uneståhl, 1994). It's progress is determined not only by the level of the development of modern neuro-science, but also by the problems that the practice brings.

In particular, one such problem is the phenomenon of altered state of consciousness (ASC) which accompanies the peak of an athlete's fitness, and which occurs in conditions of a brilliant performance at competitions (Uneståhl, 1983; Sudibio, 1994). It is quite evident that under these conditions, the subject (athlete) receives "the access" to ASC by using psychophysical reserves, and the process of "entering" the subject into ASC may be facilitated by a method of systematic psychotraining ie mental training (Uneståhl, 1992).

All the above mentioned raises for psychophysiology and sports psychology a number of theoretical and practical questions whose solution is closely connected with the brain cognition mechanisms of ASC formation. The present investigation was made as the first step on the way to ASC psychophysiological mechanism decoding for the purpose of elaborating the system for skilled athletes and for health promotion.

METHODS

Methods of Mental Relaxation and ASC

19 right-handed volunteers between the ages of 19 to 27 (10 men and 9 women) took part in the investigation. All the volunteers had a preliminary examination of the

Current Research in Sports Sciences, edited by Rogozkin and Maughan.
Plenum Press, New York, 1996

91

state of their psychophysical health on the screening diagnostic system "OFIS" and were assigned to a healthy group. For two months fifteen subjects undertook standard methods of "Mental Training for Sport and Life" by muscular relaxation, mental relaxation, teaching of using "the key", choice of own method and self-dependence training.[*]

The lessons took place three times a week. Four subjects used mental training for two years, and according to the estimation of the expert Swedish mental coaches, they possessed the ability to independently plunge into ASC in 2–4 minutes. This group (2 women and 2 men) was used as a control group. The control investigations were carried out in Central hospital Neuro-physiological clinic (Orebro, Sweden) under the guidance of professor P.Leisner. The course of the investigation of exposure to a standard audio programme took 14 min 30 s. The programme was made up of the following parts together with a musical background: muscular relaxation -4 min 30 s; submersion into so-called "mental room" -4 min; mental relaxation -4min50s; leaving the state of mental relaxation - 1 min 30 s. The multichannel recording of electroencephalogram (EEG) was made by beam methods at fixed points during the presentation the audio programme: 2 mins prior to the beginning of the audio programme, at 2.3, 6.3 and 10.3 mins of the programmes presentation, and during the first minute after leaving the state of mental relaxation. The measured fragments of EEG made up not less than 240 s.

Electro-physiological Methods of Analyses

The EEG recordings were made from 16 or 14 golden small cup electrodes connected with aural electrodes and localized strictly in accordance with the system 10/20. All the subjects reclined in a comfortable pose in an arm-chair with closed eyes. A video monitor was used for removing artifacts before processing the EEG fragments. We used the standard frequency bands for EEG analysis: delta, theta, alpha, beta-1 and beta-2. After Fourier transformation matrixes for each zone of the lead were made for each subject were made according to the following states: background, muscular relaxation, mental relaxation (ASC) and sequence. Both the individual data, by the indicated continuum of state, and the averaged data by experimental (n=15) and control (n=4) groups was analysed.

Statistical Processing of the Matrixes of Spectral EEG Analysis

The matrixes of averaged power were subjected to statistic analyses by frequent components of EEG. They were analysed according to the following parameters:

1. Changes of averaged values of the power of EEG spectral components in the continuum of state reconstruction ie calm vigilance, muscular relaxation, mental relaxation.

2. Changes of the absolute values of local activation spectral coefficients in the continuum of state reconstruction. The coefficients of local activation were calculated according to the methods of L Pavlova and A Romanenko (1988):

---

[*] In these investigations the Russian version of the Swedish system of mental training (L.-E.Uneståhl) which was elaborated by the Russian Branch of the Scandinavian International University under the leadership of P.Bundzen (S.-Petersb.) was used.

$$K \, \beta/\alpha = (\Sigma A\beta i/T\beta + 0.1)/(\Sigma A\alpha i/T\alpha + 0.1)$$

where $A\beta$ and $A\alpha$ = spectral power of the activity in β and α ranges; $T\beta$ and $T\alpha$ = time of analysis.

3. Changes of distribution in the space of brain analysing zones. The coefficients of local activation in the continuum of states and the formation of structure cortical activating patterns.

4. Changes in the values of local activation coefficients among brain zones in the continuum of studying states:

   • among symmetric zones of cortex (bilaterally);
   • among the frontal zones F3, F7, and F4, F8, the central and retrocentral zones (frontal-occipital direction).

All statistical processing was by a complex of programmes for IBM PC/AT using both parametrical (for EEG spectral matrixes) and non-parametrical analysis.

Control of ASC Development in the Course of Mental Relaxation

The control of ASC development in the subjects of the experimental group was carried out by methods previously described (Bendjukov, Emeljanov, 1993). For this, simultaneously with EEG, we measured eyes movement and made an electromyogram of the muscles of the right forearm. After completing the investigation subjects were questioned, and in some cases a neuro-linguistic test was carried out by D Spivak (1986).

RESULTS

1. Change of Averaged Values of the Power of EEG Spectral Components

The statistic analysis of the changes of the averaged values of EEG spectral components in the continuum of studied psychical states. Examination of the audio programme "Mental relaxation-2" enabled the following to be concluded:

• Firstly, in the period of mental relaxation (Fig. 1.1) there was a statistically significant intensification of theta-activity in the following zones of the cerebral cortex: F3, F4, C2, C4, P2, T6, O2 which was significantly different from all other analysing states (p<0.01). The level of significant difference in zones C3, P3, P4, is lower (p<0.05), but in all indicated zones there is some difference in the power of theta-activity, including a difference from the state of muscular relaxation;

• Secondly, in the period of mental relaxation (Fig 1.2) there was a significant decrease in the power of alpha-activity in the retrocentral zones of the cortex (T5, T6, O1 and O2 (p<0.01)) which differs from the state of muscular relaxation and from other analysing states. The statistically significant changes in activity in the band of beta-1 are found in some subjects by mental relaxation in the frontal-right zones (F3, F7) only not in all the subjects. There were no significant differences on the beta-2 band between mental relaxation and any other analysing states.

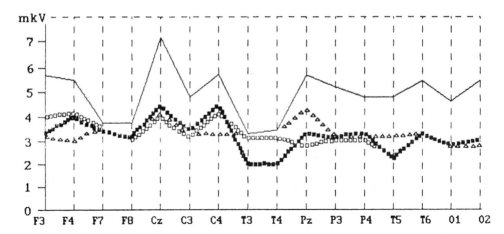

Figure 1.1. Distribution of the values of theta-activity spectral power in the analysed zones of brain in the continuum of studied psychical states.

2. Change of Absolute Values of Local Activation Spectral Coefficients (LASC) and Patterns of LASC

The calculation of LASC values in each of the studied brain zones for all the continuum of states and their averaging by separate states testify that the maximum changes of LASC are observed in the state of muscular relaxation, when their values are essentially lower (p<0.0005) than in background activity (Fig. 2.1).

Each LASC calculation is the average of 2s fragments. As shown on the brain maps of LASC, the studied states are distinguished not only by the activated level of brain cortex, but also by the patterns of maximum activation and deactivation. It should be stressed that the revealed patterns of LASC did not differ in structure by mental relaxation when

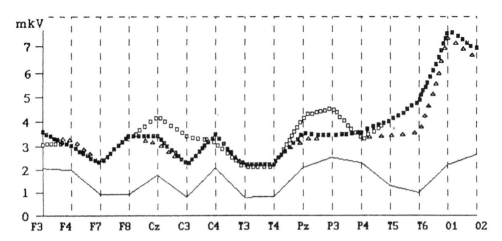

Figure 1.2. Distribution of the values of alpha-activity spectral power along the analysed zones of brain in the continuum of studied psychical states.

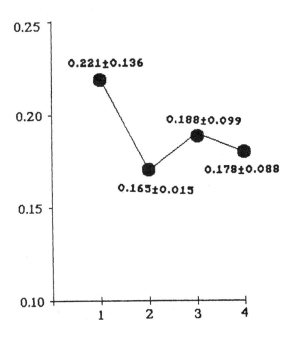

Figure 2.1. Change of averaged values of spectral coefficients of brain zone local activity in the continuum of studied psychical states. The basic conclusion from the above data is that the state of mental relaxation by the general activated level of brain cortex takes the intermediate position in the continuum of analysed states, and from the changes of EEG spectral components (see part 1) it is characterized by some specific changes, mainly over the range of theta and alpha activity. The calculation of LASC absolute values by the analysed brain zones and by each of the analysed states enabled a set pattern of cortical activated structures to be obtained. The calculation of patterns of cortical activated structures for the whole continuum of studied states of consciousness is represented in Fig. 2.2.

listening to the audio programme (the experimental group) and by ASC self-induction in the control group of subjects. There following statistically significant regularities of LASC pattern changes in the continuum of studied states of consciousness were found:

- With transition from a state of vigilance to a state of muscular relaxation there is an activation of the frontal-right and central zones of the right hemisphere (and by ASC self-induction also in the parietal zone) (Fig 2.2 1&2)

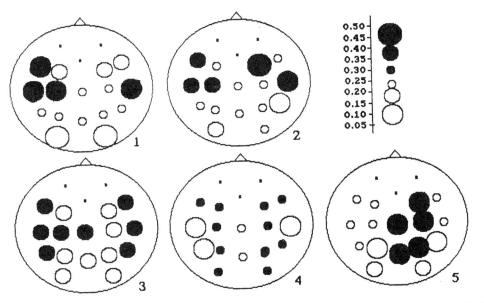

Figure 2.2. Dynamics of the averaged patterns of cortical activated structures in the continuum of studied psychical states.

- At the initial stage of mental relaxation (or ASC self induction) there is a decrease in general activated level and activation and deactivation in the right and left hemispheres increases (Fig 2.2 3)
- In the period of mental relaxation and ASC, the symmetry of the LASC pattern reaches the maximum, and it forms a smooth bilaterally activated structure which is evenly distributed towards the frontal-occipital direction[*]
- By leaving the state of mental relaxation or self-induction of ASC the reduction of the activated system with the focus of maximum activation in frontal-central-posterior sections of right hemisphere is observed (Fig 2.2 5).

The reconstruction of LASC patterns required additional analysis and statistical processing for the purpose of specifying the changes of brain bilateral functional asymmetry in the continuum of studied states of consciousness.

3. Change in the Values of Brain Zone Local Activation Coefficients in the Continuum of Studied Psychological States

Taking into consideration the peculiarities of LASC pattern reconstruction mentioned above, there were LASC bilateral differences between the symmetric zones of right (dominant) and left hemispheres (Fig. 3.1).

As is shown changes in states and mental relaxation (see curve 3) give a ($p<0.001$ for anterocentral and central zones and $p<0.05$ for retrocentral zones) decrease in functional asymmetry at the level of brain symmetrical zone local activation. Another regularity which is revealed from the present data is in the change of activation at the frontal-occipital region.

This can be seen clearer in Fig. 3.2.

As is shown in the condition of mental relaxation not only is there a decrease in the difference in the level of activation of the frontal and retrocentral zones from 0.313 ± 0.047 to 0.161 ± 0.055 (as regards F4 and F8) and from 0.212 ± 0.083 to 0.188 ± 0.043 (as regards F3 and F7), but there is also a levelling and reversal of these differences by the brain hemispheres.

Thus, under the conditions of ASC the following phenomenon are typically recognised: a smooth, isogradient activated structure, distributed both bilaterally and at the frontal-occopital direction.

DISCUSSION

The data obtained in this investigation, both in experimental and control groups, shows that the transition to ASC is accompanied by the formation in the hemispheres of the cortex a diffusely distributed activated structure which is distinguished by its activated level isogradient differences between the brain zones that are in its structure. This phe-

---

[*]
It should be noted that the activated structure is isogradient if the following relationship between separate components of the system gives a gradient of activation (GA):

$$GA = (x - y)/(x + y)$$

where x and y are the values of activation in two comparing zones (see Pavlova and Romanenko, 1988).

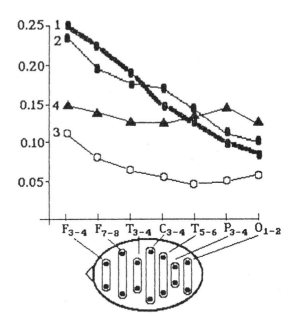

Figure 3.1. The differences of averaged values of local activation spectral coefficients between the symmetrical zones of left and right hemispheres in the continuum of studied psychical states.

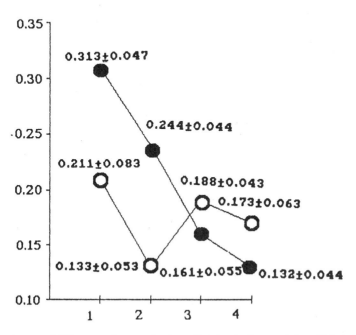

Figure 3.2. The changes of differences of averaged values of local activation spectral coefficients between frontal zones: F4 and F8 of right and F3 and F7 of left hemisphere and correspondingly central and retrocentral zones of the cortex in the continuum of studied psychical states.

nomenon is specific only to the indicated state of consciousness, and it is not reproduced in the continuum of studied psychical states.

Therefore, the described signs may be considered as a neuro-physiological marker of the transformation from a consciousness state. Taking into account neuro-psychology data (Lurija, 1973; Pribram, 1975; Kolb, Whishaw, 1990) and the results of neuro-physiological and psychophysiological mechanisms of psychical activity (Bundzen, 1978; Bekhtereva et al., 1979; Pavlova, Romanenko, 1988; Petche et al., 1992; Mishlayev, 1993), it is possible to state a number of theories concerning the brain mechanisms involved in the formation of ASC.

- The findings show that transformation of the state of consciousness is associated with some systemic reconstructions in brain activity:
  - firstly, the interaction of the thalamo-cortical and limbico-reticular systems is accompanied, according to theta activity dynamics, by an activation of the frontal-limbico complex;
  - secondly, with the reconstruction of brain functional asymmetry and the transition from the predominance of frontal-left and reflective zones and the zones of intermodal integration (C3, T3,T4) there is a phase of "non-dominance" then a change in the focus of activation to frontal-right, central zones and posterior-parietal zone of right hemisphere (T6).

These findings both in experimental and control groups enables us to suggest that the change of the state of consciousness in the course of mental relaxation is probably accompanied by a change in psychical activity brain codes (Pavio, 1978;Pavlova, Romanenko, 1988) by a simultaneous slackening of reflective processes, and a reduction in afferent synthesis. The accounts of subjects having vivid imagination in ASC which occurs under the conditions of dissociation from the environment (phenomenon of dissociative attention). Also, taking into account the information about the brain mechanisms of the activation of long-term memory matrix (Bundzen, 1981; Miller, 1989) one may suppose that the diffuse isogradient activation of brain cortex in combination with the reinforcement of theta-rhythm can create the unique conditions for the re-activation of memory engram. Whether these processes are a neuro-dynamic correlation of "top consciousness" or "insight" - it is not possible to answer at present. However, the changes of neuro-dynamics at the higher level of brain sections in ASC can explain not only the modification of the psychological state, but also the conditions of ASC and the high-effective reproduction of consolidated skills. The realization of psychomotor acts under these conditions probably depends to a lesser extent on the processes of afferent synthesis, and it is accomplished according to the type of central-determined reactions and mechanisms of shortened conditioned reflexes (Kupalov, 1963; Rabinovitch, 1975). In the theory of sports training it is probably not by chance that the given type of consolidated sports skills in motor memory appears as a motor-psychical activity whose reproduction is associated with the aim and motivation (Bulkin et al., 1994). It seems to us that the realization of sports activity in ASC which on the face of it appears to be "a paradox", becomes explicable: a combination of the peak and "tunnel consciousness", dissociation and concentration (Uneståhl, 1992).

- The reconstructions of neuro-endocrine regulation which accompanies the mental training is also essential from the point of view of the explanation of sports activity resulting in ASC. As shown, within the last few years (Isakov, 1994) mental training with the help of the programmes used in this research leads to statistically significant changes in the pain threshold, a decrease in the level of cortisol in serum and increase in the level of beta-endorphine. According to data in the literature, these neuro-chemical changes enable us not only to create the "positive" psychoemotional background and intensify the motivation for activity (Hawkes, 1992), but also to potentiate changes of hemisphere symmetry which found in conditions of mental relaxation (Silberman and Weingartner, 1988; Roschman and Wittling, 1992).

CONCLUSION

From neuro-physiological point of view, consciousness is defined as a specific state of the brain when only the accomplishment of psychological functions are possible (Sokolov, 1990). According to these findings the alternative states of consciousness whose induction is achieved by mental training, the following basic distinctive peculiarities of structural-functional relations at the levels of higher brain sections are found:

1. changes in the interaction of the thalamo-cortical and limbico-reticular structures;
2. smoothing over of the interhemispheres and frontal-occipital asymmetry of the cortical structures;
3. the formation of dominance by the level of the activation of right hemisphere as post-relaxation effect.

The findings enable us to propose that in the alternative state of consciousness the measure of brain dissipation as an information system essentially changes (Bundzen, 1978, 1981). Changes of the most essential components of consciousness are observed; there is a narrowing of the sphere of consciousness and a facilitation of inner images generation and the reproduction of central-determinant reactions associated with the activation of memory engram. Most probably, these peculiarities of psychological functions in the alternative state of consciousness enable us to ensure a highly-effective reproduction of professional and, in particular, sports skills.

In conclusion it should be stressed that the materials of the given investigation do not completely expose the brain mechanisms of mental relaxation and the formation of alternative state of consciousness. Also, the findings prove not only the adequacy of a selective approach, but also enable us to more purposefully orientate the subsequent scientific search in the developing of the methods of mental training (Uneståhl et al., 1994).

REFERENCES

Bechtereva, N. P., Bundzen, P. V., Gogolitsin, Yu. L., Malyshev, V. N., Perepelkin, P. D., 1979, Neurophysiological codes of words in subcortical structures of the human brain. Brain and Language 7:145–163.

Bendjukov, M., Emelianov, V., 1993, Diagnostics of induction of alternative state of consciousness via mental training, In: "Mental Training in Sport and Life in Russia", Orebro, Sweden :20-21.

Bulkin, V., Ershova, E., Golovin, B. et al., 1994, Motor-psychic provision of high level athletes activity, Intern. conf."Current Research into sport science", Saint-Petersburg :92–93.

Bundzen, P., 1978, Some methodological problems of the deciphering of psychical phenomena brain codes, Question of Philosophy 9:83–92.

Bundzen, P., 1981, Neuro-dynamic correlates of taking decisions. Abstract of dissertation, Leningrad, 38 p.

Bundzen, P., Uneståhl, L.-E., 1994, Thinking with image in alternative state of consciousness - neurodinamic markers. The Intern. Symposium "Psychological and Biochemical Basis of Brain Activity" June, St.Petersburg :15.

Carwford, H. J., Clark, S. W., Kitner-Triolo, M., 1989, EEG activity pattern differences in low and high hypno-
   tizables: reflection of strategy differences? Int. J. Psychophysiology 7:165–166.
Hawkes, Ch., 1992, Endorphins: The basis of pleasure? J. of Neurology, Neurosurgery and Psychiatry 55:247–250.
Isakov, V., 1994, Alteration of the level of cortisol in blood serum, pain feeling and psychoemotional status under the
   influence of mental relaxation, Abstracts of sci. conf., Saint-Petersburg, Res.Inst. of Phys. Cul.:37–38.
Kolb, B. and Whishaw, I., 1990, The Frontal Lobes. Fundamentals of Human Neuropsychology., 3rd Edn., Freeman
   and Co., New York :463–501.
Kupalov, P., 1963, Study on the reflex and reflex activity and perspectives of its development, In: Philosophical
   question of the physiology of higher nervous activity and psychology, Moscow :106.
Lebedeva, N., Dobronravova, I., 1990, Organiztion of human EEG rhytmus by special states of consciousness, J. of
   Higher Nervous Activity 40(6):951–962.
Lurija, A., 1973, Principles of neuro-psychology, Moscow.
Miller, R., 1989, Cortico-hippocampal Interplay: Set-organization Phase-locked Loops for Indexing Memory, Psy-
   chobiology 17:115–128.
Mishlayev, S., 1993, Hypnosis. Personal influence? N.Novgorod :286 p.
Pavio, A., 1978, Imagery, Language and semantic Memory. Int.J. Psycholinguistics 5:329–342.
Pavlova, L., Romanenko, A., 1988, Systemic approach to psychophysiological investigation of human brain, Lenin-
   grad, Nauka.
Petche, H., Lacroix, D., Linder, K., Rappelsberger, P., Sonmidt-Heinrich, E., 1992, Thinking With Images or Think-
   ing With Language: A Pilot EEG Probability Mapping Study, Int.J. of Psychophysiology 12:31–39.
Pribram, K., 1975, Languages of brain, Moscow.
Rabinovitch, M., 1975, Neuro-physiological mechanisms of conditioned reflex, Physiology of man and animals,
   Moscow 16:5–58.
Roschmann, R., Wittling, W., 1992, Topographic Brain Mapping of Emotionrelated Hemisphere Asymmetries, Int. J.
   Neurosciences 63:5–16.
Sabourin, M., Cutcomb, S., Drawford, G., Pribram, K., 1990, EEG Correlates of Hypnotic Susceptibility and Hyp-
   notic Trance. Int. J. Psychophysiology 10:125–142.
Silberman, E., Weingartner, H., 1986, Hemispheric Lateralization of Functions Related to Emotion, Brain and Cog-
   nition 9:322–353.
Sokolov, E., 1990, Neuro-physiological mechanisms of consciousness, J. of Higher nerv. activ. 40(6):1049–1052.
Spivak, D., 1986, Linguistics of alternative states of consciousness, Leningrad :96.
Sudibyo, S., 1994, Model of mental training based on Javanese method. Intern. conf. "Current Research into sport
   science", SaintPetersburg :17.
Uneståhl, L.-E., 1983, The Mental Aspects of Gymnastics, Veje Forlag, Orebro, Sweden.
Uneståhl, L.-E., 1992, Psychical skills for sport and life, In: "Sport and Health", Saint-Petersburg :20-28.
Uneståhl, L.-E., Bundzen, P., Bendukov, M., 1994, Psychophysiological Approaches in Mental Resource Training,
   In.: Intern. Congress on Sports, Helsinki, Finland :4.

16

LONG-ACTING REGULATORS OF MENTAL STATE IN ELITE SPORT

Control and Optimization of Their Influence

G. B. Gorskaya

Cuban State Academy of Physical Education
Budennogo St.161, 350670, Krasnodar, Russia

INTRODUCTION

Elite sport is characterized by high levels of physical and mental tension that affects performance during competition as well as the athletes mental health (Piloyan, 1985, Heyman, 1987). The most important destabilizing factors of the mental health of the elite athlete are the long acting regulators of emotional tension. These regulators can be divided into internal and external regulators. Internal regulators are those personality characteristics that increase the athletes sensitivity to emotional stress. External regulators are the stresses of life and the stress associated with training and interpersonal relations. The importance of the long acting regulators of mental state is recognized by sport psychologists (Hahn, 1983; Skanlan, Stain, Ravizza, 1991), but these regulators are not sufficiently taken into account in the psychological preparation of athletes. This situation can be attributed to the lack of information regarding the affects of long acting regulators of mental state on performance and mental health of the elite athlete. These regulators are more often discussed in connection with their influence on injury probability during sport (May, Veach, 1987; Hanson, McCullagh, Tonymon, 1992). The long acting regulators of mental state and emotional tension as well as methods to optimize their affects were investigated in the group of elite yachtsmen. Those factors influencing emotional tension and particularly those personality characteristics increasing the probability of mental stress overload were studied. These included the stability of crew position, the selection criteria for major competitions as well as national team selection and those factors influencing the competition programme. Previous investigations (Gorskaya, 1990) have shown many elite helmsmen to have a set of personality characteristics that allow rapid decision making in an ever changing situation as yacht racing. However, at the same time these personality characteristics produced tactical errors, errors in goal setting, a tendency for excessive risk making, placing of inappropriate importance to different aspects of training and a low self-control. These personality characteristics can collectively be classified as "tendency

Current Research in Sports Sciences, edited by Rogozkin and Maughan.
Plenum Press, New York, 1996

to risk and low self-control". Helmsmen with these personality characteristics often suffer from emotional overloads which can account for the reduced control over their actions. As the main goal of the elite athlete is to achieve the best result possible during competition these factors causing emotional overload have to be connected with the athlete individual style of goal setting, the probability of success and the explanation of success and failure. In present study the main focus of analysis was the emotional reaction to situations with different probability of success and to frustrating situations typical for yachtsmen with the particular personality characteristics previously mentioned. It is important for the elite athlete to have confidence in maintaining his position in the team. Uncertainty can cause increases in emotional tension and psychological overload. In order to study this problem, motivation structures in groups of yachtsmen with more or less stable positions within a crew and the analysis of emotional loads resulting from changes in position certainty need to be compared. The affects of training and competition planning on levels of emotional tension were studied in a longitudinal investigation which included the reporting of those indicators of psychological readiness for competition and the analysis of these indicators depending on the factors previously mentioned.

METHOD AND PROCEDURE

The investigation set out to analyze the emotional reactions to situations with different probability of success and to frustrating situations that are typical of elite yachtsmen with the personality characteristics described as "tendency to risk and low self-control". Yachtsmen were divided into two groups. The first group consisted of yachtsmen with the personality characteristics "tendency to risk and low self-control" while the second group consisted of yachtsmen without these personality characteristics. Differences between the two groups in emotional reaction to situations with different probability of success and to frustrating situations were studied. The emotional reaction of elite yachtsmen to situations with different probability of success was investigated using the motivation-result method (Dashkevitch, 1976). Yachtsmen were instructed to perform three series of reaction time trials with two alternative choices. Subjects were instructed that in every series their task had a different difficulty and success probability which changed from one series to another. In the first series, success probability was 90%, in the second 70% and in the third 30%. Before each trial, yachtsmen were asked to estimate the probability of success. After each trial yachtsmen were given feedback regardless of trial result but in accordance with the probability of success. Each series consisted of 20 trials. Each of two signals was presented 10 times at random. The Rosenzweig frustration test was used to determine the type of reaction to frustrating situations. The psychological load was estimated by analysing the psychological discomfort changes during three competition seasons. The subjects motivation structure was studied by the method of Piloyan (1985). Since the social position of the crew is known to be less stable than that of the helmsmen this study set out to determine the influence of social position on the level of psychological loads experienced by comparing helmsmen and crew motivation structures. In addition the subjects sensitivity to psychological discomfort and situational anxiety during competition was analyzed in situations when their social position was altered. Sensitivity to psychological discomfort was measured using the Piloyan method. The influence of training and competition planning on psychological load levels was also investigated by analysing the following indicators of mental readiness for competition: competitive motivation studied using the Kalinin questionnaire (Kalinin, 1962), sensitivity to stressors measured using the Milman ques-

Table 1. Means of reaction time and success probability value

Group of yachtsmen	Success probablity value (percent)			Reaction time (msec)		
	Series 1	Series 2	Series 3	Series 1	Series 2	Series 3
1	94.3	91.5	31.5	296.4	263.4	238.1
2	86.7	63.5	18.0	281.8	276.2	283.3
Differences significance	p<0.05	—	p<0.05	p<0.05	—	p<0.01

tionnaire (Milman, 1983), sensitivity to psychological discomfort estimated by the Piloyan method. All these indicators were registered during three competition seasons.

RESULTS

The results of the analysis of emotional reactions to situations with different probability of success revealed differences between yachtsmen with the "tendency to risk and low self-control" personality characteristics (group 1) and those without (group 2) (Table 1)

The mean values for reaction time and success probability of yachtsmen in group 1 were better at predicting success than in group 2. The greater the task difficulty the larger the difference between groups. When yachtsmen were instructed that the probability of success was 70% (series 2), yachtsmen of group 1 estimated it to be 91.5% while yachtsmen of group 2 estimated it to be 63.5%. When yachtsmen were instructed that the probability of success was 30% (series 3), yachtsmen of group 1 estimated probability of success to be 31.5% while in group 2 it was only estimated to be 18%. The prediction of success probability was found to influence reaction time. This was especially the case in series 3 which had a low probability of success. Differences between groups demonstrated in this study may also be expected during actual competition. Yachtsmen with the "tendency to risk and low self-control" personality characteristics maintained high mobilization levels in difficult situations with low probability of success. Yachtsmen without these personality characteristics often gave up competing if they perceived the situation as being too complicated. Differences in reaction to frustrating situations were found between yachtsmen of group 1 and group 2 (Table 2).

Rosenzweig Test results. Yachtsmen with the "tendency to risk and low self-control" personality characteristics differed from yachtsmen without these personality characteristics by high levels of O-D, E reactions and low levels of N-P reactions and a rather low social adaptation index. Yachtsmen without the "tendency to risk and low self-control" personality characteristics often concentrated their attention on obstacles or actions

Table 2. Rosenzweig Test results

Group of yachtsmen	Cases per cent of					
	O–D	E–D	N–P	E	I	M
1	32.8	51.4	15.8	52.7	21.9	25.4
2	25.2	49.9	24.9	36.4	27.0	31.6
Differences significance	p<0.05	—	p<0.05	p<0.05	—	p<0.05

Sensitivity to psychic discomfort

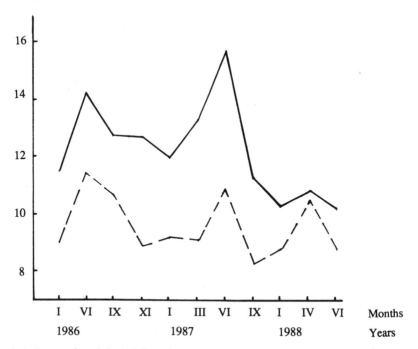

Figure 1. Indicators of psychological discomfort change in a similar fashion in the two groups of yachtsmen.

of other people that they consider as being causes of failure. These yachtsmen are less interested in analysing their own mistakes and in looking for ways to solve their problems. Their ideas regarding methods of achieving success or avoiding failure are not realistic enough. This is considered as an important issue evidenced by the fact that one of the popular topics in the yachting literature is the significance of self-control and self-analysis for yachtsmen (Fries, 1986, Walker, 1989). Yachtsmen with the "tendency to risk and low self-control" personality trait pattern have higher levels of psychological tension than athletes without this pattern (Figure 1).

Indicators of psychological discomfort change in a similar fashion in the two groups of yachtsmen. Changes in these indicators reflect competition season specificity. Psychological tension increases during the main competition periods (June 1986, 1987) and decreases during the less important periods of competition. But the indicators of psychological discomfort are constantly higher in the group of yachtsmen with the "tendency to risk and low self-control" personality trait pattern and can be considered as long acting mental state regulators. This personality trait pattern is similar to the Type A personality pattern described by Friedman and Rosenman (1977) as a health disorder risk factor associated with psychological overloads. These psychological overloads can cause an unrealistic estimation of a subject's capacity and can lead to unrealistic planning. An appropriate programme is necessary to decrease the psychological loads in the group of yachtsmen with the "tendency to risk and low self-control" pattern. A comparison of motivation indexes show differences between helmsmen and crew members. Crew members have higher indicators of three motivation components: desire for self-affirmation, desire for material security, interest in social well-being (Gorskaya, Barabanov, 1993). These

Sensitivity to psychic discomfort

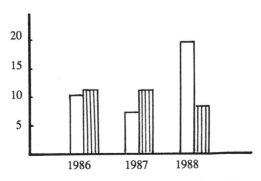

Figure 2. Mean indicators of psychological discomfort in two crew members of the same yacht during three competition seasons.

differences can be explained by differences in stability in social position between helmsmen and crew members. Crew members are usually less sure of their position. Consequently they have to be more interested in self-affirmation, material and social well-being. This conclusion is supported by a study of crew members reaction to situations where their social position changes. This investigation shows that those yachtsmen who have doubts about the stability of their social position have a higher desire of self-affirmation and interest in material and social well-being. Uncertainty about social position can increase emotional tension and cause psychological overload. The consequences of destabilization of social position is presented at figure 2.

Mean indicators of psychological discomfort in two crew members of the same yacht during three competition seasons. The social position of crew member A became less stable in the 1988 competition season and caused an increase in psychological discomfort. The data described above support the idea that social position is one of long acting mental state regulators in elite sport. The study of the affects of planning training and the competition season on psychological load levels show that the lack of information on trial criteria, insufficient exactness of their application and changing of trial criteria during the competition season resulted in increases in emotional tension. Mistakes in the planning of the competition season make the psychological rehabilitation of yachtsmen more difficult and can also increase psychological loads (Gorskaya, Barabanov, 1993).

CONCLUSION

The results of this investigation demonstrate that long acting regulators of emotional tension can significantly influence elite athletes psychological loads and consequently their mental health. Psychical overloads cause depressions and other forms of mental health disorders. There are two types of psychical characteristics that may be considered as internal long acting mental state regulators: 1) personality characteristics that decrease mental toughness (anxiety and low self-esteem); 2) personality characteristics that increase stress by unrealistic analysis of situations causing self-control to be reduced. The second personality characteristics are typical of many elite yachtsmen. A programme is necessary to decrease psychical loads. Appropriate psychological training and improvements in communication skills allow to optimize the affects of external long acting mental state regulators on the mental health of athletes.

REFERENCES

Dashkevich, O. V., 1976, Mentodi Issledovanija Emotsiy labornych uslovijach, M., VNIIFC.

Friedman, M., Rosenman, R. M., 1977, The couse-type. A behavior pattern in: Stress and coping.N.-Y.,Columbia Univ. Press : 203–212.

Fries, D., 1986, Seven ways to winning, Sail, 4: 21–24.

Gorskaya, G. B., 1990, Psychologicheskoye obespechenie podgotovki yachtsmenov, Krasnodar.

Gorskaya, G. B., Barabanov, A. G., 1993, Elite athlete mental health: factors of destabilization ans ways of maintaining, In: Sport Psychology: an integrated spproach, Procedings of 8th World Congress on Sport Psychology Lisbon : 561–565.

Hahn, E., 1983, Mesto psychologii v podgoyovke trenerov In: A.V.Rodionov, N.A.Khudalov (Eds). Psychologia i Sovremenni Sport M., FIS: 66–74.

Hanson, S. J., McCullagh, P., Tonymon, Ph., 1992, The relationship of personality characteristics, life stress and coping resources to athletic injury, Journal of Sport and Exercise Psychology 14: 262–272.

Heyman, S. R., 1987, Counselling and psychotherapy with athletes: special considerations, In: May, J. R., Asken, M. J. (Eds.) Sport Psychology, N-Y PMA Publishing Corp.: 135–136.

Kalinin, E. A., 1962, Oprosnik "motivatsiya sportivnoj deyatelnosti", M., VNIIFK.

May, J. R., Veach, T. L., 1987, The U.S. Alpine ski team psychology programm: a proposed consultation model, In: May, J. R., Asken, M. J. (Eds.) Sport Psychology, N.Y, PMA Publishing Corp.: 19–39.

Milman, V. E., 1983, Stress i lichnosthiye factori regulatsii deyatelnosti, In: Y.L.Hahn (Ed.) Stress i trevoga v sporte. M.F. i S. : 24–26.

Piloyan, R. A., 1985, Motivatsiya sportivnoy deyatelnosti, M., FIS : 91–101.

Skanlan, T. K., Stain, C. L., Ravizza, 1991, An in depth study of former elite figure skaters: III. Saurces of stress, Journal of Sport an Exercise Psychology :103–120.

Walker, S. 1989, 6 obviouse but often unrecognized truths, Sailing World, XXVIII, 7 :17–18.

17

EFFECTS OF MENTAL TRAINING ON SENSE OF RHYTHM

M. V. Yermolaeva and M. M. Chirkova

Russian Research Institute of Physical Culture
Kazakova St.18, 103064, Moscow, Russia

INTRODUCTION

Psychological preparation is an integral part of the athlete's total preparation. Through mental training athletes learn how to relax, focus attention and perform other simple and complex psychological skills. In brief, athletes learn how to use their minds to enhance their individual and collective performances. After all, the body tends to follow what the mind thinks. Mental training provides information about the particular psychological make-up of each player. This information should be used by sport psychologists when formulating mental training programs for athletes. Since athletes possess unique strengths and weaknesses, individualized mental training programs are required. In brief, sport psychologists must know athletes well before they can mentally train them. Mental training is the practising of those psychological skills that lead to performance success. Mental training is an important part of the psychological preparation of athletes and should be integrated into each stage of the training cycle. Knowing the strengths and weakness of each player is essential before designing and implementing effective mental training (MT) programs.

SENSE OF RHYTHM AND PERFORMANCE

The increase in the level of technical mastery is closely associated with the enhancement of the rhythmic structures of sporting activities. Therefore, coaches must discover and implement the psychological factors associated with motor rhythm. Our findings suggest that if athletes desire to recall the rhythmic structure of a motor movement or action, they must reproduce that action mentally. Furthermore, the recollection of rhythm is not a passive process. In contrast, recalling movements requires and demands active mental effort. Thus, the work that was done to imprint and enhance rhythm demands additional motor and mental efforts as well as the creation of a special psychological state. During the stage of specialized technical preparation, rhythm becomes the main criterion of subjec-

Current Research in Sports Sciences, edited by Rogozkin and Maughan.
Plenum Press, New York, 1996

107

tive control of the perfect sporting performance. Our data indicate that the creation of rhythm depends essentially on the development of the appropriate emotional state. In stressful situations when the level of emotional excitement is very high, athletes may temporarily lose their sense of rhythm. Since it is the sense of rhythm that determines the integrity of the sporting action, performers that are out of rhythm do not usually perform well. The optimal emotional state which promotes imprinting and enhancement of rhythm during sporting actions is characterized by moderate emotional excitement, high mental work capacity and low levels of anxiety. This is the optimum psychological state for the effective technical work at this stage. Technical work demands great emotional and physical efforts. Therefore, increments in levels of anxiety may develop if the athlete's work on rhythm enhancement is unsuccessful. Over arousal or over excitement is often the cause of poor performance. To avoid or lessen this problem, coaches must help their athletes control their emotional states while training. The creation of a safe and comfortable psychological atmosphere is very important if athletes are to perform at high levels of competition. The problems associated with rhythm structure may cause an increase in the athlete's anxiety. Anxiety management is essential because high arousal is often associated with decrements in performance. Anxiety may facilitate performance since it prepares the athlete for all-out effort. However, the increase in anxiety should not be excessive. The coach should be patient and should not place undue demands on the athlete, particularly during the early stages of learning. It is often desirable for coaches to ask their athletes to describe their own experiences and sensation following practices and actual competitions. From these descriptions, coaches learn first hand of the problems that their athletes are encountering. Therefore, the coach may adjust his/her methods to help players reach their individual and team goals. During speed-strength and technique training the athlete should be given first hand experience with mental training. Mental training is defined as the practising of the psychological skills which underlie performance success. More specifically, during this stage the athlete learns the psychological skills of relaxation, concentration, imagery, goal-setting and cognitive restructuring. Cognitive restructuring enables the athlete to get rid of negative thoughts that tend to destroy her/his performance. During mental training sessions, the athlete should prepare in the same fashion as he/she does for physical practice. Once the mental training skills are mastered, the athlete should use them during actual practice and competitions. Finally, during speed-strength and technique training, the so called "Fighting State" is introduced. The athlete must find the level of arousal or activation that will enable him/her to achieve optimum performance. The mobilization of energy is necessary if the athlete is to achieve excellence in his/her sport. Individual differences must be taken into consideration since the athlete's muscle, emotional and mental features vary from one person to another. The athlete, coach and sport psychologist must work in harmony to produce optimum arousal or activation for perfect rhythm structure. Besides the optimal emotional stage which promoted imprinting and enhancement of rhythm of sporting action, the structure of rhythm depends on proper motor control. Rhythm combines all characteristics or elements of the sporting action, e.g. strength, speed and kinematic parameters. Carrying out special investigations, we found that at the initial stage of technical work, athletes control their performance by kinematic means, i.e. the positioning of various segments of the body. During the period of technique improvement at subsequent stages, athletes control their motor movements by means of kinematic and rhythmic variables. Furthermore, on the eve of competition, athletes can control their sporting actions using the sense of rhythm. This kind of control permits athletes to direct and mange their sporting actions while performing at extremely high speeds. The basis of proper motor control is motor image and motor sets. Motor image is the most spacious and integrative aspect of self-image and is responsible, in part, for the athlete's technique. Motor space and the inner framework of actions affect the athlete's ability to reorganize his/her movements and to take into consideration an ever changing environment. The data obtained indicate that it is necessary to improve the athlete's motor image if optimal performance is desired. Improvements in motor image should take place during the initial stages of technique formation and also during the preparatory period of each annual training cycle. The adaptive role of the athlete's motor image help him/her adjust to the changing demands of technique development as the perfect or ideal image is imprinted and different modifications of competitive conditions should not influence his/her ability to perform well. The method of psychological control that is used enables the sport psychologist to determine imperfections in adaptation. He or she may also discover whether imperfections are caused by the athlete's basic personality make up or is the result of faulty practice. Deep sensations and total control of actions during the early stage of skill acquisition of each annual cycle enable the coach/sport psychologist to monitor the athlete's movements. The sport psychologist must help the coach and athlete to form a motor image that is in harmony with each other. Many errors during performances and loss of time during training is caused by a discrepancy between the motor image of the coach and the motor image of the athlete. Mental training facilitates the build-up of motor programs, control and correction of ineffective movements. Skill learning is basically a refinement process where the athlete gets rid of ineffective motor movements. Following each execution of the skill, the athlete should visualize the correct motor movements. Motor image is the most general and integrative reflection in the athlete's cognitive system. Cognition, in this context, refers to the structure of sporting actions, i.e. running, jumping, throwing and the motor space where the sporting actions take place. The motor image also includes information about the technique of the sporting movement and the possible changes in structure that need to occur under the influence of various environmental conditions, e.g. temperature and other weather conditions. The motor image is the psychological means by which the athlete forms a cognitive structure of the skill in his/her brain. The creation of the motor image is vital if the athlete desires to perform the skill well. His/her level of skill depends to a large extent on the adequacy and generalization of the motor image. Furthermore, the motor image contains information about how to execute the technique. Motor image formation is especially important for young athletes who often lack experience in executing the fundamentals of the sport. If in the very beginning they fail to form an adequate image, they will not be able to execute the skill under the stressful conditions of actual competition. It is also noted that elite athletes must also concentrate on the enhancement of clear, in focus motor images if they desire to perform well at high levels of competition. This enhancement process is accomplished during the initial stage of functional preparation which takes place in November and March. The motor image differs essentially from the motor set or aim of the skill, e.g. to put the ball into the basket. Motor set is the athlete's disposition to react in a particular way or fashion. It is also the psychological means that the athlete uses to manage or control his or her motor movements. The motor set enables coaches to monitor the movement stereotypes of athletes. If the training process is structured correctly, the motor set is formed during the nest stage of technical preparation, i.e., after the process of motor image enhancement. The adaptive role of the motor image helps athletes adjust to the changing demands of practice and competitive environments. If during the early stages of technical preparation the perfect and integrative motor image was imprinted, then the athlete's motor set will be more flexible and the athlete will be able to adapt to the ever changing competitive environment. At this stage of preparation, the coach should encourage the athlete not only to execute the skill according to the coach's directions but also to compare those directions with his own muscle sensations. In doing so the athlete should be conscious of the whole system or structure of the elements which comprise the motor movement. In other words, during the entire process of skill acquisition the athlete must not passively and unconsciously react to the directions given by the coach. Instead, the player should actively work to create and then enhance his/her motor image.

CONCLUSION

Knowledge of the psychological factors which influence motor image enhancement is of great theoretical and practical significance to players and coaches. Participation in contemporary elite level competition requires a high level of technique mastership. There is a close relationship between the enhancement of the motor rhythm, technique, and the process of motor skill formation. The results of scientific investigations have shown that if an athlete made errors in the process of technique mastery, they accompany him/her for the rest of the athlete's career. Further, mistakes tend to manifest themselves during the most stressful competitive situations. Therefore, it is only those athletes who possess the most adequate and perfect motor images who are able to reach elite levels of achievement. The possession of a clear and in focus motor rhythm enables athletes to produce integrative and vivid images of the whole structure of the sporting action and then process the psychological mechanisms that will help him/her manage motor skill execution in different environments. The coach should take into consideration the athlete's own peculiarities so that player may achieve optimum performance. Each repetition of the motor pattern may be identified as a new process of problem solving. Furthermore, coaches should be patient and should not place undue demands on the athlete, particularly during the early stage of learning. As implied from the above, the coach plays a very important role in the process of motor image enhancement. If the athlete is to form an adequate and integrative structure of rhythm of the sporting action, the coach must prepare the player for creative activity. This preparation will not only help athletes master sport techniques, but will enable them to analyze their motor sensations during recordings of training and game performances.

,

18

EXPERIENCE OF THE APPLYING APPLICATION OF MEDITATION TO SHAPING

T. V. Alfyorova-Popova and O. Votyakova

Institute of Physical Culture
Ordjonikidze St.1, 454111 Chelyabinsk, Russia

INTRODUCTION

Lately among the population there have been widely practised such mass kinds of physical culture and sport as aerobics, rhythmical gymnastics and shaping. Women of different ages and physical fitness levels go in for shaping in numerous shaping groups. Many of the women have never had any practice in sports. Therefore physical loads may also cause excessive physical stress. Hence the problem arises of helping the beginner to adapt to physical loading. Current methods often do not produce the desirable effect. As a result the beginner often complains of pains in joints and muscles and a feeling of extreme fatigue. Sometimes they refuse to continue training. The quest for a new alternative is a real one. Currently, the use of the new methods is constantly expanding in different countries of the world. Methods include yoga, meditation, psychoregulation and mental training are of great interest. Transcendental meditation is widespread in the West while Yoga and Zen Buddhism is widespread in the East. Previous results have demonstrated the effectiveness of methods such as visualization and relaxation in eliminating negative emotions in athletes during the recovery from injury (Laskowski, 1993). The results of applying meditation in shaping are presented in this paper.

METHODS

Attempts have been made to investigate the functional state of women ranging in age from 20 to 40 years old that are training twice a week according to individual computerized Federation Shaping Programme. Parameters such as electrocardiography, cardiointervalography with evaluation of statistic indices of cardiac rhythm (Baevsky et al., 1989), arterial blood pressure both before and after local loading was evaluated. Calculation of the index of double product: $DP = HR*APmax/100$ (conventional units), reflecting the functional heart resources was also carried out. Cardiac rhythm was also evaluated using the following indices: moda (Mo,ms) the most frequently registered meanings of R-R, re-

Current Research in Sports Sciences, edited by Rogozkin and Maughan.
Plenum Press, New York, 1996

111

flecting the influence of humoral-hormonal factors on regulation; variational range (ΔX, ms) the difference between maximum and minimum meanings of R-R, it reflects the activity of the parasympathetic system: amplitude of mode (AMo,%) number of cardiointervals corresponding to moda meaning, it reflects the activity of sympathetic branch of nervous system strain index (SI), conventional units calculated using the following formula: AMo/(2 ΔX*Mo) which reflects the degree of centralization of cardiac rhythm control. The bioenergetic state of the spine was determined by the number of rotations of biolocational frame ("ramka") in the region of 7 principle energetic centres ("chakras"). As a standard there was taken a normal energy distribution increasing from heart chakra towards the first and seventh chakras (Korzin, 1992). General harmony was valued at 7 marks, in disharmony the marks decreased in accordance with the number of overloaded and under loaded chakras. Meditation was used as part of complex biorelaxation fulfilled at the end of training. Meditation was carried out in Shavasana pose under the leadership of a meditation instructor, later with the help of a videocassette. Psycho-physical exercises were also used through the work with bioenergetic channels. Relaxation exercises were also applied for fingers ("husts") and arms ("Mudrs") using yoga techniques. Sportswomen were recommended to carry out independent training every day. Holotropic breathing training was also used with some of the women (Grof, 1985). The control group included women training using the shaping programme without meditation.

RESULTS

Results revealed that after 6 months of training there was marked decrease in resting heart rate ($p < 0.05$) and double product (DP) index (from 77.3 ± 2.8 to 75.5 ± 2.1, conv. units) in the experimental group. During the repeat investigation there was a decrease in heart rate and blood pressure reaction to local loading in sportswomen (Table 1).

APmax reaction to local loading and the decrease of DP at work (from 97.4 ± 3.0 to 90.3 ± 2.3 conv.units). In contrast, in the control group, there was an increase of AP max index ($p < 0.05$) and DP at rest did not change. Both in the experimental and control groups the decrease in arterial pressure reaction to local loading was evident. Cardiac rhythm structure was changed in both groups after 6 months of training (Table 2).

Heart rate changes during local loading in sportswomen of D X ($p < 0.05$) and decrease of strain index (SI) ($p < 0.05$), which demonstrates the increase of parasympathetic and autonomous mechanisms activity and decrease of central influences on the heart. During the initial investigation changes in D X, AMo and SI during local loading were vectoring differently which reveals the "strain adaptation" reaction (Alferova, 1986). This reaction was observed in 53% of subjects. Repeating the investigation revealed greater increases in central influences on the heart judging from increases in AMo and SI. Hence in the experimental group after 6 months of training there prevailed the first type of adaptation to the local loading, the so called "satisfactory adaptation" (62% of women). Both in the control and experimental groups at rest there was an increase in D X index and decreases in SI ($p < 0.05$) and AMo index ($p < 0.05$). During the initial investigation and after 6 months of shaping training the changes in cardiac rhythm during local loading corresponded to "satisfactory adaptation" but the strain of central mechanisms of regulation during the repeated investigation was less. The bioenergetic state of the spine was not different in both groups during the initial investigation (4.6 ± 0.17 and 4.4 ± 0.28 marks).

Indices of Physical Working Capacity in Sportswomen. The data obtained revealed the positive influence of shaping on physical development especially in combination with

Table 1. Heart rate and blood pressure reaction to local loading in sportswomen

Groups	Heart rate (beats/min)			Blood pressure max (mm Hg)			Blood pressure min (mm Hg)		
	1	2	3	1	2	3	1	2	3
Experimental group (n=18)	66.1 ± 2.4	76.4 ± 1.6*	67.4 ± 5.8	112.1 ± 3.0	127.1 ± 2.0*	113.5 ± 1.6	69.7 ± 1.1	82.1 ± 2.0*	73.2 ± 7.9
	59.4 ± 2.1	67.4 ± 1.9*	64.5 ± 4.9	108.2 ± 1.1	119.4 ± 1.0*	109.8 ± 2.4	68.2 ± 2.7	82.9 ± 1.8*	72.2 ± 1.8
Control group (n=19)	63.8 ± 1.1	76.7 ± 2.7*	64.5 ± 1.8	106.5 ± 2.0	138.2 ± 1.0*	118.5 ± 2.0*	66.8 ± 1.5	80.0 ± 1.8*	72.9 ± 1.6*
	61.2 ± 1.6	68.9 ± 2.5*	65.8 ± 1.4*	112.1 ± 1.4	126.4 ± 1.0*	115.0 ± 1.3	67.5 ± 1.0	81.4 ± 1.2*	71.4 ± 2.9

1 - before, 2 - during, 3 - after local loading. * - reliable differences with the initial data (p<0.05).
The first horizontal line - the initial investigation, the second horizontal line - after 6 months of training.

Table 2. Heart rate changes during local loading in sportswomen

Groups	Mo (ms)			ΔX (ms)			AMo (%)			Strain Index (SI) (conventional units)		
	1	2	3	1	2	3	1	2	3	1	2	3
Experimental group (n=18)	910 ± 30	790 ± 20*	900 ± 20	180 ± 20	200 ± 10	240 ± 10*	33.3 ± 1.4	32.9 ± 1.6	28.8 ± 2.6	92.3 ± 4.1	105.0 ± 6.2*	70.0 ± 2.1*
	1010 ± 30	840 ± 30*	930 ± 20	270 ± 20	190 ± 20*	250 ± 60	29.3 ± 2.2	41.7 ± 3.2*	32.2 ± 2.2	79.5 ± 4.4	159.0 ± 16.2*	83.0 ± 4.2
Control group (n=19)	940 ± 20	790 ± 30*	930 ± 40	185 ± 20	160 ± 10	240 ± 20*	36.2 ± 1.1	59.9 ± 3.1*	26.2 ± 0.9*	130.4 ± 6.3	239.0 ± 24.5*	108.8 ± 4.1*
	980 ± 10	870 ± 6*	920 ± 6*	250 ± 20	210 ± 10*	260 ± 10	33.0 ± 1.8	44.5 ± 0.9*	37.9 ± 1.8*	89.3 ± 3.5	117.9 ± 6.7*	87.8 ± 8.2

1 - before, 2 - during, 3 - after local loading. * - reliable differences with the initial data (p<0.05).
The first horizontal line - the initial investigation, the second horizontal line - after 6 months of training.

Experimental group

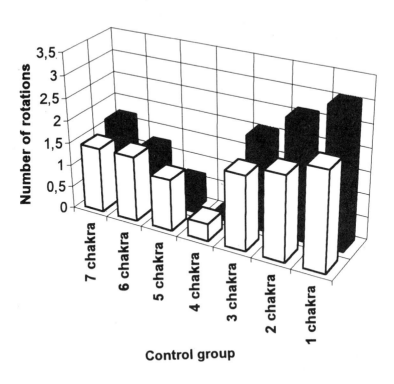

Control group

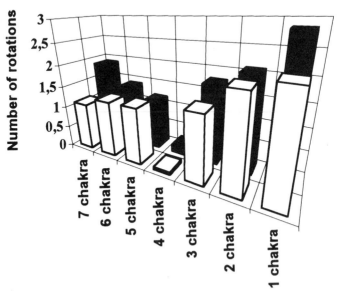

Figure 1. Changes in Bioenergetic condition of main chakras at the end of training energetic harmony index in the experimental group equaled 6.35 ± 0.15 and was less in the control group 5.7 ± 0.12 (p<0.05). The energetic loading of heart chakra decreased in the experimental group, unlike the control group, which was in accordance with the improvement of electrocardiography findings. Physical properties improved in both groups after 6 months of shaping training. In the experimental group these changes were more evident, especially in the arm muscles (Table 3).

Table 3. Indices of physical working capacity in sportswomen

Groups	VMS	VMZ	VMA	SEH
Experimental group (n=18)	43.4 ± 1.7	106.9 ± 4.0	4.8 ± 0.6	53.3 ± 2.0
	58.3 ± 2.0*	116.1 ± 2.2*	6.4 ± 0.4*	66.1 ± 2.7*
Control group (n=19)	43.1 ± 2.0	91.1 ± 4.0	5.8 ± 0.5	57.0 ± 2.1
	52.8 ± 2.0*	101.0 ± 2.6*	6.5 ± 0.5	61.5 ± 1.7

* - reliable differences with the initial data ($p < 0.05$).
The first horizontal line - the initial investigation; the second horizontal line - after 6 months of training.
VMS - volume of stomach movements (number of repetition at 4 min)
VMZ - volume of legs movements (number of repetition at 4 min)
VMA - volume of arms movements (maximum number of repetition)
SEH - static endurance of hand

meditative relaxation. Judging from heart rate and DP indices at rest heart function economization was more effective in the experimental group. The normalization of their adaptive reactions to local loading also should be noted. During the initial investigation these reactions were, as a rule, in accordance with "strain adaptation". In the experimental group the acceleration of functional indices of recovery after local loading at the end of experimental period was also found. Application of meditation in relaxation correction of the recovery period undoubtedly improves functional state and physical working capacity and increases adaptive resources. During the repeated investigation women in the experimental group noted high spirits, improvement in their general condition and wish to train. The contingent of this group was more stable than in the control group. According to recent data during meditation and other alternative states of consciousness changes at the hypothalamic-pituitary system level were found while improvements in cortical-subcortical and cerebral hemispheres reactions as well as improvements in endocrine functions were also found (Bundzen at al., 1994).

CONCLUSION

It should be emphasized that positive effects of such alternations increase adaptation to stress loading not only in sport but also in other spheres of activity and contribute to rapid functional recovery after loading. This explains the positive influence of meditation relaxation on the functional state of women trained according to the shaping programme.

REFERENCES

Alfyorova, T. V., 1986, Age pecularities of local muscle activity in sportsmen, OIPC, Omsk: 31.
Baevsky, R. M., Kirillov, O. Z., Klezkin, S. Z., 1984, Mathematical analysis of cardiac rhythm alterations in stress, Science, Moscow :221.
Bundzen, P., Unestähl, L.-E., Malinin, A., 1994, Alternative State of Consciousness: Neurodynamic Correlates and Brain Mechanisms, Mental Training for Sport and Life in Russia, Sankt Petersburg :16-20.
Grof, S., 1985, The brain: Birth, death and trancendence in psychotherapy, State University of New-YORK, Press: 274.
Korzin, O. A., 1992, Geomancial - the knowledge and skill of the ancients. Parapsychology and psychophysiology, 3:14-23.
Laskowsky, E. R., 1993, Psychology in sport medicine J.of Sports Med. and Physical Fitness 33:1029-1030.

19

ANATOMICAL DATA FOR BIOMECHANICAL CALCULATIONS

I. M. Kozlov, A. V. Samsonova, and A. B. Sinukhin

Academy of Physical Education
Dekabristov St., 35, 190121 St. Petersburg, Russia

INTRODUCTION

The human locomotor system has two main functions: moving and sensor (proprioreceptor). Both are necessary for muscle activity regulation. Knowledge about muscle morphometric characteristics (length and force arm) during motion allows analysis of: 1) regimes of muscle contraction; 2) the most effective zones of force suppling; 3) data about muscle speed ability.

There are many methods of determining muscle morphometric characteristics. They may be divided into two groups: direct measure methods and simulating. Each of these methods have positive and negative aspects. However, simulating allows more meaningful information to be obtained about the changes in muscle morphometric characteristics during movement (Kozlov et al., 1988). Therefore, it is necessary to have data about joint angles and parameters of distance between a joint centre of rotation and points of muscle attachment to bones. Determination of joint angle data may be performed with a variety of well known methods (cinematography, ciclo or video). Despite changes in joint angles during movement, parameters which characterise distance between centres of rotation and points of muscle attachment don't change. These parameters are constant and depend on the individuals characteristics. In the literature describing skeletal and muscle anatomy there are only qualitative descriptions of muscle attachment characteristics (Ivanizkij, 1938; Sinelnikov, 1972). There is not enough data about these constant parameters, and some of the literature has drawbacks: 1) sexual differences have not been taken into account; 2) subject numbers are low; 3) no examination of the relationship between anthropological characteristics of the human locomotor system and the constant parameters has been made (Kozlov and Zvenigorodskaja, 1981).

The task of this research was to quantify the numerical constants which characterise distance between centre of joint rotation and point of muscle attachment to bone and to determine the relationship between these parameters and the anthropological characteristics of the human locomotor system.

Current Research in Sports Sciences, edited by Rogozkin and Maughan.
Plenum Press, New York, 1996

METHOD FOR CALCULATION OF LENGTH AND FORCE PRODUCTION OF HUMAN LEG MUSCLES IT IS NECESSARY TO HAVE DATA OF 17 CONSTANTS

All the constants may be divided in two parts. The first group constants: a, d, e, k, l, n, p, R1, R3 may be measured just on the subject's body or on the photo (N=8). The second group constants: b, c, f, m, r, R2, u and t may be measured on cadavers. For the determination of second group constants male (N=23) and female (N=20) cadavers were measured with an age range of 25 - 45 years. The centre of joint rotation and point of a muscle attachment were marked. For the distance from points of muscle attachment and the centre of joint rotation measurements were made using Martin's anthropometer (error of measurement was 0.05 sm). If the place of muscle attachment was not precisely defined, such as with m.tibialis anterior on the tibia, the most distal point (D) and most proximal point (P) of muscle attachment were determinated. Numerical constants b, c, f, m were calculated by the following formula.

$$\text{Const} = (X1\text{-}X2)/2 + X1 \tag{1}$$

where X1 - distance from the joint's centre rotation to the point P, X2 - distance from the joint's centre of rotation to the point D.

RESULTS

To examine the numerical means of the constants and anthropological characteristics of the human locomotor system a correlation analysis was used (r - Brave-Pirson correlation coefficient). There was a weak correlation between the numerical means of the constants b, c, f, u, and R2, and the anthropological characteristics of the human locomotor system but there was a strong positive correlation between the means of constant m and femur length (n).

For constants b, c, f, u, and R2 that have weak correlations with anthropological characteristics a statistical value was calculated.

For constant m which has a strong positive correlation with femur length a regression equation was evaluated. A linear regression equation was calculated between the mean of m constant and femur length. Formulas which describe the relationship between mean value of m constant for men (2) and women (3) were obtained.

$$m = 0.560n + 0.654 \tag{2}$$

Table 1. Correlation coefficients between characteristics of the human locomotor system and measured constants

Characteristics of human locomotor system	Sex	Constants					
		b	c	f	m	R2	u
Crus length (e)	m	0.482	0.703	—	—	—	—
	f	0.606	0.404	—	—	—	—
Femur length (n)	m	—	—	0.546	0.817	0.185	0.368
	f	—	—	0.514	0.827	0.188	0.108

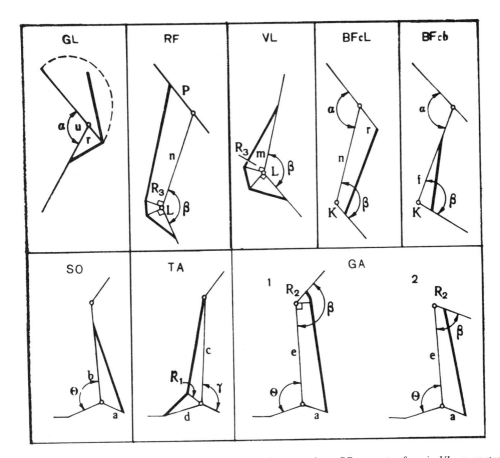

Figure 1. Models of human leg's muscles. Where: GL - m. gluteus maximus; RF - m. rectus femoris; VL - m. vastus lateralis; BFcL - m. biceps femoris caput longum; BFcb - m. biceps femoris caput breve; SO - m. soleus; TA - m. tibialis anterior; GA - m. gastrocnemius. (Designation see at the text). a - distance between the talocrural articulation centre of rotation and the point of the triceps surae attachment to the calcaneal tuber; b - distance between the talocrural articulation centre of rotation and the point of m.soleus attachment to the tibia; c - distance between the talocrural articulation centre of rotation and the point of m. tibialis anterior attachment to the tibia; d - distance between the talocrural articulation centre of rotation and the point of m. tibialis anterior attachment to the foot; e - crus length (distance from the talocrural articulation centre of rotation till the knee joint centre of rotation); f - distance from the knee joint centre of rotation till the point of m.biceps femoris caput breve attachment to the femur bone; k - distance from the knee joint centre of rotation till the point of m. biceps femoris attachment to the fibula bone; l - distance from the knee joint centre of rotation till the point of m. quadriceps femoris to the fibula bone; m - distance from the knee joint centre of rotation till the point of m. vastus lateralis to the femur bone; n - femur length (distance from the knee joint centre of rotation till the hip joint centre of rotation); p - distance from the hip joint centre of rotation till the point of m. rectus femoris attachment to the iliac bone; r - distance from the hip joint centre of rotation till the point of m. biceps femoris caput longum attachment to the tuber of the ischium; R1 - distance from the talocrural articulation centre of rotation till the superior retinaculum of the extensor muscles of the foot; R2 - distance from the knee joint centre of rotation till the point of m. gastrocnemius attachment to the femur bone; R3 - distance from the knee joint centre of rotation till the kneecap front surface. t - distance from the point of attachment to the iliac bone till the tuber of the ischium; u - distance from the hip joint centre of rotation till the point of m. gluteus maximus attachment to the femur bone.

Table 2. Mean values of constants which characterise distance
from axes of rotation to points of muscle attachment, (sm)

Constants	Sex	
	Male	Female
b	21.7 ± 0.6	20.3 ± 1.0
c	23.5 ± 0.8	21.0 ± 1.0
f	16.2 ± 1.0	15.5 ± 0.8
r	2.9 ± 1.0	2.3 ± 0.8
R2	2.2 ± 0.4	1.9 ± 0.6
t	16.9 ± 1.0	16.7 ± 0.8
u	11.9 ± 0.8	10.9 ± 0.8

$$m = 0.611n - 1.681 \tag{3}$$

where: m - distance from knee joint centre of rotation till point of m. vastus lateralis attachment to femur, sm; n - femur length, (sm).

Figure 2 shows a graphical representation of the relationship between m and femur length.

CONCLUSION

Distances from points of muscle attachment to centres of joint rotation were determined. Most of the constants had weak correlations with the anthropological charac-

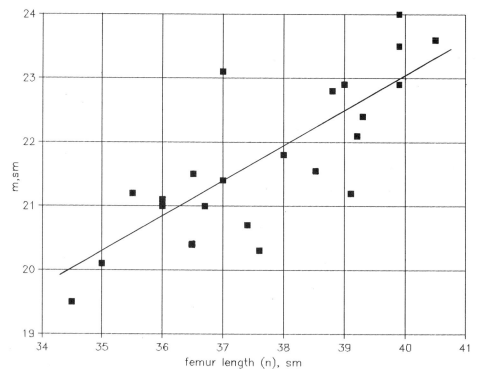

Figure 2. Relationship between values of m constant and femur length (men).

teristics of the human locomotor system. A regression equation was calculated to allow modelling of morphometric muscle characteristics during sport movements.

REFERENCES

Ivanizkij M.F.,1938., Human body movements, Moscow.
Kozlov I.M., Zvenigorodskaja A.V., 1981. Anatomic measurements data for muscle movements evaluation., in: Biomechanical factors of sport movements co-ordination, Leningrad,: 70–92.
Kozlov I.M., Samsonova À.V., Sokolov V.G., 1988, Morfometric characteristic of low limbs muscles, in: Archive of anatomy, gystology, embryology, 94: 47–52.
Sinelnikov R.D., 1972., Atlas of human anatomy, Moscow.

20

MATHEMATICAL MODELLING OF SPORTS HUMAN MOVEMENTS

L. A. Khasin and M. L. Ioffe

Moscow Regional State Institute of Physical Culture
Shosseinaya ul., 33, pos.Malahovka-2, Moscovskaya obl., 140090 Russia

ABSTRACT

The modelling of human body movements based on the analogy with movements of the n-unit mechanism has been carried out. The movement of the n-unit mechanism may be described as the system of the n-differential equation of second order. The authors have been the first to use generalized coordinates, theoretically found, in order to solve the direct, inverse and mixed problems of dynamics.

The search for generalized coordinates has been implemented in the wide class functions according to a number conditions. This approach enables the use of mathematical modelling for the solution of practical problems in sports.

INTRODUCTION

For many years, the modelling of human body movements has used the analogy of the human locomotor system with the anthropomorphous mechanism (1, 2, 3, 4). Despite the obvious resemblance of the locomotor system and the anthropomorphous mechanism and a large amount of work on this theme, the results that would let us understand better how the human being moves or at least how to make a certain movement, particularly a sporting movement more effective, have not been obtained. Nevertheless, the use of anthropomorphous mechanisms for the modelling of the human body movements appears very tempting, due to its comparative simplicity, the perfection of the mathematical apparatus, etc.

In finding the reasons of failure of the present approach we have come to the conclusion that besides the usual reasons (complexity of the task, presence of antagonist muscles, errors of computation, the converse problem of dynamics), there is one more reason that strongly restricts the investigators. When solving the direct and converse problems of dynamics (1, 2) and others have put the data obtained while registering the human body movements into the equations of movement and the unknown dynamic and kinematic

Current Research in Sports Sciences, edited by Rogozkin and Maughan.
Plenum Press, New York, 1996

123

characteristics have been computed. The characteristics obtained have an empirical character and don't allow conformity with theoretical laws. Unwarranted tie to the registration data breaks many obvious conditions to which kinematic and dynamic characteristics must comply, e.g. the position of the masses' centre, the value of the first and the second derivative of the articulation angles, and many others. Also, it is not always possible to separate the regular and occasional phenomena. And finally, in this situation it is impossible to solve the optimization tasks. Overall, we can say that these authors' methodological error is that they haven't tried to deduce the theoretical knowledge from the experience.

The approach that we have used is to treat the problem of generation of the anthropomorphous mechanism as a theoretical one i.e. the articular angles, the position of the centre of mass and other kinematic and dynamic characteristics of the anthropomorphous mechanism's movement are formulated.

SOLVING THE PROBLEM

The aim of our work is to construct a computer system modelling human body movements, and in particular, sports movements. As mentioned above, the modelling is based on the analogy of the human muscle-skeleton system and the anthropomorphous n-legs mechanism, the movement of which is described by the laws of theoretical mechanics. The process of creating the modelling system may be divided into solving two problems.

The first problem deals with the construction of the mathematical model itself, including the deduction of equations of the system's movement and the choice of the numerical algorithm for their solution. Also, the programming of the task and debugging of the software.

The second problem deals with the preparation of the basic data that are necessary for the process of modelling. The basic data necessary for the solution of the so-called converse problem, are kinematic parameters of the system i.e. angles, angular velocity and acceleration. This is a complicated, scientific and technical problem and we call this problem the problem of movement generation.

When solving the above problems the following suppositions have been made. The anthropomorphous mechanism presents n-bars sequentially jointed by spherical joint. Only movement in one plane is examined. The centres of mass are situated on the intervals that join the adjacent joints. The only outer force acting on the bodies is the force of gravity. In previous works, the mathematical modelling was based on solving the direct and converse problem of dynamics for the mechanism of interest. We would remind you that the direct problem is a problem with the known joints' moments and initial conditions, i.e. the initial values of all the angles and their derivatives.

In solving the direct problem we can find values of the angles, angular velocity and forces of reaction in joints as a function of time. In the converse problem the known parameters are the angles, angular velocity and accelerations during the whole movement. In solving the converse problem we find values of the joints' moments and forces of reaction. However, in practice there are problems which are neither direct nor converse. For instance, when we deal with the rotation of a gymnast on the horizontal bar we may probably not take into consideration the moment in the first joint. In these conditions the solution of the converse problem becomes more complicated: it is impossible to determine all the angles. Sometimes in order to construct the model it is necessary to include the additional information about the forces of reaction, e.g. if a tensometric platform is used.

In order to widen the possibilities of the modelling we have constructed the algorithm of the so-called generalized or mixed problem for the n-legs mechanism. The mechanical system in point has n+2 generalized golonomic coordinates: n angles determining the angle position of the joints in the plane of the movement and 2 linear coordinates of the end of the first leg.

The algorithm of the generalised problem as well as the algorithm of the direct and converse problems are based on the equations of movement that have been brought out by the D'Alamber method, i.e. including the forces of reaction. For each body we create equations for the projection of forces acting on the body, on the axes of the inertia system and for the moments of forces relative to the centre of mass of a body. Thus for a system consisting of n bodies we get 3n equations that are the differential equations of the second order relative to the angles and coordinates.

These equations of a movement may be written as follows:

$$
A * \begin{bmatrix} j \\ m \\ X \\ Y \\ X \\ Y \end{bmatrix} = \begin{bmatrix} 0 \\ - \\ - \\ 0 \\ L(2*n) \end{bmatrix}
$$

where A–(3nx(3n+2)) matrix depending on the values of the angles, lengths, masses, moments of inertia of the bodies and the parameters determining the position of their centre of mass,

$$
\varphi = \begin{bmatrix} \varphi(1) \\ - \\ - \\ - \\ \varphi(n) \end{bmatrix} \quad \text{(nxl) vector formed of the angles's acceleration}
$$

$$
m = \begin{bmatrix} m(1) \\ - \\ - \\ - \\ m(n) \end{bmatrix} \quad \text{(nxl) vector formed of the moments acting on the link}
$$

$$
X = \begin{bmatrix} X_{01} \\ - \\ - \\ - \\ X_{n-1n} \end{bmatrix} \quad \text{(nxl) vector formed of the projections of the forces of reaction on the horizontal axis}
$$

$$
Y = \begin{bmatrix} Y_{01} \\ - \\ - \\ - \\ Y_{n-1n} \end{bmatrix} \quad \text{(nxl) vector formed of the projections of the forces of reaction on the vertical axis}
$$

X,Y—scalar quantity equal to the second derivative coodinates L(2xn)-(2nxl) vector of the centrifugal forces and the forces of gravity.

Let's introduce vector Z with a dimension 4xn+2Z = (φ,m,X,Y,X,Y). Depending on the type of problem to be solved (direct, converse, mixed) n+2 components of vector Z are the known quantity while the rest, 3xn, the unknown ones. For example, for the direct problem in a support movement the known quantities are n moments and two projections of the acceleration X and Y which probably equal to zero. For the same problem without support movement, the known quantities, besides the moments, are two projections of the force of reaction in the first joint which are probably equal to zero. The remaining unknown 3xn quantities can be found by solving the respective linear system of the equations directly deducible from (1). Depending on the type of problem among 3xn quantities there will be a certain amount of second derivatives of the angles' coordinates. For example, for the direct problem there will be the following second derivatives: n for the support movement and n+2 for the others, while there will be none for the converse problem. Thus, the solution of the mixed problem is reduced to the problem of the integration of the system of the differential equations, the order of which varies from 2x(n+2) to the zero. In the last case the calculations deduce to algebraic ones.

GENERATION OF MOVEMENTS

Our approach to the generation of movements is based on setting the theoretical conditions to which kinematic and dynamic characteristics of the movement must satisfy. Besides this, an expert sets the characteristic phases of the investigated movement. "The phase" means the period of monotonous changing of the angle's coordinate. Usually, a small number of such phases is chosen. Since for the modelling of the movement continuous recording is needed, i.e. it is necessary to know the angles' value, the angular velocities and the angles accelerations in whatever moment of time from the beginning of the movement, we need to link the characteristic points (phases) by the smooth curve, i.e. by a curve with continuous first and second derivatives.

Following from the above, the following mathematical problem has arisen. It is necessary to construct a set of functions, in particular, polynoms that have the following characteristics on the interval (xl, x2): a) on the ends of the interval the function takes the set meanings and has the first derivatives equal to zero; b) inside the interval the function is monotonous, i.e. its first derivative retains the sign; c) in the set interval the function has not more than one point of inflection, i.e. inside the interval the second derivative has no more than one zero; d) on the ends of the interval the second derivative turns into zero.

With the help of homographic correspondence the problem may be deduced to the problem of finding the function y(t) on the interval (0,1) with the following characteristics:

$$y(0) = 0, \ y(1) = 1, y'(0) = y'(1) = y''(0) = y''(1) = 0, \quad y'(t) \geq 0 \text{ for } 0 \leq t \leq 1 \qquad (1)$$

Inside the interval (0,1) Y''(t) has no more than one zero. The relation of the functions F(x) and Y(t) is defined by the formula:

$$F(x) = F(x)_1 + (F(x_2) - F(x_1)) * y((x - x_1)/(x_2 - x_1)) \qquad (2)$$

Further the problem of constructing a set of functions on the interval (0,1) is examined. Our approach to the solution of the raised problem of finding the set of functions with the certain attributes is as follows. Suppose the second derivative of the function that is being found is defined by the formula:

$$F''(x) = \gamma * x * (x-1) * (x-\alpha) * x^n \tag{3}$$

This function depends on three parameters: γ, α, n. It is obvious that $F''(x)$ becomes zero when $x = 0$, $x = \alpha$, $x = 1$. Thus from equation (3), the only point of inflection and the second derivative found for the function $F(x)$ is the point $x = \alpha$. Three parameters γ, α, n are linked by two dependencies, brought from the conditions: $F''(1) = 0$, $F(1) = 1$. Thus,

$$\alpha = (n+2)/(n+4) \tag{4}$$

$$\gamma = 0.5 * (n+3) * (n+4)^2 * (n+5) \tag{5}$$

Since n>0 it follows from (4) that

$$5 \leq \alpha \leq 1 \tag{6}$$

Thus the formulas (3), (4) and (5) define the function

$$F(x) = \gamma * x^{n+3} * (x^2/((n+4)*(n+5)) - (\alpha+1)*x/((n+3)*(n+4)) + +\alpha/((n+2)*(n+4))),$$

that satisfies the set conditions and the point of inflection of which is situated in the right half of the interval (0,1).

MODELLING SOME SPORTS MOVEMENTS (GIANT SWING AND PRESSES)

The movement of a gymnast accomplishing a giant swing has been modeled by the movement of a three-links joint mechanism with known mass-inertial characteristics: l(i) - length of links, m(i) - masses of links, in(i) - moments of inertia of links relative to the centre of mass, al(i) - coefficients determining the position of the centre of mass of links. Formerly, it was considered that a horizontal bar is absolutely stiff i.e. it supported movement (the beginning of the first link was regarded as a fixed point—the beginning of the system of coordinates with the horizontal and vertical axes)

It has been considered that the moment of the forces of the interaction of a gymnast with a horizontal bar may be disregarded, ie. mo1 = 0. The problem has been solved in relative coordinates. A mixed problem has been put, i.e. the unknown parameters are: angle φ_1 of a turn of the first link relative the immobile (fixed) system of coordinates, joint moments in joints, and also the forces of reaction.

Thus, the angles of turn of the second link relative to the first one (φ_{12}) and the third link relative to the second one (φ_{23}) are considered to be known. The known angles have been set by their characteristic values (phases) at known periods of time.

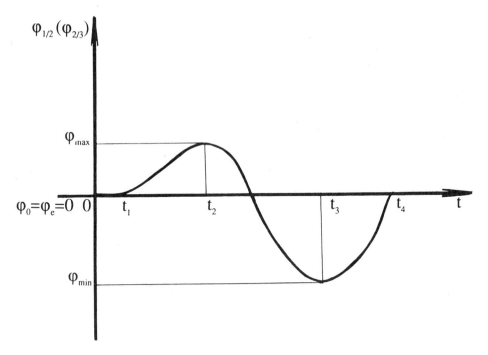

Figure 1. The structure of the execution of a giant swing.

Further, the values are approximated at corresponding intervals by functions examined above. It has been considered that in all functions, the point of inflection is situated in the middle of an interval, i.e. $\alpha = 0.5$. That is,

$$n = 0; \gamma = 120; f(x) = 6*x^5 - 15*x^4 + 10*x^3 \tag{7}$$

$$f'(x) = 30*x^4 - 60*x^3 + 30*x^2 \tag{8}$$

$$f''(x) = 120*x^3 - 180*x^2 + 60*x \tag{9}$$

The algorithm of modelling of the giant swings has been created on the base algorithm of numerical solution with the help of the Runge-Kutt method on a given interval of a differential equation of the second order for an unknown angle of turn of the first link.

In accordance with the described numerical algorithms the following programs have been made: a) for the modelling of the giant swings on an absolutely stiff horizontal bar; b) for the modelling of the giant swings on an elastic horizontal bar. The basic data for the modelling in both programs are: a) mass-inertia characteristics involving one-dimensional array of link mass $m(i)$, link length $l(i)$, coefficients that determine the position of the centre of mass of links $al(i)$ and the moments of inertia relative to the centre of mass $in(i)$; b) the initial conditions necessary for the integration of the differential equations include the value of an angle φ and of an angle's velocity φ, and the values of the coordinates X, Y and their derivatives XT, YT; c) the array of the moments of time $x1(i)$, $x2(i)$ and the corresponding angles' arrays φ_{12}, φ_{23} values $y1(i)$, $y2(i)$ necessary for the calculation of the angles, their first and second derivatives at any moment of time according to the ap-

Table 1. Initial data for two variants of movement execution

		Variant 1					Variant 2		
t_s	t_1	t_2	t_3	t_4	t_s	t_1	t_2	t_3	t_4
	0	0.975	1.275	1.5		0	0.97	1.23	1.57
$\varphi 1/2$	$\varphi 0$	φmax	φmin	φe	$\varphi 1/2$	$\varphi 0$	φmax	φmin	φe
	0	15°	−90°	0		0	20°	−60°	0
t_s	t_1	t_2	t_3	t_4	t_s	t_1	t_2	t_3	t_4
	0	0.975	1.175	1.5		0	0.92	1.23	1.57
$\varphi 2/3$	$\varphi 0$	φmax	φmin	φe	$\varphi 2/3$	$\varphi 0$	φmax	φmin	φe
	0	30°	−60°	0		0	40°	−50°	0

proximating formulas (see above); d) the interval of time during which an exercise is accomplished (time of modelling) and the number of steps.

For the modelling of giant swings on an elastic horizontal bar the coefficients of rigidness (Cl and C2) also need to be set. It should be noted that the program of modelling of the giant swings on an elastic horizontal bar also allows modelling of a nonsupport movement. In this case C1 = C2 = 0.

The output parameters of the model are: a) the angle values φ_1 and its first and second derivatives; together with the known angle values it allows the position of a mechanism during the whole period of movement to be demonstrated; b) the value of a kinetic moment (angular momentum) and the moment of inertia of a system (mechanism) relative to its centre of mass that are calculated on the basis of known angles and angular velocities at any moment of time, including the moment of the ending of a movement; c) joint moments m_{12}, m_{23}; d) forces of reaction in the joints; e) for an elastic horizontal bar the values of the coordinates of a point of support X, Y and their derivatives XT, YT.

Eight variations of the giant swing have been modeled. The structure of a movement (fig. 1, table 1) as well as the approximate angles' values and the time needed to execute them have been set by the expert N.G. Suchilin who has used his own experience and the registration data of a movement.

These values have been refined according to the results from the mixed problem of dynamics. The angular velocity of a gymnast in a vertical position has been taken as a criterion of the quality of the execution of an exercise. Moreover, the relative angular velocities and the joints angles in vertical position are equal to zero. The initial data (moments of time and the angles φ values) for two versions of the giant swing are shown in table 1. The results of the modelling show that for version 1 (which is correct, according to the experts) there is an increase in angular velocity of 3.4 rad/sec, while for version 2 (which has

Table 2. Results of modelling of the bar press

			Moment max (NM)	
N	Time(s)	Force max (N)	1	2
1	0.5	1800	600	1100
2	1	900	140	480
3	2	685	50	200
4	3	645	10	170
5	5	625	7	158

been executed so that the flexion in pelvis-thigh joint and in shoulder joint has been late) there is an angular velocity of 1.7 rad/sec.

The modelling of the press of a bar has been made in an analogous way. The results of modelling are shown in table 2.

The dynamic characteristics of a movement greatly depend on the speed of its execution. The generation of a movement (press) is based on solving the problem of optimization. Besides the above conditions as functions of the joints' angles, the maximum deviation of the trajectory of a bar from the vertical line has been minimized.

The approach we offer allows a comparison of variations in the execution of sporting movements and gives theoretical laws that can be used in sports science and practice.

REFERENCES

1 Aleshinsky S. G., Zasccozky, V. M. Modelizovanic prostzanstvennogo dvigenia cheloveka // Biophisica. 1975: 20 1121–1126.
2 Aleshinsky S. G., Zasccozky, V. M. Opzedelenie megzvennix momentov vnutzennix sil voznikaushix pridvigenii cheloveka // teorici i praktexa phizicheskoy kuetuzi. 1974: 11.
3 Marshall R.N., Sensen R.K., Wood G.A. A General Newtonian Simulation of an N-segment Open Chain Model // J. Biomechanics. 1985: 18(5) 359–367.
4 Pandy M., Zajak F., Sim F., Levine W. An optimal Control Model for Maximum Height of Human Jumping // J. Biomechanics. 1990:23(12) 1185–1198.

21

THE CONTROL OF MOVEMENT

S. P. Evseev

Academy of Physical Education
Dekabristov St. 35, 190121, St. Petersburg, Russia

INTRODUCTION

This work proposes a structure for developing motor actions which differs from the traditional ones in many respects. Firstly, starting from initial stages, the structure orients both an athlete and a coach to obtaining a near perfect movement, and, secondly, it requires minimal changes from the standard optimum technique for a given task and therefore guarantees the best technical result.

The movements must start to be developed from the very beginning of training in a laboratory setting using training devices that imitate the necessary joints movements of an athlete. Once the athlete is competent at the movements the controlled training can be stopped and all the technical requirements of the natural movement should remain. The devices which manipulate an athlete's joints movements are called imperative devices (ID). On the one hand they provide conditions which allow athletes to perform certain actions with a planned level of quality and, on the other hand, they restrict the possible deviations of his body segments from the technically correct positions. This method also makes it possible to prepare an athlete (using the mentioned device) to meet typical critical situations which are not uncommon in natural conditions. It also helps to avoid learning possible technical faults in a movement.

Designing, construction and control of the training devices governing joints movements can help to create the necessary prerequisites to form a special structure for developing motor actions. Several new stages are incorporated with this type of learning:

- the development of orientational base of a motor action and sensor-perceptive training at the imitation of performance stage of the action with a training device;
- the development of neuro-muscular coordinations and the corresponding muscular sensations using orientational stage of the action imitated by the device;
- the development of specific qualities and abilities necessary for the action with the device;
- the development of self-controlled skills, prevention and correction of mistakes;
- the transitions to independent motor activity exercises execution and the demonstration of the achieved result in natural conditions. The defined stages correspond to the primary and advanced stages of traditional motor activity training.

Current Research in Sports Sciences, edited by Rogozkin and Maughan.
Plenum Press, New York, 1996

RESULTS

I.

The main tasks of the first stage are:
- to form visionary logical perception about the optimum program and admitted deviations;
- to study the instruction of the basic activity;
- to form sight, vestibulatory, statodynamic, joint and other perceptions, the perception of the exercise as whole as well as the conditions in which it is performed; the execution of the main part which the help of the training device.

The activity of the coach during this stage includes the next operations:
- Demonstrates the optimum program and admitted deviations with the help of a device, video, multiplicated films, cinemagrams and others. The main and corrective directing jolts movements are defined; makes the analysis of the shift of separate parts of the device. The coach controls the activity of his trances, stimulates them asking questions;
- Using the device the coach demonstrates all the reciting movements and dynamic stance which are to be performed by the trained; explains the kinematic and dynamic character of all operations, gives the instructions. Describes feelings and perceptions which appear during the exercise execution. Performs the exercise, defines when to switch of the sight control from body rings to exterior conditions and vice versa.
- Performs the same activity as in the second task, but not with a model but on the device with the trance inside. Makes the fixation of separate body positions at variable speed, asks the trainee about his feelings and thus corrects the of the device.

Athlete's activity during the first stage includes the next operations:
- Precepts the optimum pose program and admitted deviations. Analyzes the elements of directing movements and dynamic stance. Works with device moves its parts with optimum speed, slows down their movements, loudly pronounces the components of the technique, stops and the most important moments. Comprehends, summarizes know ledge and perception of optimum pose program and admitted deviations.
- Precepts the instruction, contents. Comprehends and consolidates them, manipulating the device. Memorizes full and restricted verbal formulae of the studied action.
- Analyses, puts into system and comprehends the born perceptions, connects them with the contents of verbal formulae, strives to achieve the coincidence of real perceptions with the described in the instruction. When they do not coincide he informs the coach. Defines consecutive row of external movements.

Functions of the training device are demonstration of the optimum posture program and reproduction of optimum directing joint movements instead of the trainee as well as the elements of dynamic stance.

The expected pedagogical effects for main tasks of the first stage are:

- the understanding of the motor task, the formation of logical perception end verbal logical constituent of the studied image, the formation of associative connections;
- the understanding of the essence of the motor task, the formation of the verbal logical constituent of the studied action;

• the formation of sensory perceptive constituent of the mode of action, the way of its execution and conditions in which it is performed.

II.

The task of the second stage is to create neuromuscular perceptions, corresponding to the part of the action modeled by the device which the help of the basics studied at the previous stage.

The activity of the coach. Observes the trainee s activity and indicators of immediate information on technical aids. Correct the athlete's activity, asks him about his muscular perceptions, tries to make them coincide with those in the instruction. In some cases allows to correct instruction. Performs the activation or switch off of some analysing systems (for example, to better muscular perception understanding, recommends to perform the action with closed eyes).

Athlete's activity. Performs the necessary directing movements and elements of dynamic stage using previously learned basic standard actions. At the beginning with the minimum intensity of efforts, paying attention to the training device. Then gradually increases the intensity of muscular efforts up to necessary magnitudes. Doing it analyzes, comprehends muscular efforts, combines them with verbal instruction formula. The athlete constantly gets information whether his movements are correct (from motor and tacitly analyzers, vision and audio).

Functions of the training device. Imitation of the athletes optimum directing movements in joints and stance elements independent of his activity. Bringing information to an athlete through motor and tacitly analyzes (when there are transformers of deformations in some parts of the device, sight and sound analyzers are used).

The expected pedagogical effect. The end of the approximate basics formation. Concretization of the standard image (at all levels-sensory, perceptive, imaginary, verbal, logical) of motor action and conditions of its execution. The end of the studied action execution.

III.

The task of the third stage is to develop qualities and abilities necessary for the athlete to perform the action in natural conditions and to create the stock of durability while performing simultaneous (parallel) action.

The activity of the coach. Defines the quantity of aggravation and exercise dosage. Observes the trainee s activity and the indicators of technical aids, corrects the athlete's activity. Talks with the trainee, controls his feelings and perceptions which appear when he is tired.

Athlete's activity. Performs the action with maximum speed and strength and a grant number of repetitions etc., parallel to the training device. Thus alternates the active work with passive pauses during which the body ring direction is fully performed by the training device. Performs the action with supplementary loading (weighted belts, waste-coats, wrist-bands, shock-absorbers, etc.) fixed on the athlete's body.

Functions of the training device. The guaranteed reproduction instead of the trainee, the optimum joint movements bringing the athlete to necessary results and informing the athlete in case of any ring deviations from the optimum trajectories. The expected pedagogical effect. The end of the execution part formation (with standard technique and the given result). The development of excessive stability to concrete loading with the aim of increasing the reliability of the performance.

IV.

The main tasks of this stage are:

- to form feelings and perceptions characteristic for incorrect action options and for critical zones when deviations cause mistakes;
- to form skills and abilities which allow to correct and prevent possible typical mistakes.

The activity of the coach. Changing the device construction leads to alternative "conduction" with optimum pose program. Brings out athlete's body rings outside the border of accessible deviations and stimulates the trainee s actions when returning into the frames of permitted "corridor". Forms the skills of typical mistakes prevention.

Athlete's activity. Analyses senses and perceptions that arise at reproduction of an incorrect posture and compares it with a standard one. Works at the actions preventing mistakes and ensuring the return of body rings from the "zone of mistakes" to the zone of admissible deviations.

Function of the training device. Contrasted reproduction of optimum and incorrect program of the posture. The modelling of conditions when the athlete's body rings return from the "zone of mistakes" to the zone of admissible deviations.

The expected pedagogical effect. The development of perceptions of incorrect options and critical zones of standard posture programs. The development of abilities and skills of self-control as well as mistakes prevention and correction.

V.

The task of fifth stage is to perform gradual transition to the execution of the planned result in natural conditions and to form abilities and skills of self-support.

The activity of the coach. Performs gradual "release" of the athlete from device. If necessary uses traditional support devices (safety belts, foam pits, et.). Teaches the skills of self support in critical situations.

Athlete's activity. Having mastered the actions on the device the athlete starts to perform them in natural conditions. Works out abilities and skills of self-support. In case of unsuccessful exercise execution uses support device if necessary.

Functions of the training device. Gradual decrease of the influence on the trainee s body rings.

The expected pedagogical effect. The execution of the motor action with given properties, demonstration of the planned result in natural conditions.

Despite of the great importance of first stage of the oriented basics of motor activity formation as well as sensory perceptive training, it must be recommended not to be so long. According to the pedagogical experiments it is sufficient to perform 4–6 attempts with 3–5 partitions of the action. Depending on coordinative complexity of the action these indices can vary.

The second stage is considered to be the maim in the given method. Here the athlete masters the necessary directing joint movements using ID mastered during the first stage. It is vividly seen that the athlete's activity during the second stage principally differs from the traditional motor activity training. It should be remembered that the first stage of action development in traditional methods is characterized by low speed, tension and vague movements. It is explained by the necessity to block the redundant biokinematic chains, as it is impossible to organize the necessary motor action in natural conditions and the motor task can't be solved.

Imperative training devices having the functions of motor activities organization give the possibility to execute motor tasks both when the athlete is passive and at different levels of his activity (from minimum up to the necessary ones) and even when he is unable to solve the task. It allows the athlete to seek for the necessary governing movements and to correct them during their execution.

The work on the second stage comes to an end when the athlete is able to perform the program governing joint movements consciously irrespective of the travelling device when the latter does not influence his body rings.

If the athlete is unable to perform the action after he had been trained on the device, the device can be used for the increase of strength and functional abilities of different organs and systems.

To raise motor potential the third stage is introduced to develop special qualities necessary for the action execution with the help of the device.

During this stage motor qualities and abilities are trained in order to create a stock of durability. It is efficient to perform exercises using ID with weights supplementary amortizators, limiting the amplitude of different joint movements and other complex movements. The possibility to show maximum tension without any deviation of the optimum pose program which is controlled by the training device allows to exclude many negative features and mistakes. ID allows to organize new conditions for the realization of all methods of motor perfection. All training methods can be differentiated according to strength and other parameters, body rings being transferred in the typical gymnastic manner. For instance, ID can help to model the conditions of statodynamic, isokinetic and some other most effective regimes of muscular work.

The fourth stage is introduced to create the mobile skills for functioning in changing surrounding conditions. It is used to create skills of self-control, prevention and correction of mistakes with the help of training device.

Unlike the traditional training process self-control is performed more purposefully after the action has been mastered. A special method—contrast reproduction of optimal program of pose change and deviated (mistaken) program is used here. This stage gives opportunity to prevent and correct mistakes. For this purpose athlete's body rings are excluded from the zone of accessible deviations with ID and it works at the process, hindering the body rings exit from the zone of deviations, returning them from "zone of mistakes" to the zone of the permissible deviations.

The fifth stage is the transition to self execution of motor action and demonstration of the planned result in natural competitive conditions. There occurs either an increase of the "clearance" between athlete s body rings and the parts of the device or some parts of the device are taken away. In the first case it is efficient to use unequal "clearance" for different phases of motor activity. To find the individual athlete's technique it is efficient to increase the "clearance" between body rings and parts of the device in the phases not effecting the final result of the action and to keep the program unchanged for the phases directly influencing the final result. Similar method can be used during gradual "clearance" introduction. At the beginning "clearance" is introduced for the rings, performing governing joint movements, and then for the rings performing the main directing movements.

In conclusion the method of controlling movement to aid learning of correct technique using the ID makes it possible to create, from the very first attempts of the exercise execution, the correct motor actions, to form the correct neural pathways, to develop the necessary motor qualities and abilities, and to correct technical errors.

22

THE APPLICATION OF COMPUTER TECHNOLOGIES TO THE MANAGEMENT OF SPORT SPECIFIC TRAINING IN RHYTHMIC SPORTS

V. Kleshnev

Research Institute of Physical Culture
Dynamo 2, St. Petersburg, 197042, Russia

INTRODUCTION

It is well known that results in sports such as rowing and most other rhythmic sports are determined by the quality of the athlete's sport-specific training. There are two main parts of this special training that have a close interrelationship with each other, but are also to some extent independent of each other: the athlete's sport-specific fitness or work capacity and technique development routines. The application of computer technologies could increase the efficiency of the management of this special training in cyclic sports by allowing simultaneous testing and training of both the special work capacity and the exercise technique. The purpose of this study was to develop a computer-aided testing and training system for cyclic sports and then to demonstrate its use in rowing.

This study was designed to address the following questions:

- to find a way of computer-aided analysis of rowing technique and assessment of the work capacity of rowers
- to determine the main biomechanical features of rowing using computer-aided methods of data processing;
- to define interrelations between a rower's technique and his work capacity.

There are a number of studies devoted to the analysis of rowing technique using racing boats and improved means of filming (Martin and Bernfield, 1980; Nelson and Widule, 1983; Zatsiorsky and Yakunin, 1991). A common disadvantage of these methods is that they take no account of the forces applied by rowers to the oar and to the footrest and therefore can not evaluate precisely the rower's power output and work capacity. Some direct measurements of force have been made during real boat rowing (Dal-Monte et al, 1985; Fukunaga et al, 1986), but due to difficulties in the standardisation of exercise

Current Research in Sports Sciences, edited by Rogozkin and Maughan.
Plenum Press, New York, 1996

conditions and the evaluation of a sufficient number of subjects, these studies have generally failed to produce clear results.

The main principles of an effective rowing technique were reported by Nolte (1991) who determined the four most important principles:

1. the greatest possible transfer of physiological capacity into boat propulsion
2. a long stroke to maximise power output
3. the movement of rower must be as horizontal as possible
4. the horizontal velocity of the rower relative to the boat should be as small as possible.

It is clearly necessary that more information on the relevance of these principles to rowers should be obtained.

Some researchers have used rowing ergometers with the aim of measuring mechanical parameters of rowing and the work capacity of rowers of different skill levels (Kleshnev, 1992, Hartmann et al, 1993). At the present time, a variety of off-water devices (including the "Concept II "and others) are widely used by rowers in training and also for testing purposes. More than 100 different competitions are held each year on rowing ergometers, including World and continental championships. Martindale and Robertson (1984) and Lamb (1989) have shown that the biomechanics of ergometer rowing are different from those in a real boat. Ergometer rowing can not therefore be used for the analysis of rowing technique.

METHODS and SUBJECTS

The problem of combined investigation of a rower's technique and work capacity were solved by the design of a special-purpose rowing ergometer "IGL-1" with a moveable seat, which simulates real boat rowing in off-water conditions. This new development became possible because the rower performs exercise in a moving unit. External forces applied to the unit are similar to the real boat external forces. Six mechanical parameters were recorded by the attachment of appropriate recording apparatus to the rowing simulator. Handle force and the force applied to the horizontal part of the foot rest force were measured by tensiometer gauges. Movements of the handle, the seat and the athlete's back were measured by electrical resistance gauges. Horizontal acceleration of the moveable unit was measured by an electric resistance accelerometer.

An analogue-to-digital converter (12 bit, 16 channels, 50 Hz) with amplifiers provided data transformation from the recording apparatus to a personal computer, which performed the data processing, visualisation and storage of the results.

The total work power of the rower was estimated by calculation of the energy dissipated by rotation of the ergometer flywheel. A performance monitor taken from the Concept 2 rowing ergometer was used for this purpose.

A software program was designed to provide an original algorithm for information handling and evaluate patterns of each of the mechanical parameters during the stroke cycle. The main advantage of this approach consists in resolving the problem of selection from a lot of cycles a typical one that can be used for evaluation of the quality and stability of the athlete's rowing action. The algorithm also removes any electrical or movement artifacts. Testing of the algorithm validity showed that it removes random variability without compromising the integrity of the data.

A group of male rowers of different levels of achievement took part in the study (n = 28, age 19.3 ± 6.2 years, height 1.87±0.04 m, weight 78.9±7.0 kg, mean±S.D.). On the CARE ergometer, each oarsman performed two rowing exercises: a 10-stroke maximal test (T10) and a 6-min test that imitated a standard 2000 m rowing race (T6).

Then mechanical power and work of the arms, trunk and legs were calculated. Also ratios of segment powers to the total one were calculated which we mean as part performances of body segments.

RESULTS

An example of the primary information received using CARE from gauges and inputed to the computer is shown in Figure 1.

As it can be seen, There are disturbances due to vibration and electric-magnetic interference. The curve shapes of the parameters change from cycle to cycle during the exercise and therefore, it is quite difficult to select from these the most typical one for further analysis. The primary data were processed using the above mentioned algorithm. From the pattern of each primary parameter received, derived parameters were calculated. The typical patterns of primary and derived parameters of skilled rower received in T6 are shown on Figure 2.

Checking of the data processing by algorithm shows that it was quite accurate. Evidence for this fact was the correlation factor of 0.95 (p<0.001) between total power received by measuring the flywheel rotation and total power received using the derived pattern of total power. The patterns show the structure of exercise movements of each rower. They show in great detail, mistakes and clear differences between skilled rowers and beginners. Moreover, it was possible to define average patterns of a group of rowers and therefore allows a typical rowing style of each coach and crew to be determined.

Quantity criteria of the received patterns were also calculated (Table 1.).

Six groups of criteria were studied: times, displacements, velocities, accelerations, forces and powers. Some differences were seen comparing the quantity criteria of rowing with maximal and race intensity. As expected, the most significant differences were seen in rowing rate, time of drive phase, total velocities of the handle (maximal and average), handle and footstretcher forces, positive device acceleration, total power and powers of legs and arms. There were no differences in longitudinal criteria, legs and trunk velocities and trunk power, though they could have been expected. Unexpected differences found were in force decreasing time, ratio of average to maximal footstretcher force and in the ratio hand power to total power. Relative work power (a ratio of total power to the athlete's body weight in Wt/kg) was accepted as a measure of efficiency of rowing movements. Values of Pearson's correlation factor of relative work power with quantity criteria of simulated rowing are shown in the Table 2.

In T6 the relative power has a significant correlation with: stroke rate, increasing and decreasing force in the drive phase, maximal and average velocities of handle drive, maximal and average velocities of arm contraction, maximal negative and positive acceleration of device, work power of trunk. On the other hand, the relative power in T10 has a significant correlation only with: stroke rate, average velocities of handle drive, maximal negative and positive acceleration of device, maximal and average forces on the handle.

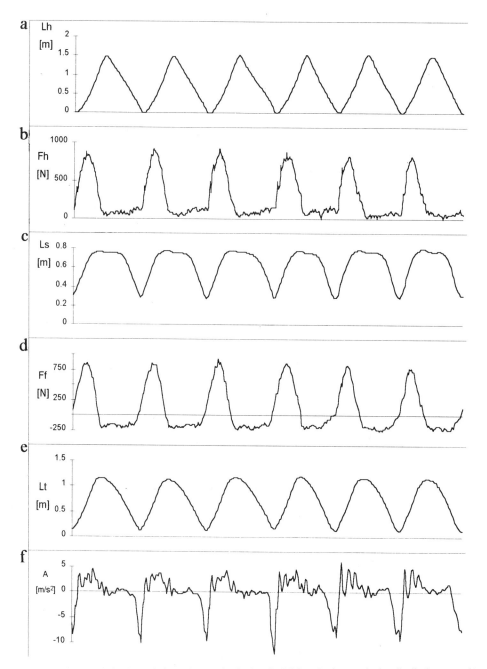

Figure 1. An example of primary information received using CARE in a 6 min test: a) - handle displacement; b) - handle force; c) - seat displacement; d) - horizontal footstretcher force; e) - trunk displacement; f) - device acceleration.

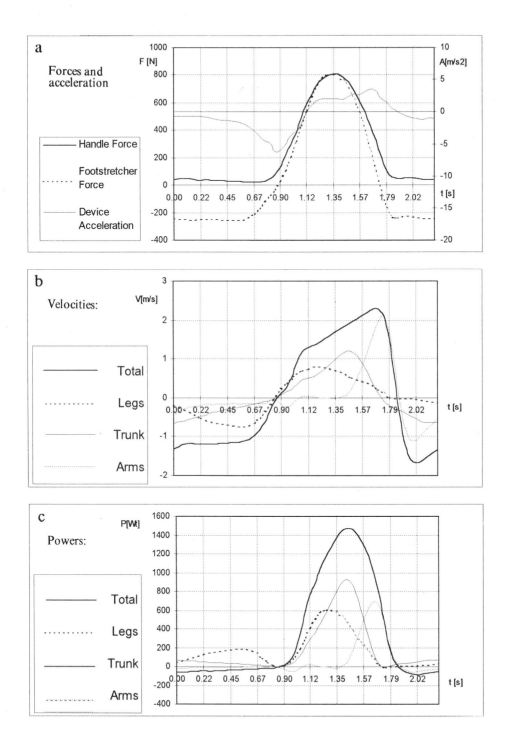

Figure 2. Typical patterns of primary and derived parameters of skilled rowers during a 6 min test: a) - forces and device acceleration; b) - body segments and total velocities; c) - body segments and total powers.

142

V. Kleshnev

Table 1. The quantity criteria of simulated rowing with different intensity (P = degree of significance between T10 and T6)

Criteria	T10	±S.D.	T6	±S.D.	P
1. Time criteria					
Stroke time [s]	1.46	0.22	2.07	0.23	0.03
Drive phase [s]	0.84	0.10	1.08	0.10	0.05
Force increasing [s]	0.20	0.06	0.27	0.07	0.25
Force holding [s]	0.34	0.05	0.37	0.04	0.35
Force decreasing [s]	0.30	0.6	0.44	0.06	0.05
Stroke rate [1/min]	42.0	6.6	29.2	3.2	0.04
Drive / Stroke [%]	57.9	4.0	52.7	6.4	0.25
Increasing / Drive [%]	24.3	6.1	24.8	5.3	0.48
Holding / Drive [%]	40.6	4.8	34.1	3.6	0.14
Decreasing / Drive [%]	35.1	6.6	40.9	5.4	0.25
2. Longitudinal criteria					
Stroke length [m]	1.47	0.16	1.43	0.12	0.43
Legs displacement [m]	0.40	0.07	0.46	0.04	0.25
Trunk displacement [m]	0.51	0.10	0.50	0.07	0.48
Arms displacement [m]	0.55	0.08	0.46	0.06	0.21
Stroke length / Height	0.78	0.08	0.76	0.06	0.42
Legs / Stroke Length [%]	27.7	4.3	32.2	2.6	0.19
Trunk / Stroke Length [%]	34.7	5.4	35.3	3.18	0.46
Arms / Stroke Length [%]	37.4	3.9	32.4	3.47	0.17
3. Velocity criteria					
V max. Total [m/s]	2.57	0.26	1.93	0.15	0.02
V max. Legs [m/s]	1.10	0.24	0.80	0.08	0.12
V max. Trunk [m/s]	1.48	0.59	1.08	0.15	0.26
V max. Arms [m/s]	2.26	0.31	1.68	0.24	0.07
V average Total [m/s]	1.82	0.14	1.38	0.10	0.01
V average Legs [m/s]	0.72	0.13	0.51	0.05	0.09
V average Trunk [m/s]	0.84	0.28	0.60	0.06	0.20
V average Arms [m/s]	0.95	0.26	0.97	0.16	0.47
4. Force criteria					
F max. Handle [N]	829	109	642	86.3	0.09
F average. Handle [N]	530	63.7	385	49.8	0.04
Handle F aver./F max. [%]	64.1	2.91	60.0	2.69	0.15
F max. Footstretcher [N]	822	105	641	86.3	0.09
F aver. Footstretcher [N]	453	56.7	303	42.4	0.02
Footstretcher F aver./F max. [%]	55.3	3.55	47.4	2.92	0.04
5. Device accelerations					
A max. negative	11.2	3.27	6.61	1.62	0.10
A max. positive	4.61	1.15	2.59	0.40	0.05
A positive - A negative	15.8	4.17	9.19	1.72	0.07
6. Power criteria					
Total Power [Wt]	566	124	282	48.1	0.02
Legs Power [Wt]	188	43.3	101	19.7	0.03
Trunk Power [Wt]	213	75.6	118	26.0	0.12
Arms Power [Wt]	164	34.3	62.8	12.2	0.01
Power / Weight [Wt/kg]	7.16	1.20	3.56	0.44	0.01
Legs / Total [%]	33.6	5.06	36.0	3.69	0.35
Trunk / Total [%]	37.1	7.03	41.6	4.57	0.29
Arms / Total [%]	29.1	3.54	22.3	3.33	0.08

Table 2. Values of correlation of relative work power (in Wt/kg) with quantity criteria of simulated rowing

	T10	T6		T10	T6
1. Time criteria					
Stroke time [s]	-0.56^b	-0.56^b	Stroke rate [1/min]	0.49^b	0.56^b
Drive phase [s]	-0.38^c	-0.01^{ns}	Drive/Stroke [%]	0.56^b	0.45^c
Force increasing [s]	-0.16^{ns}	0.45^c	Increasing/Stroke [%]	-0.02^{ns}	0.58^b
Force holding [s]	-0.17^{ns}	0.03^{ns}	Holding/Stroke [%]	0.17^{ns}	0.07^{ns}
Force decreasing [s]	0.20^{ns}	-0.54^b	Decreasing/Stroke [%]	0.30^{ns}	-0.60^a
2. Longitudinal criteria					
Stroke length [m]	0.31^{ns}	0.44^c	Stroke length / Height	0.30^{ns}	0.42^c
Legs displacement [m]	0.07^{ns}	0.46^c	Legs/Stroke Length [%]	-0.12^{ns}	0.16^{ns}
Trunk displacement [m]	0.17^{ns}	0.36^{ns}	Trunk/Stroke Length [%]	-0.01^{ns}	0.13^{ns}
Arms displacement [m]	0.34^{ns}	0.09^{ns}	Arms/Stroke Length [%]	0.14^{ns}	-0.24^{ns}
3. Velocity criteria					
V max. Total [m/s]	0.27^{ns}	0.62^a	V average Total [m/s]	0.82^a	0.70^a
V max. Legs [m/s]	0.06^{ns}	-0.08^{ns}	V average Legs [m/s]	0.18^{ns}	0.03^{ns}
V max. Trunk [m/s]	0.16^{ns}	0.40^b	V average Trunk [m/s]	0.13^{ns}	0.54^b
V max. Arms [m/s]	0.46^c	0.77^a	V average Arms [m/s]	0.09^{ns}	0.78^a
4. Force criteria					
F max. Handle [N]	0.61^a	0.40^c	F max. Footstretcher [N]	0.61^b	0.40^c
F average. Handle [N]	0.64^a	0.37^{ns}	F aver. Footstretcher [N]	0.65^a	0.33^{ns}
Handle F aver./F max. [%]	-0.08^{ns}	-0.18^{ns}	Footstretcher F aver./F max. [%]	0.03^{ns}	-0.16^{ns}
5. Device accelerations					
A max. negative	0.84^a	0.58^b	A positive - A negative	0.82^a	0.66^a
A max. positive	0.62^a	0.50^b			
6. Power criteria					
Total Power [Wt]	0.92^a	0.85^a	Power / Weight [Wt/kg]	1.00^a	1.00^a
Legs Power [Wt]	0.75^a	0.62^a	Legs / Total [%]	-0.21^{ns}	-0.29^{ns}
Trunk Power [Wt]	0.69^a	0.84^a	Trunk / Total [%]	0.23^{ns}	0.49^b
Arms Power [Wt]	0.85^a	0.57^b	Arms / Total [%]	-0.16^{ns}	-0.30^{ns}

Degree of significance marked as follows: [a]P < 0.01; [b]P , 0.05; [c]P , 0.1; [ns]not significant.

DISCUSSION

As shown above, the typical stroke cycle of the skilled rower has the following features:

- at the beginning of the drive phase, mainly the legs work and produce acceleration of rower's body mass;
- the trunk then joins in the drive movement and produces, with the legs, the highest total power on the handle;
- at the end of the drive phase, the arms perform a fast but not very powerful movement and slow down the movement of the rower body mass.

Unskilled rowers have sharp differences in the drive movement structure in comparison to more experienced athletes. The typical patterns of drive structure of an unskilled rower are shown in Figure 3.

The most typical mistakes of unskilled rowers consist of the following:

- when the trunk begins its drive movement it decreases the acceleration of the rower's body mass;

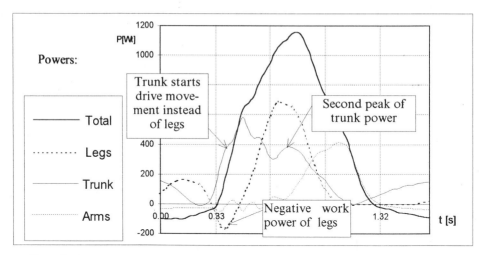

Figure 3. Patterns of body segment and total handle powers of an unskilled rower during a 6 min test.

- the legs have negative values of work power in the first stage of the drive phase, i.e. they work as a power receiver but not as a supplier;
- the trunk has two peaks of work power but both are smaller than the leg power peak.

One reason for these mistakes is incorrect coaching which encourages rowers to increase the handle force as soon as possible without applying a high force to the footstretcher at the same time. Using CARE helps to detect these mistakes and then to look after their correction. The criteria of simulated rowing are show that the following feature are required for effective movement structure.

- The handle force and power peak must be achieved as late as possible during drive phase that there is a positive relationship between relative power and the time of increasing force (r=0.58, p<0.01) and a negative relationship with the decreasing force (r=-0.60, p<0.01). This fact means that a rowers body mass must be accelerated for as long as possible.
- A long acceleration of body mass impossible without a rapid slowing down by means of a fast and powerful arm movement. This was confirmed by a high correlation of relative power with maximal and average velocities of arm contraction (r=0.77 and r=0.78 respectively, p<0.01 both).
- Increasing the relative power leads to permanent increase of both negative and positive acceleration (r=0.58, p<0.01 and r=0,50, p<0.05 respectively).
- A high level of relative power could be achieved only with powerful trunk movement during the drive phase (r=0.45, p<0.05).

These features are mostly important during long rowing races (T6) while during short exercises with a high intensity (T10) more important parameters are: rate, drive force, velocity and power. Certainly, the above features are concerned only with the first of Nolte's principles of effective rowing as they are concerned only with transferring physiological performance into mechanical power onto the handle. An additional experiment should be conducted with the purpose of correlating these new facts with boat propulsion power. However, these findings give a more detailed view of rowing technique and could help to increase the efficiency of special training of rowers.

CONCLUSION

A computer-aided rowing exerciser allows investigation of an athlete's work capacity and technique during exercise. The using of the CARE enables us to determine both quantity and quality criteria of a rower movements and brings new knowledge about the principles of rowing efficiency. Furthermore, it is possible to improve the quality and stability of rowing motions more effectively by using CARE, when the rowers can see on a computer display the level of rowing motion stability, force and power curves for each stroke and can compare them with patterns of skilled rowers.

REFERENCES

Dal-Monte A., Faina M., Cecioni N., Leonardi L., 1985, Analysis of the inertial forces in rowing using a force platform. In: Winter D.A. (ed.) et al., Biomechanics IX-B, Champaign, Ill., Human Kinetics Publishers, 481–485.

Hartmann U., Mader A., Wasser K,. Klauer I., 1993, Peak Force, Velocity, and Power During Five and Ten Maximal Rowing Ergometer Strokes by World Class Female and Male Rowers. Int J Sports Med, 14: S42-S45.

Fukunaga T., Matsuo A., Yamamoto K., Asami T., 1986, Mechanical efficiency in rowing, Eur. J. Appl. Physiol. 55: 471–475.

Kleshnev V., Kleshneva E., 1992, Work performance of different body segments of rowers. Biology of sport, 9: 127–133.

Lamb D.H., 1989, A kinematic comparison of ergometer and on-water rowing Am. J. Sports Med. 17: 367–373.

Martin T.P., Bernfield J.S., 1980, Effect of stroke rate on velocity of a rowing shell. Med. Sci. Sports Exer. 12: 250–256.

Martindale W.O., Robertson D.G.E., 1984, Mechanical energy in sculling and in rowing an ergometer. Can. J. Appl. Sport Sci. 9: 153–163.

Nelson W.N., Widule C.J., 1983, Kinematic analysis and efficiency estimate of intercollegiate female rowers. Med. Sci. Sports Exer. 15: 535–541

Nolte V., 1991, Introduction to the biomechanics of rowing. FISA-coach, 2: 1–6.

Zatsiorsky V.M., Yakunin N.I., 1991, Mechanics and biomechanics of rowing: a review. Int. J. Sport Biomech. 7: 229–281.

23

COMPUTER SOFTWARE FOR CANOEING SIMULATION

General Opportunities

I. Sharobaiko, P. Temnov, A. Korotkin, and T. Gladilova

Saint-Petersburg Research Institute of Physical Culture
Dynamo Ave. 2, 197042 St. Petersburg, Russia

GENERAL APPROACH

Basing upon the biomechanical principles of paddling (V. Issurin, 1986) product on IBM PC AT was developed for evaluation of paddling efficiency in different technical variants of canoeing and kayaking. It was developed in the canoeing department of the Research Institute of Physical Culture in Saint-Petersburg under the leadership of Prof Issurin. The program permits operative solving of problems of investigators and coaches and helps to find parameters for optimal technical performance. It estimates the influence of different factors upon the system conoe–canoeist during movement along the water. These factors are:

- characteristics of the boat (its resistance, mass (together with the canoeist) and steering parameters),
- strength and direction of wind,
- different elements of paddling technique (rate, duration of underwater and air transfer phases of the stroke, configuration of force function) and their variations in the process of paddling.

All factors seem to be significant for the coach and canoeist. Modelling of canoe's movement is based upon differential equation solving. The initial data are collected in dialogue form. They are:

- type of sport boat;
- length of race;
- mass of the boat with paddles;
- mass of the canoeist;
- coefficient of hydrodynamic resistance;
- duration of quasi-stationary period of racing;
- length of finishing section of the race;

Current Research in Sports Sciences, edited by Rogozkin and Maughan.
Plenum Press, New York, 1996

- duration of underwater phase of paddle's movement at the start, middle and end section of the race;
- duration of airtransfer phase of paddle's movement at the start, middle and end section of the race;
- number of points to determinate the force impulse;
- values of force on the paddle at the declared points of impulse;
- wind speed;
- wind direction.

The dynamic parameters of paddling can be obtained from tenziodynamograms when the paddler covers the distance using special instrumentation and devices. The program collects results for calculation any time needed:

- boat's speed;
- distance covered;
- power of paddling;
- time-result.

Because paddling can be modelled by PC there is the unique opportunity to separate the clear effect of differing factors. In order to test the program the following problems were solved.

EVALUATION OF RESULT IN CANOEING

Olympic Champion M.Slivinsky covered 500m at competitive speed in a single canoe. The parameters of his movement were registered by means of the tenziodynamographic system. His individual data was input to the program together with the real weather conditions (Fig. 1).

The PC modelled his strokes at the starting section, middle of the race and its finish. The modelled result in 500m canoeing was 124.66 s., the real one, 125.00 s which shows the adequate accuracy of the program.

General Reflection of Force-Function Influence

To evaluate the possibilities in kayaking there was input of 2 different force curves (Fig. 2).

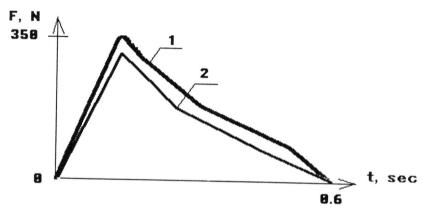

Figure 1. Force-impulse of Olympic champion M. Slivinsky.

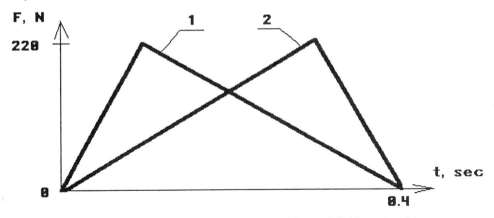

Figure 2. Force impulses 1 and 2 with different maximum positions modelled for testing of the program.

The impulse values were identical, but the location of the maximum point was shifted to the beginning of underwater phase (stroke 1) and to its end (stroke 2). The results show that under equal conditions the paddler that used stroke 1 won by 6.73 seconds over 500m of racing. This shows that the program reflects the influence of the force-curve upon the result.

DETAILED REFLECTION OF FORCE-CURVE INFLUENCE

In order to estimate the level of influence, the input data (mass of boat and paddles, mass of sportsman, duration of underwater and airtransfer phases of stroke, value of power-impulse, duration of quasi-stationary period of boat movement, duration of finishing part of distance) was standardised; only the values of force on the blade corresponded with 25 fixed points of underwater phase of stroke. One by one we modelled strokes of the same force impulse with different depths in the curve (Fig. 3).

The analysed stroke variants could be divided into of paddlers of identical anthropometric characteristics and physical fitness, but different technical abilities or into the same paddler whose dynamic characteristics of stroke were influenced by certain circumstances. The results show that under equal conditions the presence of curvity break in force-impulse can cause disadvantage in the result of up to 7 seconds (500 m).

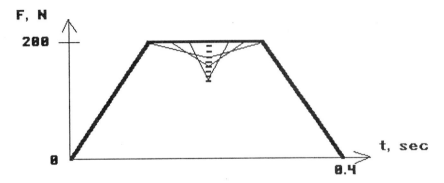

Figure 3. Alterations of model impulse shape.

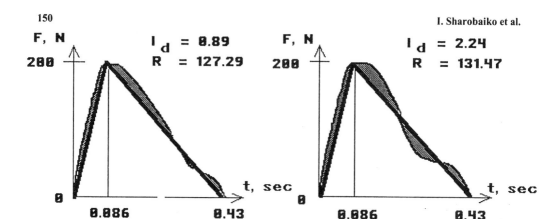

Figure 4. Variants of investigated strokes.

INDEX OF DISCREPANCY

Under the same input conditions we modelled force-impulses similar to those in ca-noeing. The optimal impulse form is known (V. Issurin, et al, 1983), but real impulse forms differ slightly or greatly from it. As a measure of the difference a special index of discrepancy was chosen (square root of mean square deviation of curve related to a time unit). The closer the real stroke is to the optimal one, the lower is the index value (Fig. 4).

The results show the correlation between the index and boat speed; it is an inverse relationship. Therefore, the program permits us to follow the influence of individual vari-ation upon the result.

INFLUENCE OF WIND

The next series of problems was studied in order to investigate the influence of wind upon the race time in canoeing and to examine the program's possibilities in this field. The wind direction was set by positive values of angle from 0 to 360 degrees (0 degrees coincides with a head wind (Fig. 5).

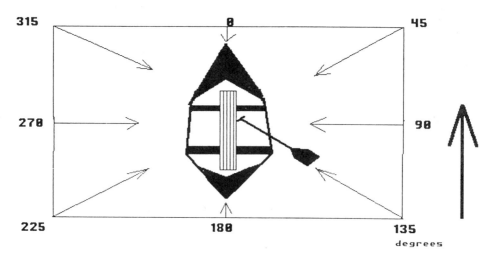

Figure 5. Coordinative system for estimation of wind influence upon the movement of canoe.

Table 1. Initial data for modelling the process of canoeing

1.	Duration of quasi-stationary period of racing, sec	15
2.	Length of finishing section of distance, m	50
3.	Mass of the boat with paddles, kg	18
4.	Mass of the paddler, kg	83
5.	Coefficient of hydrodynamic resistance	4.2
6.	Duration of underwater/airtransfer phase of paddle's movement in the starting section of distance, sec	0.6/0.47
7.	Duration of underwater/airtransfer phase of paddle's movement in the middle section of distance, sec	0.62/0.57
8.	Duration of underwater/airtransfer phase of paddle's movement in the finishing section of distance, sec	0.62/0.53
9.	Number of points to determinate the force impulse	7

There were 34 variation in 500m distance canoeing and the same number in 1000 m. The one chosen was that used by top-level Soviet canoeists. The initial data were standardised: the kinematic (Table 1) and dynamic (Table 2) parameters of paddling allowed the canoeist on PC display to get a result 123.26 s for 500 m, 248.17 s for 1000 m.

The only varying parameters were wind characteristics: from 2 to 5 m/sec and from 0 to 360 degrees. Changes from the result without any wind were calculated. Analysis of the results showed that the program generally reflects the influence of wind both in terms of its strength and direction. To find the level of correspondence between the real and modelled results an examination test was offered to 13 top-class Russian, Belorussian and Ukrainian canoeists (members of the former USSR National canoeing team) and among them there were Olympic champions M.Slivinsky and S.Postrechin and also the coach of M.Slivinsky, P.Bratash.

The difference between the experts and modelled curves are of principal consideration. The modelling program observes the moving canoeist as the generator of the changing resistance to the wind taking into consideration the size of his surface facing the wind. It is accompanied by the idea that the force on the paddle is set once and remains constant under the changing circumstances. However, the system canoe–canoeist reflects the slightest alterations in the surroundings. Wind is a powerful factor which can force the paddler

Table 2. Values of force on the paddle in the 7 declared points of impulse

		$F^{\dagger}$, N					
		Part of 1000 m distance			Part of 500 m distance		
N	Relative T-I*	Start	Middle	Finish	Start	Middle	Finish
1.	0.00	0.00	0.00	0.00	0.00	0.00	0.00
2.	0.17	360.00	340.00	340.00	370.00	350.00	350.00
3.	0.33	310.00	240.00	240.00	320.00	250.00	250.00
4.	0.50	205.00	155.00	155.00	205.00	155.00	155.00
5.	0.67	130.00	110.00	110.00	130.00	110.00	110.00
6.	0.83	80.00	60.00	60.00	80.00	60.00	60.00
7.	1.00	0.00	0.00	0.00	0.00	0.00	0.00

*Relative T-1–relative time-interval from the beginning of underwater phase of stroke
†F–current values of force on the paddle determinated with the help of autonomic tenziodynamography.

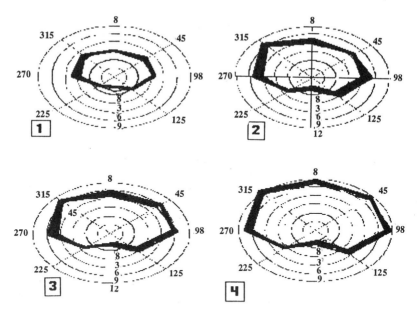

Figure 6. Alterations in 1000 m result (Δt sec) in canoeing depending on the direction of wind. (1) 2 m/s, (2) 3 m/s, (3) 4 m/s, (4) 5 m/s deviation of the program 2 results from the expert opinion.

to desperate shifts.[*] The above mentioned considerations can be estimated only by top-class canoeists. In order to compensate for this we decided to put a correcting coefficient into the main equation of the program. When this was done, the value of impulse became dependent upon the wind characteristics. The results calculated with the new coefficients are shown in Fig. 6.

By the statistical program Statgraphics (v. 2.1) on IBM PC AT the results were calculated by multiple linear regression. The correlation coefficients obtained confirm the close connection of the experts curve with results of the new program compared to the original one (Table 3).

CONCLUSION

The above mentioned investigations improved the program. Now it can help both the coach and sportsman to do the following:

- to investigate the individual impulse-form correspondence with model one;
- to follow the influence of different technical parameters upon canoeing;
- to search possible ways to decrease the value of individual technical mistakes;
- to foresee the result of alterations in training sessions;
- to estimate the effect of external factors upon the individual paddling technical variants;
- to investigate characteristics of new boat construction.

---

[*] Inconvenient wind prevents realization of individual optimal dynamic and kinematic characteristics of stroke.

Table 3. Correlation coefficients of experts graph

Speed of wind (m/s)	500 m		1000 m	
	With prog1	With prog2	With prog1	With prog2
2	0.79	0.84	0.52	0.93
3	0.75	0.86	0.68	0.91
4	0.71	0.86	0.66	0.91
5	0.76	0.87	0.69	0.91

ACKNOWLEDGMENTS

The authors highly appreciate the help of Dr. Timofeev V. D. in collecting material for the investigation.

REFERENCES

Issurin, V. B., 1986, Biomechanics of canoeing, Ed. M. Zatsiorsky, Physical culture and Sport, Moscow - 112.
Issurin, V. B., Begak, M. V., Krasnov, E. A., Razumov, G. G., 1983, Comparative effectiveness of different technical variants in paddling (biomechanical modelling), Theory and Practice of Physical Culture 9:11–13.

24

NUTRITIONAL PREPARATION FOR SPORTS PERFORMANCE

The Elite Performer

R. J. Maughan[1] and Louise Burke[2]

[1] University Medical School
   Aberdeen AB9 2ZD, Scotland
[2] Australian Institute of Sport
   PO Box 176, Belconnen, ACT 2616, Australia

INTRODUCTION

At the highest level of competition in sport, where opposing competitors are predisposed to success by genetic endowment, have undergone the most rigorous training, and are equally matched in skill and motivation, nutritional intervention may make the difference between success and failure. It is not surprising therefore that sportsmen and women generally are concerned about their diet, although this concern is not always matched by a knowledge of basic nutrition. Although it is clear that the ability to train and compete will be impaired if the diet is inadequate, the concept of dietary inadequacy may be quite different for the elite athlete compared with the sedentary individual. Conversely performance may be improved by dietary manipulation, but we still have an incomplete understanding of how best to control diet to optimise sports performance. Some of the dietary practices followed by athletes in pursuit of success are sound, but others have no beneficial effect and may even be harmful. As in other areas of nutrition, these ideas are often encouraged by those who stand to gain financially from sales of dietary supplements.

Two distinct aspects of the athlete's diet must be considered; the first is the diet in training which must be consumed on a daily basis for a large part of the year, and the second is the diet in the immediate pre-competition period and during competition itself. The most important aspect of the athlete's diet is that it allows consistent hard training to be performed, because it is only from such training that improvements in performance result. In most sports, the number of major competitions in the year is small, and peak performance must be achieved on only a few times each year. In some team games (eg, football, basketball) competition is frequent, but peak effort is not produced on every occasion. Considering the range of activities encompassed by the term sport and the variation in the

Current Research in Sports Sciences, edited by Rogozkin and Maughan.
Plenum Press, New York, 1996

155

characteristics of the individuals taking part, it is not surprising that the nutritional requirements vary. For non-competitive activities, and for the individual who exercises for recreational and health reasons, the daily diet forms part of a lifestyle which may be quite different from that of the competitive athlete, but the nutritional implications of exercise participation apply equally, albeit to different degrees.

NUTRITION FOR TRAINING

The primary need for the diet of the athlete in training is to meet the additional nutrient requirement imposed by the training load. In sports involving prolonged strenuous exercise on a regular basis, participation has a significant effect on energy balance. Metabolic rate during running or cycling, for example, may be 15–20 times the resting rate, and such levels of activity may be sustained for several hours by trained athletes (Maughan and Leiper, 1983). Even the sprinter, whose event lasts only a few seconds, may spend several hours per day in training, resulting in very high levels of energy expenditure

Evidence suggests that the metabolic rate may remain elevated for at least 12 and possibly up to 24 hours if the exercise is prolonged and close to the maximum intensity that can be sustained (Bahr, 1992), and this will further increase the energy cost of training. The recreational exerciser, whose aim is often to lose weight, is unlikely to benefit from this effect because the duration and intensity of exercise will be too short, but the elite athlete training at the limits of the tolerable load will incur an additional energy cost. If body weight and performance levels are to be maintained, the high rate of energy expenditure must be matched by a high energy intake. Available data for most athletes suggest that they are in energy balance within the limits of the techniques used for measuring intake and expenditure (Westerterp and Saris, 1992). This is to be expected as a chronic deficit in energy intake would lead to a progressive loss of body mass and a limited capacity to tolerate training. If body weight and performance levels are to be maintained, the high rate of energy expenditure must be matched by a high energy intake. However, data for women engaged in sports where a low body weight, and especially a low body fat content, are important, including events such as gymnastics, distance running and ballet, consistently show a lower than expected energy intake (Stanton, 1994). There is no obvious physiological explanation for this finding other than methodological errors in the calculation of energy intake and expenditure, but it seems odd that these should apply specifically to this group of athletes. There are, of course, major difficulties in training at high intensity on a regular basis and maintaining body fat content at a level well below that which would be considered normal. Many of these women athletes have a very low body fat content: a total fat content of less than 10% of body weight is not uncommon in female long distance runners. Secondary amenorrhoea, possibly related more to the training regimen than to the low body fat content, is common in these women, but is usually reversed when training stops (Drinkwater et al, 1986).

Athletes engaged in strength and power events have traditionally been concerned with achieving a high dietary protein intake in the belief that this is necessary for muscle hypertrophy. In a survey of American college athletes, 98% believed that a high protein diet would improve performance. While it is undoubtedly true that a diet deficient in protein will lead to loss of muscle tissue, there is no evidence to support the idea that excess dietary protein will drive the system in favour of protein synthesis (Lemon, 1991). Excess protein will simply be used as a substrate for oxidative metabolism, either directly or as a precursor of glucose, and the excess nitrogen will be lost in the urine. Exercise, whether it is long distance running, aerobics or weight training, will cause an increased protein oxidation compared with the resting state. Although the contribution of protein oxidation to energy production during the exercise period may decrease to about 5% of the total energy requirement, compared with about 10–15% (ie the normal fraction of protein in the diet) at rest, the absolute rate of protein degradation is increased during exercise (Dohm, 1986). This leads to an increase in the minimum daily protein requirement, but this will be met if a normal mixed diet adequate to meet the increased energy expenditure is consumed. In spite of this, however, many athletes ingest large quantities of protein-containing foods and expensive protein supplements; daily protein intakes of 3–4 g/kg body weight are not unknown in some sports, especially in body building and other strength training events (Burke and Inge, 1994). Disposal of the excess nitrogen is theoretically a problem if renal function is compromised, but there does not appear to be any evidence that excessive protein intake among athletes is in any way damaging to health (Lemon, 1991). The recommended diet for athletes, especially those in endurance events, may even contain a lower than normal proportion of protein on account of the fact that total energy demand is increased to a greater extent than is protein requirement.

The energy requirements of training are largely met by oxidation of fat and carbohydrate, with only a very small contribution from protein oxidation. The higher the intensity of exercise, the greater the reliance on carbohydrate as a fuel: at an exercise intensity corresponding to about 50% of an individual's maximum oxygen uptake (VO2max), approximately two thirds of the total energy requirement is met by fat oxidation, with carbohydrate oxidation supplying about one third. if the exercise intensity is increased to about 75% of VO2max, the total energy expenditure is increased, and carbohydrate is now the major fuel. If carbohydrate is not available, or is available in only a limited amount, the intensity of the exercise must be reduced to a level where the energy requirement can be met by fat oxidation.

The primary need, therefore, is for the carbohydrate intake to be sufficient to enable the training load to be sustained at the high level necessary to produce a response. During each strenuous training session, substantial depletion of the glycogen stores in the exercising muscles (Bergstrom and Hultman, 1967) and in the liver (Hultman and Nilsson, 1971) takes place. If this carbohydrate reserve is not replenished before the next training session, training intensity must be reduced, leading to corresponding decrements in the training response. Any athlete training hard on a daily basis can readily observe this; if a low carbohydrate diet, consisting mostly of fat and protein, is consumed after a day's training, it will be difficult to repeat the same training load on the following day.

Recovery of the muscle and liver glycogen stores after exercise is a rather slow process, and will normally require at least 24–48 hours for complete recovery (Piehl, 1974). The rate of glycogen resynthesis after exercise is determined largely by the amount of carbohydrate supplied by the diet (Ivy et al, 1988a), and the amount of carbohydrate consumed is of far greater importance for this process than the type of carbohydrate. If the diet is deficient in carbohydrate, little restoration of muscle glycogen will occur (Bergstrom et al, 1967), as most of the dietary carbohydrate is then used by nervous tissue and red blood cells.

Feeding a high-fat, low-carbohydrate diet for prolonged periods has been shown to increase the capacity of muscle to oxidise fat and hence improve endurance capacity in the rat. There have been suggestions than man can also adapt by training on a low carbohydrate diet, but same effect appears not to be observed. Similarly, short term fasting increases endurance capacity in the rat, but results in a decreased exercise tolerance in man

(Gleeson et al, 1988). The training diet, therefore should be high in carbohydrate, with perhaps 60% or more of total energy intake coming from carbohydrate (Williams and Devlin, 1992). This suggestion conforms with the recommendations of Expert Committees such as NACNE that carbohydrates provide more than 50% of dietary energy intake. It has been shown that a high carbohydrate diet (70% of energy intake as carbohydrate) enabled runners who were training for 2h per day to maintain muscle glycogen levels, whereas if the carbohydrate content was only 40%, a progressive fall in muscle glycogen content was observed (Costill, 1988). A daily dietary carbohydrate intake of 500–600g may be neces- sary to ensure adequate glycogen resynthesis during periods of intensive training, and for some athletes, the amount of carbohydrate that must be consumed on a daily basis is even greater (Coyle, 1992). The amount that is necessary will be determined primarily by the training volume and intensity, but will also be influenced to a large degree by body size. These high levels of intake are difficult to achieve without consuming large amounts of simple sugars and other compact forms of carbohydrate, as well as increasing the fre- quency of meals and snacks towards a "grazing" eating pattern. Athletes may find that sugar, jam, honey and high sugar foods such as confectionery, as well as carbohydrate- containing drinks, such as soft drinks, fruit juices and specialist sports drinks, can provide a low-bulk, convenient addition of carbohydrate to the nutritious food base (Clark, 1994).

For the athlete training at least once and perhaps even two or three times per day, and who also has to work or study, practical difficulties arise in achieving the necessary energy and carbohydrate intake (Clark, 1994). Most athletes find it difficult to train hard for at least 3 hours after food intake, and the appetite is also suppressed for a time after hard exercise. In this situation, it is particularly important to focus on ensuring a rapid re- covery of the glycogen stores between training sessions. This seems to be best achieved when carbohydrate is consumed as soon as possible after the end of training, as the rate of glycogen synthesis if most rapid at this time (Ivy et al, 1988a). At least 50–100 g should be consumed at this time, and a high carbohydrate intake continued thereafter (Ivy et al, 1988b). There is clearly a maximum rate at which muscle glycogen resynthesis can occur, and there appears to be no benefit in increasing the carbohydrate intake to levels in excess of 100 g every 2 hours (Blom, 1987). Although it was clearly demonstrated by Ivy et al (1988a) that the type of carbohydrate is less crucial than the amount consumed, there may be some benefit from ingesting high glycaemic index foods at this time if rapid restoration of muscle glycogen is a critical issue (Coyle, 1992).

With regular strenuous training, there must be an increased total intake to balance the increased energy expenditure. Provided that a reasonably normal diet is consumed, this will supply more than adequate amounts of protein, minerals, vitamins and other die- tary requirements. There is no good evidence to suggest that specific supplementation with any of these dietary components is necessary or that it will improve performance. A diet which may be considered inadequate for a sedentary individual consuming 4MJ per day, may meet the requirements of an athlete taking 12–15 MJ/day. Indeed without resort- ing to sweets, snacks and convenience foods, such a high intake may be difficult to achieve. There is, however, no evidence that this pattern of eating is harmful; for the indi- vidual who has to fit an exercise programme into a busy day, it is inevitable that changes to eating patterns must be made, but these need not compromise the quality of the diet. When the energy expenditure is very high, carbohydrate-rich drinks and snacks become an essential part of the diet (Brouns et al, 1989).

The only exceptions to the generalisation about dietary supplements may be iron (Eichner, 1986) and, in the case of very active women, calcium (Clarkson, 1991). Highly trained endurance athletes commonly have low circulating haemoglobin levels, although total red cell mass may be elevated due to an increased blood volume. This may be considered to be an adaptation to the trained state, but hard training may result in an increased iron requirement and exercise tolerance is impaired in the presence of anaemia. Low serum folate and serum ferritin levels are not associated with impaired performance, however, and correction of these deficiencies does not influence indices of fitness in trained athletes. Moderate exercise has been reported to increase bone mineral density in women, and this may be a significant benefit of exercise for most women: hard training, however, may reduce circulating oestrogen levels and hence accelerate bone loss. For these athletes, an adequate calcium intake should be ensured, although calcium supplements themselves will not reverse bone loss while oestrogen levels remain low: restoration of normal menstrual function is, however, associated with a gain of bone mass (Drinkwater et al, 1986).

NUTRITION FOR COMPETITION

There is no doubt that the ability to perform prolonged exercise can be substantially modified by dietary intake in the pre-exercise period, and this becomes important for the individual aiming to produce peak performance on a specific day. The pre-exercise period can conveniently be divided into two phases - the few days prior to the exercise task, and the day of exercise itself.

Dietary manipulation to increase muscle glycogen content in the few days prior to exercise has been extensively recommended for endurance athletes following observations that these procedures were effective in increasing endurance capacity in cycle ergometer exercise lasting about 1½-2h. The suggested procedure was to deplete muscle glycogen by prolonged exercise about one week prior to competition and to prevent resynthesis by consuming a low-carbohydrate diet for 2–3 days before changing to a high-carbohydrate diet for the last 3 days during which little or no exercise was performed. This procedure can double the muscle glycogen content and is effective in increasing cycling or running performance, measured as the time for which a given workload can be sustained (Coyle, 1991).

There is now a considerable amount of evidence that it is not necessary to include the low-carbohydrate glycogen depletion phase of the diet for endurance athletes. All that is necessary is to reduce the training load over the last 5 or 6 days before competition and to simultaneously increase the dietary carbohydrate intake. This avoids many of the problems associated with the more extreme forms of the diet. Although an increased pre-competition muscle glycogen content is undoubtedly beneficial, there is a faster rate of muscle glycogen utilisation when the glycogen content itself is increased, thus nullifying some of the advantage gained.

Consumption of a high carbohydrate diet in the days prior to competition may also benefit competitors in games such as rugby, soccer or hockey, although it appears not to be usual for these players to pay attention to this aspect of their diet. Karlsson and Saltin showed that players starting a soccer game with low muscle glycogen content did less running, and much less running at high speed, than those players who began the game with a normal muscle glycogen content. It is common for players to have one game in midweek as well as one at the weekend, and it is likely that full restoration of the muscle glycogen content will not occur between games unless a conscious effort is made to achieve a high carbohydrate intake.

Although this glycogen-loading procedure is generally restricted to use by athletes engaged in endurance events, there is some evidence that the muscle glycogen content may influence performance in events lasting only a few minutes (Maughan and Greenhaff, 1991). A high muscle glycogen content may be particularly important when repeated sprints at near maximum speed have to be made: at major athletics championships, the sprinter who competes in the 100 and 200m as well as in the relay may be required to run as many as 8 or 9 races within a rather short space of time. Short term high-intensity exercise can also be improved by ingestion of alkaline salts prior to exercise to enhance the buffering of the protons produced by anaerobic glycolysis (Maughan and Greenhaff, 1991).

There is scope for nutritional intervention during exercise only when the duration of events is sufficient to allow absorption of drinks or foods ingested and where the rules of the sport permit. The primary aims must be to ingest a source of energy, usually in the form of carbohydrate, and fluid for replacement of water lost as sweat (Maughan, 1994). High rates of sweat secretion are necessary during hard exercise in order to limit the rise in body temperature which would otherwise occur. If the exercise is prolonged, this leads to progressive dehydration and loss of electrolytes. Fatigue towards the end of a prolonged event may result as much from the effects of dehydration as from substrate depletion. It is often reported that exercise performance is impaired when an individual is dehydrated by as little as 2% of body weight, and that losses in excess of 5% of body weight can decrease the capacity for work by about 30% (Maughan, 1991). Sprint athletes are generally less concerned about the effects of dehydration than are the endurance athletes, but the capacity to perform high intensity exercise which results in exhaustion within only a few minutes has been shown to be reduced by as much as 45% by prior prolonged exercise which resulted in a loss of water corresponding to only 2.5% of body weight: smaller, but substantial, reductions in performance occurred after administration of diuretics or after sweat loss in a sauna. Although there is little opportunity for sweat loss during sprint events, athletes who travel to hot climates are likely to experience acute dehydration, which will persist for several days and may well be of sufficient magnitude to have a detrimental effect on performance in competition.

The composition of drinks to be taken during exercise should be chosen to suit individual circumstances (Maughan, 1994). During exercise in the cold, fluid replacement may not be necessary as sweat rates will be low, but there is still a need to supply additional glucose to the exercising muscles. Although consumption of a high-carbohydrate diet in the days prior to exercise should reduce the need for carbohydrate ingestion during exercise in events lasting less than about 2 hours, it is not always possible to achieve this; competition on successive days, for example, may prevent adequate glycogen replacement between exercise periods. In this situation, more concentrated glucose drinks are to be preferred. These will supply more glucose thus sparing the limited glycogen stores in the muscles and liver without overloading the body with fluid. In many sports there is little provision for fluid replacement: participants in games such as football or hockey can lose large amounts of fluid, but replacement is possible only at the half-time interval. Until very recently the opportunities for drinking during long road races were severely restricted, but the rules have now been relaxed to allow more frequent intake.

In the post-exercise period, replacement of fluid and electrolytes can usually be achieved through the normal dietary intake. Restoration of water balance is not achieved, however, unless the electrolytes lost in sweat (especially sodium) are also replaced (Ladell, 1955). If there is a need to ensure adequate replacement before exercise is repeated, extra fluids should be taken and additional salt (sodium chloride) might usefully be added to food. The other major electrolytes, particularly potassium, magnesium and calcium, are present in abundance in fruit and fruit juices. Mineral supplements are not normally necessary.

NUTRITIONAL SUPPLEMENTS: CREATINE

The effects of several different nutritional supplements are dealt with elsewhere in these proceedings, illustrating the widespread interest in any component which has the potential for improving exercise performance. Nutritional supplements used by elite athletes generally fall into one of two categories: those that are banned by the governing bodies of sport, and those that have no effect on performance. One exception to this generalisation is creatine: this is now widely used by athletes in many different sports. There is no doubt as to its beneficial effects on high intensity exercise (Balsom et al, 1994), and its use is not against the rules of any sport. The normal daily dietary intake of creatine is about 1–2 grams: the major sources are meat and fish, so the intake in vegetarians is minimal. Daily creatine turnover is about 2 g, so any shortfall in the dietary intake is met by endogenous synthesis (Walker, 1979). Harris et al (1992) showed that ingestion of small amounts of creatine (1 g or less) had a negligible effect on the circulating creatine concentration, whereas feeding higher doses (5 g) resulted in an approximately 15-fold increase. Repeated feeding of creatine (5 g four times per day) over a period of 4–5 days resulted in a marked increase in the total creatine content of the quadriceps femoris muscle: in some cases an increase of 50% was observed. Approximately 20% of the increase in total muscle creatine content was accounted for by creatine phosphate.

There appears to be no beneficial effect of creatine supplementation on the peak power output that can be achieved in a range of tests, but the balance of the available evidence suggests that performance is improved in high intensity exercise tasks, especially where repeated exercise bouts are carried out. Greenhaff et al (1993) measured the ability of two groups of subjects to complete five bouts of 30 maximal voluntary contractions on an isokinetic dynamometer: half the group then received a placebo, and the others received 24 g of creatine per day for 5 days. Peak torque was not different after the placebo treatment compared with the initial measurements, but the creatine-supplemented group showed significantly improved peak torque production. Other recent studies (Balsom et al, 1993; Harris et al, 1993; Soderlund et al, 1994) have confirmed the beneficial effect of creatine supplementation in high doses (20 g/d for 6 d) on high intensity exercise performance.

CONCLUSIONS

There is no doubt that exercise performance is influenced by the composition of the diet. In spite of this, however, many elite athletes have no clear idea of the nutritional demands of their sport, and often pursue practices which are unhelpful. The main requirements are to ensure that energy intake is appropriate for the level of energy expenditure, after taking account of requirements for growth and of and need to change body mass or fat content. A high carbohydrate intake is necessary to meet the demands of intensive training, and carbohydrate supplementation prior to competition may also be beneficial. If a varied diet is eaten in an adequate amount, it is extremely unlikely that deficiencies of protein or of any of the micronutrients will arise. There may be a need to pay attention to the intake of iron, and possibly also of calcium, especially in female athletes and anyone on an energy-restricted diet. Water intake must be increased to meet the increased losses which occur through sweating, and carbohydrate-electrolyte drinks may be beneficial if consumed during events lasting longer than about 40 min. Dietary supplements are not recommended, although information is emerging to suggest that creatine supplementation may enhance performance in sprint events.

REFERENCES

Bahr R (1992) Excess postexercise oxygen consumption - magnitude, mechanisms and practical implications. Acta Physiol Scand 144 (Suppl 605) 1–70

Balsom, PD, B Ekblom, K Soderlund, B Sjodin, E Hultman. Creatine supplementation and dynamic high-intensity intermittent exercise. Scand. J. Med. Sci. Sports 3: 143–149, 1993.

Balsom PD, K Soderlund, B Ekblom (1994) Creatine in humans with special referenec to creatine supplementation. Sports Med 18; 268–280

Bergstrom L and E Hultman (1967) A study of the glycogen metabolism during exercise in man. Scand J Clin Lab Invest 19 218–228

Bergstrom J, L Hermansen, E Hultman and B Saltin. Diet, muscle glycogen and physical performance. Acta Physiol Scand 1967; 71: 140–150

Blom PC, AT Hostmark, O Vaage, KR Vardal, S Maehlum. Effect of different post- exercise sugar diets on the arte of muscle glycogen synthesis. Med Sci Sports Ex 1987; 19: 491–496

Brouns F, WHM Saris, J Stroecken, E Beckers, R Thijssen, NJ Rehrer, F ten Hoor (1989) Eating, drinking and cycling. A controlled Tour de France simulation study. Int J Sports Med 10 S41-S48

Burke LM, K Inge. Protein requirements for training and "bulking up". In: Burke LM, V Deakin (eds) Clinical Sports Nutrition. McGraw-Hill, Sydney 1994, pp 124–150

Clark K. Nutritional guidance to soccer players for training and competition. J Sports Sci 1994; 12 (Special Issue): S43-S50

Clarkson PM (1991) Minerals: exercise performance and supplementation. J Sports Sci 9 (Special Issue) 91–116

Costill DL. Carbohydrates for exercise: dietary demands for optimal performance. Int J Sports Med 1988; 9: 1–18

Costill DL, Miller JM. Nutrition for endurance sport: Carbohydrate and fluid balance. Int J Sports Med 1980; 1:2–14.

Coyle EF (1991) Timing and method of increased carbohydrate intake to cope with heavy training, competition and recovery. J Sports Sci 9 (Special Issue) 29–52

Dohm GL. Protein as a fuel for endurance exercise. Ex Sport Sci Rev 1986; 14: 143–173.

Drinkwater BL, K Nilson, S Ott, CH Chestnet (1986) Bone meineral density after resumption of menses in amenorrheic athletes. JAMA 256 380–382

Eichner ER (1986) The anemias of athletes. Physician Sportsmed 14 122–130

Gleeson M, PL Greenhaff, RJ Maughan. Influence of a 24-hour fast on high-intensity cycle exercise performance in man. Eur J Appl Physiol 1988; 57:653–659

Greenhaff, PL, A Casey, AH Short, RC Harris, K Soderlund, E Hultman. Influence of oral creatine supplementation on muscle torque during repeated bouts of maximal voluntary exercise in man. Clin Sci 84: 565–571, 1993.

Harris, RC, K Soderlund and E Hultman. Elevation of creatine in resting and exercised muscle of normal subjects by creatine supplementation. Clin. Sci. 83: 367–374, 1992.

Harris, RC, M Viru, PL Greenhaff and E Hultman. The effect of oral creatine supplementation on running performance during maximal short term exercise in man. J. Physiol. 467: 74P, 1993.

Hultman E and LH Nilsson (1971) Liver glycogen in man Adv Exp Biol Med 11 143–151

Ivy, JL, AL Katz, CL Cutler, WM Sherman, EF Coyle. Muscle glycogen synthesis after exercise: effect of time on carbohydrate ingestion. J Appl Physiol 1988a; 65: 1480–1485

Ivy, JL, MC Lee, JT Brozinick, MJ Reed. Muscle glycogen storage after different amounts of carbohydrate ingestion. J Appl Physiol 1988b; 65: 2018–2023

Lemon PWR (1991) Effect of exercise on protein requirements. J Sports Sci 9 (Special Issue) 53–70

Maughan RJ (1991) Fluid and electrolyte loss and replacement in exercise. J Sports Sci 9 (Special Issue) 117–142

Maughan RJ (1994) Fluid and electrolyte loss and replacement in exercise. In: Oxford Textbook of Sports Medicine. Ed Harries, Williams, Stanish & Micheli. Oxford University press, New York. pp 82–93

Maughan RJ, PL Greenhaff. High intensity exercise and acid-base balance: the influence of diet and induced metabolic alkalosis on performance. In: Brouns F, (Ed) Advances in Nutrition and Top Sport. Karger, Basel. (1991) pp147–165

Maughan RJ, JB Leiper. Aerobic capacity and fractional utilisation of aerobic capacity in elite and non-elite male and female marathon runners. Eur J Appl Physiol 1983; 52: 80–87

Murray R. The effects of consuming carbohydrate-electrolyte beverages on gastric emptying and fluid absorption during and following exercise. Sports Medicine 1987; 4: 322–351

Piehl K. Time course of refilling of glycogen stores in human muscle fibres following exercise-induced glycogen depletion. Acta Physiol Scand. 1974; 90: 297–302

Soderlund, K, PD Balsom, B Ekblom. Creatine supplementation and high intensity exercise: influence on performance and muscle metabolism. Clin Sci. 87, Suppl: 120–121, 1994.

Stanton, R. Dietary extremism and eating disorders in athletes. In: Burke LM, V Deakin (eds) Clinical Sports Nutrition. McGraw-Hill, Sydney 1994, pp 285–306

Walker, JB. Creatine biosynthesis, regulation and function. Adv. Enzymol. 50: 117–142, 1979.

Westerterp KR, WHM Saris. Limits of energy turnover in relation to physical performance; achievement of energy balance on a daily basis. In: Williams C, JT Devlin (eds). Foods, Nutrition and Sports Performance. London, E & FN Spon. 1992. pp 1–16

Williams C, JT Devlin. Foods, Nutrition and Sports Performance. London, E & FN Spon. 1992

Williams MH. Nutritional aspects of human physical and athletic performance. Springfield, Charles C Thomas, 1985.

25

ERGOGENIC EFFECTS OF THE CREATINE SUPPLEMENTATION DURING THE TRAINING OF TOP-CLASS ATHLETES

Nikolai I. Volkov, Egon R. Andris, Yurij M. Saveljev, and
Vladimir I. Olejnikov

Department of Biochemistry
Russian Academy of Physical Education
Moscow, Russia

INTRODUCTION

Application of the creatine and creatine phosphate in the preparation of athletes causes marked ergogenic and anabolic effect (2, 13, 16, 18, 20, 23, 24, 25). The exogenic creatine moves rather easily across the cell membrane (7, 8) and can change markedly the intracellular creatine pool. The permeability of cell membranes for the exogenic creatine phosphate is considerably lower but it can cause an increase in the premembrane pool of macroergs that in its turn contributes to maintaining a high bioelectric activity of the excited cells under conditions of hypoxia (9, 17).

Ergogenic effects of intaking the creatine preparations and creatine phosphate are manifested in (1) enhancing a maximal anaerobic power and alactic anaerobic capacity (18, 23, 24), (2) improving the indices of aerobic efficiency, reflecting a rise in the speed of energy transportation from the mitochondrial centers of oxidation to myofibrils (4, 5, 11, 23) and also raising the intramuscular buffer capacity (10, 16, 23). These effects, produced in the laboratory trials and model experiments, are differently manifested under revile conditions of sports training and so they are the subject of our further research in real efficiency which could be applying the creatine preparations in the process of preparing top-class athletes.

SUBJECTS AND METHODS

Acute effects of applying the creatine preparations were studied on the experimental group of 5 top-class track-and-road riders-cyclists. The experiment program consisted in taking the physiological and biochemical measurements with athletes at rest and in the

Current Research in Sports Sciences, edited by Rogozkin and Maughan.
Plenum Press, New York, 1996

course of conducting the standardized laboratory tests on (1) graded increase in load (1, 15, 19), (2) maintenance of critical power (22, 26), (3) repeated maximal efforts (12, 14) and (4) repeated utmost one-minute exercises (20, 27). This program in testing the physical workcapacity was repeated three times: under control conditions without taking the creatine preparations, with intaking the creatine preparations and with combined intake of the creatine preparations and amino acid mixtures.

Creatine preparations were intaken in a kind of drink cooked on the fruit juice base with creatine sulphate supplement with a dosage of 125 milligrams per a kg of body weight. The combined intake of creatine preparations and amino acid mixtures was used in the same way with supplementing 52 grams of amino acid mixture, cooked with autolys of baker yeast with supplementing some separate amino acids. This drink of 200 milligrams, containing the mentioned above preparations with the given dosages, was cooked "ex tempore" and was intaken 90 minutes prior to the beginning of tests.

At rest and at work, including the testing loads the gasometric measurements were taken with using the MMC "Beckman" monitor system. The lactic acid concentration in blood was determined with the calorimetric method of Barker S. and Summerson W. (3). The indices of acid-base equilibrium were measured with IL-213 microanalyses manufactured by the firm "Instrument Laboratory" and BMS-2M microanalyses of the "Radiometer" firm.

Cumulative effects on intaking the creatine preparations for a long period of training were studied on the base of two arrangements of our experiment. The work on the 1st. arrangement consisted of the following: 16 national sprinters-runners intake regularly creatine preparations for one year, on those days when they had a large volume of speed-strengthening exercises. Daily dosed was about 5 grams, but first 2.5 grams were taken before training session and the second ones-after training session.

Second arrangement of our experiment : seven athletes-students of the Russian State Academy of Physical Culture in Moscow had to fulfill a definite training program during the two following sports seasons (two years). During the first year of experimental training these athletes did not intake any ergogenic aids, during the second one they in took regularly the creatine preparations according to the instruction described above. At the beginning and at the end of each stage of the season preparation they were examined according to the full program of standardized laboratory and special tests.

RESULTS

Results of the laboratory researches in estimating the acute effect on intaking the creatine preparations were illustrated in table 1.

Results of the conducted research have established that intake of creatine preparations and aminoacid mixtures under conditions of hard interval work enhances maximal power, increases the organism's buffer reserves and heightens efficiency of the aerobic energy creation. To a considerable extent these effects are manifested during the repeated performance of exercises of a maximal power without wore intense glycolysis and marked acidity of the organism's internal medium. In the other studied kinds of exercises where a higher degree of acidity with considerable reduction of the anaerobic capacity occurred the effects of intaking the creatine preparations were of less appreciation. Proceeding from the achieved results much benefit from applying the creatine preparations and amino acid mixtures can be got in those sports events where short term efforts of maximal power pre-

Table 1. Influence of creatine application on indices of sportsmen work capacity in standardized laboratory tests

No	Indices	Control	Creatin	Creatin + autolysin
1	Maximum power (Wmax), kpm/min	5238 ± 411	6380 ± 901	6378 ± 534
2	Power of exhaustion (Wexh), kpm/min	2768 ± 128	2750 ± 165	2886 ± 259
3	Critical power (Wcr), kpm/min	2068 ± 110	2037 ± 70	2080 ± 167
4	Marginal work in test of repeated maximum efforts, kpm	940 ± 68	1096 ± 35	1105 ± 66
5	Limited work in test of repeated maximum efforts (3×1' after 1' rest), kpm	8309 ± 385	8249 ± 423	8658 ± 778
6	Limited work in test on holding of critical power, kpm	10994 ± 1171	11212 ± 1204	15239 ± 864
7	Value of O_2 request in test 3×1' after 1' rest	13.7 ± 0.9	13.7 ± 1.8	14.0 ± 0.9
8	Value of O_2 request in test on holding W_{cr}	20.0 ± 2.3	20.2 ± 2.6	28.7 ± 3.0
9	Value of pH in test MAM, conv. units	7.046 ± 0.015	7.087 ± 0.036	7.120 ± 0.009
10	Value of pH in test 3×1' after 1' rest, conv. units	7.089 ± 0.033	7.065 ± 0.023	7.065 ± 0.031
11	Value of pH in test on holding W_{cr}, conv. units	7.104 ± 0.019	7.102 ± 0.025	7.115 ± 0.002

dominate, here belong short distance running, jumps, throwing events, cycling track races and so on.

For estimating the cumulative effect on intaking the creatine preparation special experiment on seven top-class sprinters-runners was conducted. The plan of this experiment included the fulfillment of a definite training program with a large volume of interval work during the two sports seasons. In the first year of experimental training athletes did not apply any ergogenic aids. In the second year of experimental training they applied regularly the creatine preparations and amino acid mixtures. In the beginning and at the end of each stage of the season preparation athletes underwent examination according to the program on standardized laboratory and special tests.

Data on performing the volumes of training loads of different directness and an increment in work capacity indices in the experimental group of sprinters-runners, who trained for two sports seasons with application of the creatine preparations and without it was given in table 2.

Table 2. Data of training loads volumes of various orientation and of increase of efficiency indices for runners on short distances under influence of one year training with and without application of creatine

No	Indices without creatine	Training with creatine	Training
1	Volume of loads of aerobic action, min	2891 ± 246	2797 ± 366
2	Volume of loads of combined aerobic/anaerobic action, min	3620 ± 816	3965 ± 412
3	Volume of loads of glycolitic anaerobic action, min	210 ± 40	215 ± 22
4	Volume of loads of alactate anaerobic action, min	3487 ± 521	3472 ± 488
5	Total volume of loads per year, min	10226 ± 570	10272 ± 710
6	Result in ranning on 100 m, s	10.62 ± 0.23	10.51 ± 0.22
7	Result in running on 200 m, s	22.07 ± 0.35	21.81 ± 0.31
8	Increase of result in standing jump, cm	24.5 ± 2.6	37.0 ± 3.4
9	Increase of result in running on 20 m, s	0.07 ± 0.01	0.11 ± 0.02
10	Increase of result in running on 60 m, s	0.18 ± 0.04	0.257 ± 0.06
11	Increase of result in running on 150 m, s	0.56 ± 0.11	0.8 ± 0.10
12	Increase of result in running on 300 m, s	1.2 ± 0.21	1.97 ± 0.24
13	Increase of O_2 maximum consumption index, ml/kg·min	-2.85 ± 0.64	-1.043 ± 0.06

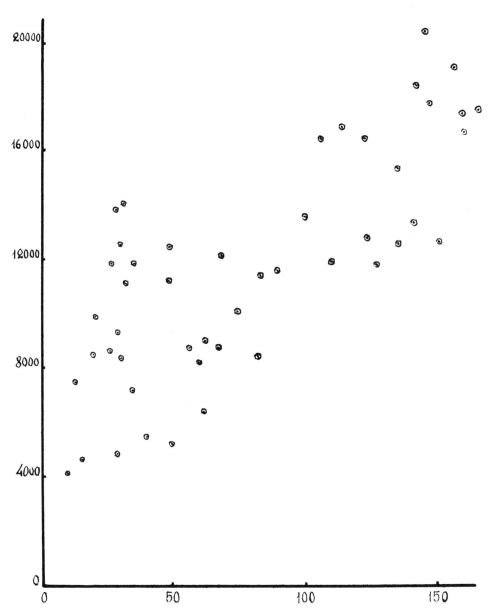

Figure 1. Dependence of total volume of training loads on quantity of consumed creatine: ordinates—value of total volume of loads, min; abscises—quantity of consumed creatine, g.

Given at shows the absence of significant difference in the indices of volume of the training loads performed by the examines of our experimental group for both two years under supervision. However , sports results in 100, 200 m running and special workcapacity indices improved significantly in the second year of our experiment, when creatine preparations were applied.

Cumulative effect of applying the creatine preparations was prominently manifested in the improvement at those functional and physical qualities at which training means, used in the period of applying the creatine preparations, were directed.

Table 3. Matrix of intercorrelations of indices of relative increase of sporting results, volfilled training loads and quantity of consumed creatine

No	Function	Independent variables H_1	H_2	H_3	H_4	Cr
1	Volume of loads of aerobic orientation (H_1)	X				
2	Volume of loads of combined aerobic/anaerobic orientation (H_2)	0.75	X			
3	Volume of loads of glycolitic anaerobic orientation (H_3)	−0.11	−0.53			
4	Volume of loads of alactate anaerobic orientation (H_4)	0.40	0.13	0.53	X	
5	Quantity of consumped creatin (Cr)	0.51	0.47	0.11	0.80	X
6	Relative increase of sporting results (ΔR)	−0.49	−0.41	0.05	−0.31	−0.88

Close relationship between a volume of work done at training's and an amount of the creatine preparations, intake at separate studies of the preparation, was established in the national short distance running team where we studied the cumulative effect as a result of applying the creatine preparations for one year of the training (Fig.1).

Table 3 illustrates the data of intercorrelations between the indices of a volume of the performed training loads of different divectness, an amount of the intaken creatine and an increment in sports results in the short distance running.

Data, given in this table, shows that the rate of an increment in sports results in the short distance running is significantly correlated with an amount of the performed training loads of alactic-anaerobic influence.

Table 4 illustrates the results of done calculations at optimal values of a volume of the training loads and an amount of the intaken creatine due to which a maximal increment in sports results in the short distance running is achieved with applying the square approximation method.

This table illustrates average values of the performed training loads and magnitudes of the standard deviations including also the values of an optimal volume of loads due of which greater increments in sports results are achieved (localization of extreme). Due to the used form of the square approximation for the loads of combined aerobic-anaerobic influence we managed to establish only general character of the relationship: with an increase in a volume of loads the value of an increment in sports result heightens too (sign "+" was written in the appropriate column of the table). The calculations of the data concerning volumes of the loads of glycolytic -anaerobic influence did not produced signifi-

Table 4. Optimum values of training volume load of various orientation and quantities of consumed creatine

No	Training action	Average error	Standard error	Local extreme	Value of error	Maximum increase of sport result, %
1.	Load of aerobic action, min	4813	1783	7765	3125	1.29
2.	Loads of combined aerobic/ anaerobic action, min	2804	1822	+	?	2.8
3.	Loads of glycolitic anaerobic action, min	683	602	?	?	?
4.	Loads of alactate anaerobic action, min	2486	1501	4582	2800	1.14
5.	Quantity of consumed creatine, gram	127.5	258.0	270.0	190.0	1.98

cant results (sign "?" was written in the appropriate column of the table). Our calculations show that "pure" effect, produced as a result of applying the creatine preparations, can better sports results only approximately by 2 %.

In the course of approbation the creatine preparations by national top-class sprinters-runners in the preparation and participation in international competitions total increment in results in 100 m running was equal to 2.1 % and 2.7 % in 200 m running.

Thus the conducted research has shown that the application of creatine preparations enables us to heighten considerably efficiency of using the training loads of speed-strengthening character and it is associated with improving the indices of special work capacity and also an increment in results in the short distance running .

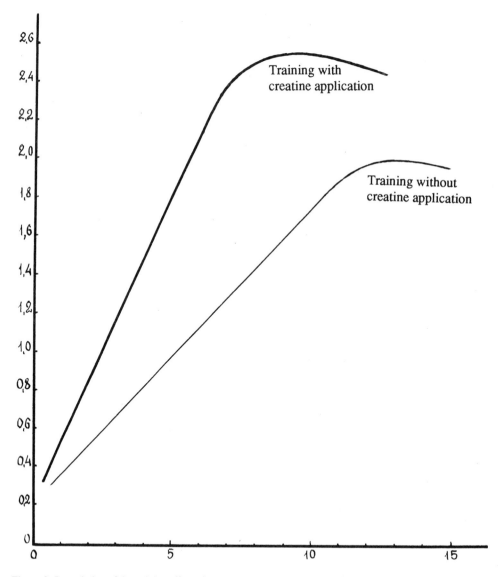

Figure 2. Potentiating of the training effect of physical loads by intaking the creatine preparations in the process of season sports preparation of short distance-runners.

DISCUSSION

Results of the conducted research with the participation of top-class athletes testify in full measure the theoretical suggestions stated previously on possible mechanisms of ergogenic action of the creatine preparations (10, 13, 17, 24, 25). Improvement of the indices of maximal anaerobic power and some enhancement of the buffer capacity can be considered as an evidence of the possibility of increasing the ultramuscular creatine phosphate stores under conditions of performing the loads of maximal power. The results of recent experimental research of the Swedish authors point out also such a possibility (10). Enhancement of the buffer capacity, achieved at the expense of applying an extra amount of creatine, is relatively slight and it does not produce a somewhat marked positive effect under conditions of considerable acidity of the intramuscular medium in the course of performing the loads of glycolytic-anaerobic influence. We also observed an insignificant influence of the exogenic creatine on improving the indices of aerobic metabolism that could be connected with intensifying the activity of creatine–creatine phosphate energy transport shuttle (4, 5, 17). At the same time marked anabolic influence of the creatine preparations at the loads of speed-strengthening character can be considered to have been proved. This has been ascertained by the results of estimating the cumulative effect of training in the prolonged periods of regular application of creatine in the combination with loads of alactic-anaerobic influence. Total result of this research is graphically illustrated on fig.2.

ACKNOWLEDGMENTS

We wish to thank Dr.Bobkov Yu. and Dr.Saks V. for their serious discussions to thank Cirkov V. and Barinova Y. for their invaluable technical assistance during the experiments and Remizov L. for the mathematical calculations.

We thank Dr.Yu.Y.Bobkov and Dr.V.A.Saks for helpful discussions and encouragement. We also thank V.S.Cirkov and Y.V.Barinova for their skilled technical assistance during the experiments and L.P.Remizov for mathematical calculations.

REFERENCES

1. Andersen K.L., R.J.Shephard, H.Derolin, E.Varrauskas, R.Masironi. Fundamentals of exercise testing. Geneva: WHO, 1971, 133 pp.
2. Andris E.R., Yu.M.Saveljev, N.I.Volkov Use of creatine preparations for the enhancing of physical work capacity of sportsmen and potentiating training effect of exercise. In: Pharmacological correction of hypoxic states. Abstr. of the All-Union Conf., January 24–28, 1988, Moscow, p.6–7.
3. Barker S.B., W.H.Summerson. The colorimetric determination of lactic acid in biological material. J.Biol. Chem. 138: 535–542, 1941.
4. Bessman S.P., C.L.Carpenter. The creatine–creatine phosphate energy shuttle. Ann. Rev. Biochem. 54: 831–862, 1985.
5. Bessman S.P., F.Savabi. The vole of the phosphocreatine energy shuttle in exercise and muscle hypertrophy. In: International Series on Sports Sciences, edited by A.W.Tylor, P.D.Gollnick, H.J.Green, C.D.Januzzo, E.Y.Noble, Y.Metivier, J.R.Sutton. Champaign, In.: Human Kinetics, 1990, v.21, p.167–178.
6. Crim M.C., D.H.Calloway, S.Margen. Creatine metabolism in men: creatin pool size and turnover in relation to creatine intake. J.Nutr. 106: 371–381, 1976.
7. Fitch C.D., R.P.Shields. Creatine metabolism in skeletal muscle F. Creatine movement across muscle membranes. J.Biol. Chem. 241: 3611–3614, 1966.

8. Fitch C.D., R.P.Shields, W.Payne, J.M.Dacus. Creatine metabolism in skeletal muscle. III. Specificity of the creatine entry process. J.Biol.Chem. 243: 2024–2027, 1968.

9. Gattuso C. Fosfocreatine ed eccitabilita neuromuscolare. Minerva medica Siciliana, 1965, 12 pp.

10. Harris R., K.Soderlund, E.Hultman. Elevation of creatine in resting and exercised muscle of normal subjects by creatine supplementation. Clinical Science, 83: 367–374, 1992.

11. Kammermeier H. Why do cells need phosphocreatine and a phosphocreatine shuttle? J.Mol.Cell.Cardiol. 19: 115–118, 1987.

12. Katch V., A.Weltman, R.Martin, L.Grey. Optimal test characteristics for maximal anaerobic work on the biajcle ergometer. Res.Quart. No 2: 263–276, 1977.

13. Marcenaro A., R.Balestreri, F.Basso. L'influenza del trattamento confosfocreatine sul reudimento energetics della respiratzione e sul lavono aerobico massimo. Arch., "E.Maragliano" patol. e clin. 18: 155–163, 1962.

14. Margaria R. An outline for setting significant tests of mascular performance. Arch. fisiol. 64: 37–44, 1965.

15. Mellerowicz H., V.N.Smodlaka (Editors). Ergometry: Basics of Medical Exercise Testing. Berlin: Urban u. Schwarzenberg Med. Publ., 1981, 432 pp.

16. Olejnikov V.I. Efficiency of application physical loads in training of short distance runners with using special aids. Dissertation Abstr., Moscow, 1989, 13 pp.

17. Saks V.A., Yu.G.Bobkov, E.Strumia (Editors). Creatine phosphate: biochemistry, pharmacology and clinical efficiency . Torino: Ed.Minerva medica, 1987, 270 pp.

18. Saveljev Yu.M. Efficiency of strength and power training of athletes by application special aids. Dissertation Abstr., Omsk. 1993, 18 pp.

19. Shephard R.J. Standard tests of aerobic power. In: Frontiers of fitness, edited by R.J.Shphard—Springfield: Ch.Thomas, 1978, p.233–264.

20. Silver M.L., V.N.Litvinova, B.I.Morosov, A.V.Pliskin, A.I.Pshendin, V.A.Rogozkin. Effect of creatine on protein and relonucleic acid synthesis in embryonic chicken myoblasts. In: Materials All-Union Symposium "Biochemical Pathways of Enhancing Sports Training Efficiency". Leningrad; Research Institute of Physical Culture, 1974, p.110–116.

21. Szogy A., G.Cherebetiu. Minutentest auf dem Fakrradergometen zur Bestimmung der anaeroben Karazitat. Europ. J.appl. Physical. 33: 171–176, 1974.

22. Volkov N.I. Tests and criterion's for assessment of endures. Moscow, RIO GCOLIFK, 1989, 44 p.

23. Volkov N.I. Use of creatine preparations and amino acid mixtures for potentiating the training effect of exercise. In: Materials All-Union Symposium "Hutrition and physical work capacity". Leningrad: Research Institute of Physical Culture, 1991, p.126–134.

24. Volkov N.I., V.I.Olejnikov. Use of creatine preparations for enhancing the training efficiency in short distance running. In: Factors limiting and enhancing physical work capacity of elite sportsmen, edited by M.A.Godik, A.N.Korobov. Moscow, GCOLIFK, 1985, p.204–210.

25. Volkov N.I., Yu.M.Saveljev, E.R.Andris, N.D.Altukhov. Application if creatine and amino acid for enhancing physical work capacity and training efficiency in sports. Moscow, GCOLIFK, 34 pp.

26. Volkov N.I., Ye.A. Shirkovetz. About the energetical criterions of physical work capacity of athletes. In: Bioenergetics . Leningrad, 1973, p.18–30.

27. Wolkow, N.I. Bioenergetyczne podstawy i' ocena wytrzymatosci. Sport Wyczynowy, 27, Nr.7–8, s.7–18, 1989.

26

EFFECT OF PRODUCTS OF ENHANCED BIOLOGICAL VALUE ON THE PERFORMANCE OF ATHLETES

A. Pshendin, N. Shishina, and V. Zagrantsev

Research Institute of Physical Culture
Dynamo Ave. 2, 197042 St. Petersburg, Russia

INTRODUCTION

The nutrition of an athlete, as of any person systematically performing physical exercise, is considered to be an important factor influencing the organism's metabolic adaptation to physical and nervous-psychical loads (Rogozkin et al, 1985; Williams, 1991). Nutritional factors can actively influence metabolic processes and increase physical efficiency; they can also speed up or slow down the processes of recovery in the period of rest after training and competition (Berlutti, 1991).

The possibilities and peculiarities of metabolic processes derived from the direct action of low-molecular food components during physical exercise and subsequent rest enabled us to formulate certain theoretical principles for developing product compositions of high biological value (PHBV). The leading principle was correspondence of PHBV's physico-chemical and nutritional-physiological properties to the metabolic specificity of an organism (Rogozkin, 1976; Korovnikov and Letshyk, 1989). The main aim of PHBV usage is to enhance adaptability to muscular activities of varied duration and intensity. PHBV introduction into athletic practice may lead to improvement of certain procedures, such as feeding between events, in heavy exercise and in the recovery period, regulation of body weight, water and salt metabolism, reduction in the volume of food ingested on the days of competitions, and an increase of the number of meals etc. (Rogozkin and Pshendin, 1989).

The aim of this paper is to examine the influence of MIXOVIT-FORTE PHBV systematic administration on the development of specific work-capacity of athletes within a training program.

METHODS

21 athletes volunteered to take part in the research program. The participants' average age was 15.2 ± 1.1 years, height 168.0 ± 3.2 cm and body mass 58.0 ± 2.6 kg. The ath-

Current Research in Sports Sciences, edited by Rogozkin and Maughan.
Plenum Press, New York, 1996

Table 1. Food and chemical composition MIXOVIT-FORTE

	Components	g/100g
Proteins:	Caseins, Lactoalbumins, Lactoglobulins, Proteosopeptons, Immunoglobulins	65
Vitamins:	B_1, B_2, B_6, C, PP, B_{12}, K and other	+
Minerals:	K, Ma, Ca, Mg, Co, Mn, Zn, Cu, Fe, P, Ce, S	+
Carbohydrate:	Lactose	175
	Glucose	6
Organic acids		+
Lecithine		+
Kcal		360

letes (Alpine combination) were undertaking a 12-day training period for an international competition. The experimental group (n=10) received 75 g of MIXOVIT-FORTE per person in the form of a water cocktail during the period of recovery after all physical exercise, the control group (n=11) received placebo. MIXOVIT-FORTE is a mixture of lactic proteins obtained by means of membraneless osmoses technology, basic vitamins, macro- and microelements, organic acids, flavouring supplements (Table 1).

The energy value of 75 g MIXOVIT-FORTE is 200 kcal. The athletes received a diet (energy value of 4318 ± 130 kcal) which corresponded to the recommended standards.

The maximal physical work-capacity level was determined by an aerobic productivity in short-term exercises of maximal power. The lactate power (ALP) of work was judged from the maximal value of mechanical productivity on a bicycle ergometer for 6 seconds, followed by 5–10 minutes rest; later, the lactate-glycolic power (LGP) was determined during 20 s of work (Volkov and Jaruzni, 1984).

Biochemical criteria of lactate-glycolytic working power and capacity Creatine (Cr) and inorganic phosphate (Pi)) were determined before and after the testing on a bicycle ergometer.

Blood urea and serum cholesterol content, as well as creatininase activity in blood were determined by REFLOTRON system (Boehringer Mannheim GmbH, 1985).

Examinations were performed before and after the 12-day administration of PHBV, in the morning after a standard breakfast. Vitamin status was analysed by gram/hour content method (Pokrovsky, 1969). The chemical composition of diets was calculated by table (Skurihin, 1984). The statistical treatment of the data was carried out according to Student's criterion.

RESULTS

An experiment involving evaluation of nutritional status must be preceded by analysis of the normal diet which forms the background of investigation. For data analysis on the habitual diet of athletes participating in the experiment see Table 2.

According to the data obtained the caloric value of the athletes diet (4318 ± 132 kcal) was within recommended standards. Dietary protein value ($147 \pm 5,8$ g) covered the physiological needs, while the fat value (189 ± 31.3 g) exceeded the standard, with a complete absence of vegetable fats; the carbohydrate value (506 ± 26.7 g), on the contrary, was always insufficient.

Table 2. Total daily nutrient intake

	Requirements	Intake
Kcal	4000–4500	4318 ± 132
Protein, g	140–151	147 ± 5.8
Fat, g	133–150	189 ± 31.3
Carbohydrate, g	597–674	506 ± 26.7
Vitamins, mg/day:		
C	160–180	131.3 ± 43.8
B_1	2.6–2.9	2.4 ± 0.19
B_2	2.6–2.9	2.6 ± 0.23
B_6	4.8–3.4	4.9 ± 0.11
PP	44–49	46 ± 9.6
A	3.0–3.4	3.1 ± 0.2
Minerals, mg/day:		
Ca	1200	1206 ± 7
P	1800	2101 ± 36
Fe	20	25 ± 36
Mg	600–900	564 ± 23
K	5000	5467 ± 234

We also noted the limited and monotonous character of the diets; insufficient consumption of fruit, vegetables, juices, greenery, cereals and whole meal bread, resulted in mineral and vitamin unbalance and deficiency. Under such conditions, systematic administration of a vitamin- and salt-enriched PHBV induced positive changes in the athletes, especially in vitamins C and B_1. A second study (on termination of PHBV administration) revealed that in the experimental group there was a significant decrease, at rest on an empty stomach, of blood urea (t=2.55), cholesterol (t=2.47), similar to the decrease in Creatine kinase activity (t=2.5), which might imply acceleration of restitution processes in general, and muscular restitution in particular, similar to a positive shift in lipid metabolism, which is particularly important in speed-strength and strength physical loads (Table 3).

The results of investigation into physical work-capacity before and after administration of MIXOVIT-FORTE are presented in Table 4.

Table 4 demonstrates that the control group which took the placebo displayed an alactate power change from 683 ± 22.3 to 723 ± 20.3W, with the specific power change from 11.6 ± 0.3 to 12.3 ± 0.27W/kg, but these increases are insignificant (p>0.05). The alactate power increase of 40W can be put down as a result of the speed-strength training.

Table 3. Clinical-biochemical data (n=10)

	Urea mmol/l	Cholesterol mmol/l	Creatinekinase μ/l
Placebo			
Before	6.71 ± 0.30	3.10 ± 0.15	297 ± 46
After	7.20 ± 0.20	3.17 ± 0.20	302 ± 37
Mixovit-Forte			
Before	7.01 ± 0.39	3.23 ± 0.15	269 ± 21
After	6.50 ± 0.18[a]	2.81 ± 18[b]	196 ± 16[a]

Mean and S.D.
[a] p<0.05 after vs before.

Table 4. Alactate power and alactate-glycolytic power before and after placebo and
MIXOVIT-FORTE (n=10)

Variables	Alactate power			Alactate glycolytic power		
	kP	watt	w/kg	kP	watt	w/kg
Placebo						
Before	5.4 ± 0.08	688 ± 22.3	11.6 ± 0.30	3.6 ± 0.09	447 ± 15.4	7.6 ± 0.25
After	5.7 ± 0.09	723 ± 20.3	12.3 ± 0.271	3.7 ± 0.08	506 ± 14.51	8.6 ± 0.371
Mixovit-Forte						
Before	5.3 ± 0.10	656 ± 19.1	11.5 ± 0.36	3.6 ± 0.18	461 ± 19.5	8.16 ± 0.30
After	5.5 ± 0.11	$759 ± 19.1^{a}$	$13.3 ± 0.43^{a,b}$	3.9 ± 0.07	$555 ± 12.4^{a,b}$	$9.6 ± 0.30^{a,b}$

mean and S.D.

[a] p<0.05 after vs before.

[b] p<0.05 Mixovit-Forte vs placebo.

Table 4 presents changes of alactate glycolytic power; the power went up from 447 ± 15.4 to 506 ± 14.5W i.e. by 59W. Specific alactate glycolytic power changed from 7.6 ± 0.25 to 8.6 ± 0.37W/kg. The increases in power are insignificant (p>0.05). The tendency for the increase seems to be the result of intensive regular training.

We then investigated physiological changes in the experimental group that received the supplement MIXOVIT-FORTE for 12 days. Table 4 demonstrates the alactate power increase from 656 ± 23.9 to 759 ± 19.1W i.e. by 103W. Specific alactate power changed from 11.5 ± 0.36 to 13.3 ± 0.42W/kg i.e. by 1.8W/kg. The increase in alactate power was significant (p<0.05). Alactate glycolytic power in the experimental group also changed. Its power went up from 461 ± 19.5 to 555 ± 12.4W, i.e. by 94W. Specific alactate glycolytic power went up from 8.1 ± 0.3 to 9.6 ± 0.3W/kg i.e. by 1.5W/kg. Only in the case of the training program Cr and Pi increased 57 ± 8 mmol/l and 0.19 ± 0.03 mmol/l after testing on the bicycle ergometer.

The systematically application of MIXOVIT-FORTE led to more reliable alteration of biochemical indexes of work capacity. So the changes of Cr was 83 ± 10 mmol/l and Pi 0.30 ± 0.03 mmol/l.

Physical work-capacity indexes in the experimental group increased considerably (p<0.05). Positive changes of alactate power by 103W and alactate glycolytic power by 94W in the experimental group must be ascribed to the introduction of the high alimentary density due to the athletes' diet. Evaluation of MIXOVIT-FORTE showed it to have a high percentage of protein; the percentage of protein in 100g of the product is 65%, 96% of this being casein and the rest serum proteins lactoglobulines, lactoalbumines, proteosopeptones, immunoglobulins (ca. 3–4%). It is general known that for amino acids, milk proteins are of the most value; their digestibility is 96–98% and pure utilization index is 82%. Caseines are food proteins of mammals and perform alimentary function, developed in the process of evolution. Casein contains amino acid sequences that are released in proteolise as physiologically active peptides. Active casein peptides that participate in saturation processes, influence the functions of the alimentary canal and change the organism's metabolic status. They seem to be able to act as local "food hormones", executing heterogeneous influence on the organism. Their possible influence exceeds nutritional function, involving also the organism's general adaptation capacities; they can be regarded as antistress supplements.

Thus, apart from being used to the extreme for plastic needs, casein acts as a source of physiological active peptides that display hormonal activity, similar to phosphorus and calcium ions. MIXOVIT-FORTE contains 17% lactose, which is an important component of the PHBV. Unlike other sugar products, lactose is poorly dissolved in water, is slowly absorbed into the intestine and thus stimulates the development of lactic acid bacteria that, by forming lactic acid, suppress putrefactive flora, ensuring better absorbtion of calcium and phosphorus.

PHBV influence on adaptation to muscular activities and neuro-endocrine mechanisms of training was investigated. The data obtained allow us to suggest that the introduction of MIXOVIT-FORTE into a diet (10% of the daily protein ration) facilitates adaptation to muscular loads in the training process. Food with MIXOVIT-FORTE stimulate an increase in heart and skeletal muscles weight, formation of haemoglobin, executes a marked influence on lipid metabolism, on the metabolic intensity and oxidating processes in liver and thymus.

CONCLUSION

MIXOVIT-FORTE posses active physiological and food properties that allow effective application to athletes. The physiological and biochemical analysis of work power in experimental and control groups permits us to make conclusion about the substantial influence of PHBV on the work ability of athletes.

Systematic taking of MIXOVIT-FORTE for two weeks reveal a positive increase of alactate and alactate-glycolytic power.

REFERENCES

Berlutti, G., Giampetro, M., 1991, Sport a sforze breve; nutrizione e potensiamento muscolare, Scuola dello sport. 21:7–12.

Boehringer-Mannheim GmbH, 1985, Reflotron zur quontitativen Besimmung von Parametern der klinischen Chemia.

Korovnikov, K. A., Letshyk, Y. D., 1989, Nutrition and the sports working capacity, Theory and practice of physical culture, 11:9–12.

Pokrovsky, A. A., 1969, Biochemical methods in clinic, Moscow, Medicine, 367.

Rogozkin, V. A., 1976, The role of low molecular weight compounds in the regulation of skeletal muscle genom activity during exercise, Med. sci. sports. 8:223–225.

Rogozkin, V. A., Pshendin , A. I., Shishina, N. N., 1985, Nutrition of athletes, Moscow, Physculture and Sport, 198.

Rogozkin., V. A., Pshendin, A. I., 1989, The utilization of high biological value products for nutrition of athletes, Theory and practice of physical culture, 11:13–15.

Scurihin, I. M., 1984, Chemical composition of food productes, Moscow, Agropromisdat. 1:1–223.

Volkov, N. I., Jaruzni, N .I., 1984, The limit factors of athletes performance, Moscow, Physcultura and Sport, 52–56.

Williams, C., 1991, Foods, Nutrition and Sports performance, London.

27

THE METABOLIC BASIS OF ROWING

Variability of Energy Supply and Possibilities of Interpretation

U. Hartmann and A. Mader

Deutsche Sporthochschule Koln
50927, Koln, Germany

INTRODUCTION

The majority of the test and interpretation methods used in the framework of performance-diagnostic investigations in rowing are purely descriptive procedures based on lactate performance diagnosis (5, 12, 14). These investigations are mostly restricted to the determination of:

- the anaerobic threshold,
- the performance and lactate values measured during the last stage of the multi-stage test, or
- the maximal performance during the 6 min maximal test (6MMT).

In this context, the anaerobic threshold is frequently equated with the endurance performance ability (12, 14, 15), and an attempt is made at measuring the limits of maximal performance via the criterion of "maximally tolerable lactate acidosis" during maximal exercise (8, 16).

The results or the energy provision are not interpreted in such a way that the physiological conditions and activities of metabolism are explained. The question as to whether and how the optimal use of the available metabolic capacities could be achieved can not be answered by routine examinations. However, this problem could be solved by simulating the results obtained during the experiment by using a corresponding simulation model (9, 11). Such a simulation model provides a description and representation of the energy-providing processes in the working muscles. This description and representation are sufficiently correct.

THE METABOLIC BASE OF ROWING

The physiological rowing-specific performance ability is determined to a great extent by metabolic performance ability (2, 11). According to our own studies, the work out-

Current Research in Sports Sciences, edited by Rogozkin and Maughan.
Plenum Press, New York, 1996

put during the 6MMT on a GJESSING rowing ergometer is between 410 to 430 watt in men and about 300 watt in women; the maximal value in men is 470 watt, whereas the corresponding value in women is 330 watt. An oxygen uptake (VO2) of up to 6,000 ml/min in athletes of average performance ability and of up to 6,500 ml/min in elite athletes is to be expected. As far as women are concerned, the corresponding values are about 4000 ml/min and 4,300 ml/min, respectively.

The responses of the parameters which are routinely measured during rowing ergometer tests have been shown by various authors (2, 3, 12). Here, in general, the behaviour of corresponding parameters of rowers in qualitatively different groups is identical (3). The literature as a whole does not include much information about corresponding guideline values (2, 3, 5).

The most simple presentation of a model of metabolic performance in rowing is the presentation of the three energy-providing reactions in the form of an area pattern illustrating the proportional energy contributions (3). A simple distribution of the energy requirement as proportions of the three existing energy resources (anaerobic alactic, anaerobic lactic and aerobic) can be calculated using the area pattern of proportional energy contributions (3). 80% of the energy recruited during the competition (of about 5:30 to 7:00 min duration) are produced by the aerobic energy system, 11% are provided by the anaerobic lactic system, and 9% are of anaerobic alactic nature (3, 4) . In the same literature, the data available from the references of other authors of recent years is quantified.

From the metabolic performance available no direct inferences to competition velocity can be made, not even if a corresponding motivation during competition is taken into account. Elite athletes can use an excellent rowing technique to compensate for a low metabolic performance only to some extent. Therefore, although a high metabolic performance is necessary, this is not a sufficient prerequisite for competition success.

However, using the area calculation (3), statements about the development or the type of recruitment of the relevant proportions of energy provision are not possible. The reason for this is that, beyond the findings obtained, the really existent maximal aerobic, performance (VO2 max) or the maximal lactacid performance (glycolytic performance = maximal velocity of lactate development (VLAmax)) cannot be determined.

In the following, certain aspects of this problem will be discussed in more detail.

MODELLING AND INTERPRETATION OF CONDITIONS IN ROWING THROUGH POST-EXERCISE SIMULATION

Theoretical Base of Modelling

The dynamics of the energy metabolism of the working muscles must be regarded as a consequence of the energy demanded for contraction (9, 11). The metabolic dynamics can be calculated using a system of differential equations (9). Each of the calculations is related to a kilogram (kg) of muscle mass.

As far as the maximal oxidative metabolic performance is concerned, the following assumptions can be made:

The formation of 1 mmol of ATP requires approx. 3.95 ml of O2 (1). The mitochondrial volume included in a normal red muscle fiber is about 3%, which corresponds to an O2 uptake of the muscle of approx. 120 ml/min*kg (120*9/3.0 (see below) = 3600 ml/min of measured VO2). In the case of a mitochondrial volume of approx. 5% in an endurance-trained rower (6, 10) this corresponds to an O2 uptake of approx. 200 ml/min*kg (200*90/3.0 (see below) = 6000 ml/min of measured VO2). The share of the working muscles in the body mass (active mass) is approx. 33%, while the total fluid volume (active and passive) is approx. 68–70% of the body mass, or about 47% as lactate distribution space (9, 11). Taking into consideration these assumptions, there is a conversion factor between active, muscle mass and body mass of approx. 3.0–3.1 for a rower weighing 90 kg and having a VO2 max of approx. 6000 ml/min (6, 10).

The estimation of VLAmax is considerably more difficult. If the increase of post-exercise blood lactate concentration (PBLC) by 21 mmol/1 after a 6MMT is taken as the basis of the measurement, the set lactate formation is 0.06 mmol/1*s (21 mmol/1 net PBLC/(360s exercise duration - 3s lactate-free interval)). However, it would be wrong to assume that this very low rate of lactate formation calculated in the 6MMT corresponds to a rower's VLAmax. The VLAmax of a sprinter, for example, is 1.0–1.4 mmol/1*s (11).

A rowing performance consisting of 10 to 15 maximal rowing strokes is comparable to a sprint performance, which can be performed in approx. 9 to 12s. Here, the maximally determinable PBLC is between 4 to 12 mmol/1, and there is a 3s interval without measurable increase of lactate. From this a VLAmax of 0.4 up to more than 1.4 mmol/1*s can be calculated. As even the lowest value of 9.1 mmol/l*s is about 6 times as high as the VLA during the 6MMT, the maximal VLA of rowers occurring during a competition cannot be utilised in a direct way.

In the case of 5.0 mmol/kg ATP, the store of energy-rich phosphates during rest prior to the start of the respective exercise is 21.0 mmol/kg CRPH and 7.0 mmol CR/kg muscle with a pH value of 7.0. All data correspond to the average values found in muscle biopsy examinations (7, 13). Explanations and purely theoretical foundations are to be found in other publications (9, 11).

Results of Simulation of the 6 Min Maximal Test

The post-exercise simulation of the 6MMT (Figure 1) shows significant parallels to the parameters measured in the practical experiments (3), too. For the absolute O2-uptake of 620 ml/min during a 490 watt maximal performance (420 watt plus 70 watt body shift performance) of an elite athlete there is a gross O2-uptake of 210 ml/kg for actively used muscle mass (MM) . This corresponds to a net VO2 (VO2 max. minus VO2-rest (400 ml/min)) of 197 ml/kg per kg active MM. According to the parameters mentioned above, there is a mean performance of 16.5 watt/kg, which is also related to active MM (Figure 1).

Because of the high initial performance (here 1 to 5 s approx. 1000 watt, 6th-10th s 800 watt) the CRPH content of 21 mmol/kg decreases by 3 mmol/kg or 15% within about 25s. During this period the decrease of ATP is about 0.7 mmol/kg (17%), VLA reaches about 40%, after about 70s it decreases by about 15% to 12% and then hardly changes until the end of the test. Because of the initially short-term activation, muscle lactate concentration rises to about 7 mmol/kg, which corresponds to about 30% of the final concentration. The increase of blood lactate follows the rise of muscle lactate after a short delay. After an initial short-term delay (dead time of about 4s), there is a steep rise which comes to an end after about 90s. The slight decrease of VO2 during the remaining time of the test results from the slight decrease of CRPH, which, especially during the final time of the load, leads to a further decrease of lactate concentration.

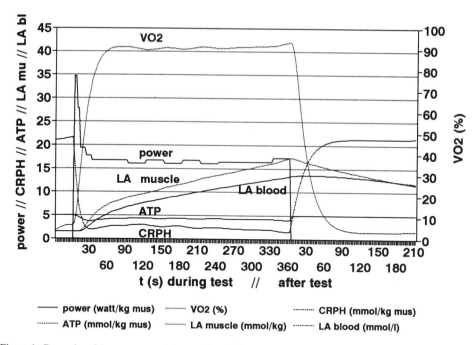

Figure 1. Dynamics of the energy-supplying reactions during maximal exercise; for further information see text.

As can be seen in Figure 1 the maximally possible performance was not fully exploited under the given circumstances as neither the VO2 (only by 90%) nor the VLA (16 mmol/1 PBLC) were completely utilised.

Figures 2 and 3 show the result of the simulation as far as the 6MMT is concerned. The same VO2 as above was taken as a basis. However, a low (0.35 mmol/1*s) and high VLA (1.40 mmol/1*s) were assumed.

It can he seen that the exploitation of the theoretically available VO2 max could still be slightly increased by a decrease of the VLAmax (Figure 2); however, here the CRPH concentration, which is almost exhausted after only 40s because of the high initial performance, is a limiting factor. A further significant decrease would make it impossible to continue an exercise of the same intensity (13).

The assumption and utilisation of a higher VLA (18 mmol/1 instead of 16.0 mmol/1 PBLC) alone would. lead to a decrease of the utilisation of the theoretical VO2max to approx. below 90% (Figure 3).

Results of Simulation Under the Assumption of Different Mechanisms of Energy Supply

By way of summary, Table 1 shows the results of the simulation on the basis of the assumption of different metabolic conditions. So there is the possibility of quantifying the influence of the existing or variable VLA on the result of the 6MMT.

As Figures 1 to 3 or especially the 2nd block of Table 1 show, the maximally possible performance ability was not fully exploited under the given circumstances (see above).

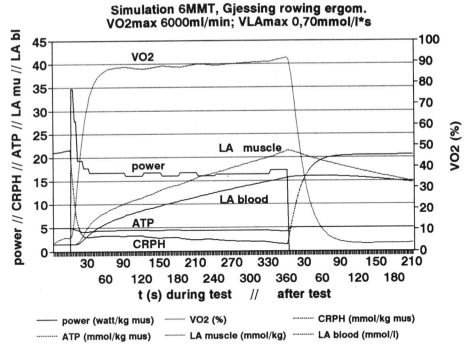

Figure 2. Dynamics of the energy-supplying reactions during maximal exercise; for further information see text.

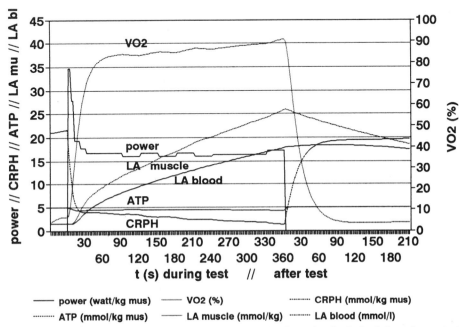

Figure 3. Dynamics of the energy-supplying reactions during maximal exercise; for further information see text.

Table 1. Summary of the results of. the simulation on the basis, of a high (6000 ml/min), very high (6300 ml/min) and a low (5400 ml/min) VO2 or a high (1.40 mmol/1*s), medium (0.70 mmol/1*s) and low (0.35 mmol/1*s) VLA (with regard to rowers). At the same time the calculated proportions of energy provision (%) as well as the cellular pH values and PBLC are presented. The assumed performance is 430 watt Gjessing

Metabolic conditions		Share of energy production			Remarks		
VO2max (ml/min)	VLAmax (mmol/l*s)	aerob. (%)	aner.lact. (%)	aner.alact. (%)	VO2(%) / ph / LA(mmol/l)		
6000	1.40	80.0	15.6	4.4	<90	6.4	18.0
6000	0.70	82.6	13.1	4.3	=90	6.6	16.0
6000	0.35	85.0	10.8	4.2	>90	6.7	13.0
6300	1.40	82.5	13.4	4.1	<85	6.6	16.0
6300	0.70	85.1	10.9	4.0	=85	6.7	13.5
6300	0.35	87.5	8.6	3.9	>85	6.8	11.0
5400	1.40	76.2	18.9	4.9	>95	<6.40	23.0
					ATP depleted; breakd.		
5400	0.70	78.3	16.7	5.0	>95	<6.45	21.0
					ATP nearly dep.; limit		
5400	0.35	80.1	14.8	5.1	>95	=6.55	18.0
					glyc.cap.too low; breakd.		

If a high VO2 max of 6300 ml/min and the different assumptions of VLAmax, which can be seen in Table 1, are taken as a basis, the partial or complete utilisation of individual proportions of energy metabolism is not necessary.

If an 02 uptake below 5400 ml/min is assumed as a basic condition, the given performance of 430 watt Giessing - can just be realised if the VO2 max is utilised to a high degree and using a medium maximal VLA of 0.7 mmol/1*s. However, a high VLA of 1.4 mmol/1*s and the maximal PBLC or pH values connected to this would not allow the complete utilisation of the VO2. If a low VLA (0.35 mmol/1*s) were assumed, the PBLC and the pH value would be within a tolerable range; however, in this case the capacitative proportions of VO2 as well as of VLA would be too low for producing the performance required in the simulation.

CONCLUSION AND PROSPECTS

The possibility presented here by way of example shows to what extent further findings and interpretations are made possible by the post-simulation of given metabolic circumstances. The information, which is only obtainable this way, goes considerably beyond the current performance diagnosis in the laboratory, which is primarily of a descriptive character. Furthermore, there is the possibility of simulating the metabolic changes and their effects, which otherwise would only he possible via training experiments, without incurring the corresponding long-lasting, mostly negative, effects on the athlete. The post-simulation of given metabolic circumstances should also be preferred for reasons of the objectivation of performance-diagnostic findings and interpretations.

REFERENCES

1. Di Prampero, P.E., 1981,Energetics of muscular exercise. Rev. Physiol.. Biochem. Pharmacol. 89: 144–222.
2. Hagerman, F.C., 1984, Applied physiology of rowing. Sports Med, 1: 303–326.
3. Hartmann, U., 1987, Querschnittuntersuchungen an Leistungsruderern im Flachland und Längsschnittunter-suchungen an Eliteruderern in der Höhe mittels eines zweistufigen Tests auf einem GIESSING-Ruderer-gometer. Konstanz; Hartung-Gorre.
4. Hartmann, U. and A. Mader., 1993, Modelling of metabolic conditions in rowing through post-exercise simu-lation. Coach 4: 1–15.
5. Hartmann, U. and A. Mader, 1995, Zur Einordnung ruderergometrischer Befunde verschiedener Alters- und Zielgruppen in das Spektrum der Leistungsfähigkeit, In, Deutscher Ruderverband (ed). Rudern - lehren, lernen, trainieren. Wiesbaden; Limpert.
6. Hoppeler, H., O. Mathien-Costello and S. R. Kayat., 1991, Mitochoridria and microvascular design. In: Crys-tal, R.G. J.B. West, P.J. Barnes, N.S. Cherniak and H.R. Weibel (eds). The Lung; Raven Press; New York; 1991, 1467–1477.
7. Karlsson. J,. Lactate and phosphagen concentration in working muscle of man. Acta Physiol. Scand. Suppl. 358: 7–72.
8. Kindermann, W., 1984, Grundlagen der aeroben und anaeroben Leistungsdiagnostik. Schweiz, Zt. Sportmed. 32: 69–74.
9. Mader A., 1994, Aussagekraft der Laktatleistungskurve in Kombination mit anaeroben Tests zur Bestimmung der Stoffwechselkapazitat. In: Clasing, D., H, Weicker and D. Böning (eds). Stellenwert der Laktatbestim-mung in der Leistungsdiagnostik. Stuttgart, Jena, New York; Fischer; 133–152.
10. Mader, A., U. Hartmann and W. Hollmann., 1988, Der Einfluß der Ausdauer auf die 6minutige maximale an-aerobe upd aerobe Arbeitskapazität eines Eliteruderers. In; Steinacker, J.M. (ed) . Rudern. Berlin, Heidel-berg, New York, London, Paris, Tokyo; Springer; 62–78
11. Mader, A. and H. Heck., 1991, Möglichkeiten und Aufgaben in der Forschung und Praxis der Humanleis-tungsphysiologie. Spectrum der Sportwissenschaften 3: 5 - 54.
12. Steinacker, J.M., W. Lormes and M. Stauch, 1994, Leistungsdiagnostik und Trainingssteuerung im Rudern. In: Clasing, D H. Weicker, D. Boning (eds). Stellenwert der Laktatbestimmung in der Leistungsdiagnostik. Stuttgart, Jena, New York;- Fischer; 105–110.
13. Spriet, L.L., K. Söderlund, M. Bergström and E. Hultmann, 1987, Anaerobic energy release in skeletal mus-cle during electrical stimulation in men, J. Appl. Physiol. 62: 611–615.
14. Urhausen, A., B. Coen, B. Weiler and W. Kindermann, 1994, Individuelle anaerobe Schwelle und Laktat steady state bei Ausdauerbelastungen. In: Clasing, D., H. Weicker, D. Böning (eds). Stellenwert der Laktat-bestimmung in der Leistungsdiagnostik. Stuttgart, Jena, New York; Fischer; 37–46.
15. Urhausen, A., B. Weiler and W. Kindermann, 1987, Laktat- und Katecholaminverhalten bei unterschiedlichen ruderergometrischen Testverfahren. Dt. Zt. Sportmed. 38: 11–19.
16. Zinner, J., B. Pansold and R. Buckwitz, 1993, Computergestützte Auswertung von Stufentests in der Leis-tungsdiagnostik. Leistungssport 23: 21–26.

28

RELATIONSHIPS BETWEEN FATIGUE AND REHABILITATION

V. Volkov

State Institute of Physical Culture
Gagarin St.,23, 214018 Smolensk, Russia

INTRODUCTION

Fatigue and rehabilitation processes should be considered together with the personal adaptation acquired in the course of an individuals development and training (phenotypical adaptation). It is known that the correlation between functional and genetic apparatus in cells is the most important link of the adaptational mechanism. As a result, a physical load forms the mechanism for increasing the capacity of the adaptational functional systems (Edington and Edgerton, 1976). The compensatory processes are dependent to a major degree upon the processes of fatigue which stimulate the physiological processes of rehabilitation (Hollmann and Hettinger, 1980) fatigue and recovery not only have a functional, but also a structural aspect. Molecular biology has proved the existence of intramolecular regeneration as the basis of muscle cell regeneration. Each organism, depending on its genetic program, has its own pattern of cellular regeneration. Rational coaching intensifies the intracellular regenerative rhythm and helps to use the functional reserves of an organism to a full degree. The causes of fatigue and the mechanisms of the compensatory processes vary according to the different muscles, the energy direction of the loads and the muscular working regime.

Nowadays the following possible mechanisms of fatigue are proposed:

- Central nervous and hormone-humoral mechanisms;
- Oxygen-transmission;
- Local mechanisms of the nervous-muscular system;
- Depletion of the energy sources such as glucose.

Recently more attention has been paid to the role of intracellular factors. Various mechanisms of fatigue lead to different types of compensatory processes after physical exercise (Margaria, 1976). Because of the interdependence of fatigue and recovery, it is necessary to take into account the phase of weariness at which exercise is stopped.

The processes of recovery are characterized by a number of specific features dependent on fatigue (Dubrovskiy, 1991; Volkov et al.,1994).

Current Research in Sports Sciences, edited by Rogozkin and Maughan.
Plenum Press, New York, 1996

AEROBIC CHARACTER OF ENERGY RESOURCES

Under heavy loads, energy is released at different phases of the aerobic and anaerobic process. During the period of recovery after exercise, energy metabolism is mainly aerobic. After hard training and competition, resulting in fatigue, the increased O_2 consumption lasts for 24–36 hours or more during the rest period. With exhaustion there can be an oppression of the compensatory mechanisms as a result of the reduced activity of enzymes.

INTENSIFICATION OF HORMONE-HUMORAL MECHANISMS OF REGULATIVE FUNCTION

After exercise, the role of humoral mechanisms in regulative function is intensified. Muscular metabolites, physico-chemical blood changes and hormones determine the processes of restoration. Exercises which result in an accumulation of metabolites and hormones turn out to be the most effective.

UNEVENNESS OF THE PERIOD OF REHABILITATION

Recovery is conventionally divided into immediate and delayed. For example, after a 30km ski-race the O_2 debt is restored quickly in the 1st 2–5h after exercise but after 10–12 hours it is slowed down (Fig. 1).

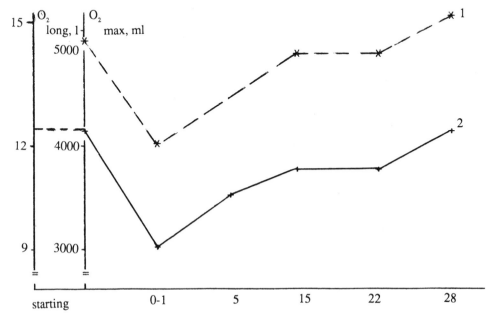

Figure 1. Change of O_2 max and O_2 long in ski racers after a 30 km race. The harder the physical exercise the more time is required for the postponed rehabilitation.

Table 1. Muscular capacity changes of weight-lifters after hard coaching loads (K9, X ± M)

Muscular groups coaching	Before	After coaching (hours)				
		0–1	12	20	28	36
Extensors of shank	72 ± 3	53 ± 3	57 ± 2	62 ± 3	65 ± 3	70 ± 3
Extensors of trunk	259 ± 9	216 ± 9	230 ± 8	235 ± 8	247 ± 8	254 ± 8
Extensors of forearm	40 ± 2	32 ± 1	35 ± 1	36 ± 1	40 ± 1	40 ± 2
Flexors of forearm	38 ± 1	39 ± 2	37 ± 2	39 ± 2	38 ± 2	39 ± 2
Flexors of foot	106 ± 6	86 ± 5	91 ± 6	96 ± 6	100 ± 6	106 ± 6
Flexors of hand	72 ± 6	73 ± 6	70 ± 7	70 ± 6	70 ± 7	70 ± 6

Select Processes of Recovery

In weight-lifting there are differences in muscles time and capacity. The longest rehabilitation is found in major groups of muscles such as the extensors of the trunk and the leg. The strength of the minor muscular groups recovery more quickly (Table 1).

The process of recovery is also related to different muscular working regimes. The biggest diversity of changes occurred during rehabilitation and after isometric training and lesser changes after high speed-strength training.

CONSTRUCTIVE CHARACTER OF REHABILITATION

During the period of recovery after the heavy loads there is a so-called supercompensatory phase. The characteristics of this phase are dependent upon fatigue.

- The supercompensatory phase is most marked after exhausting muscular activity ie after considerable lessening of the functional abilities of an organism.
- The supercompensatory phase does not always occur. After excessive heavy loads leading to a high energy consumption it occurs much later or is absent completely.
- The process of supercompensation is selective. It occurs in the muscular structures which undergo the most intensive loads.
- The processes of supercompensation can be influenced. For example, consumption of glucose provides good conditions for glucose supercompensation.

In order to increase the effect of supercompensation it is necessary to have increments in the physical load. The employment of equal loads leads to adaption to them and lessens the supercompensation. A factor closely related to fatigue is the length of the period of rehabilitation after training and competition. In the early 1970's it was believed that after heavy exercise, the working capacity was restored by the 5th-7th day. Research carried out in the 1980's showed the supercompensatory phase occurring at 2–3 days. Reducing the period of rehabilitation is related to superior coaching methods and better adaptation to loads. While evaluating the duration of the period of recovery, respiratory and blood circulation functions and the processes of metabolism, should be taken into consideration. It is also connected with the activated oxidizing processes necessary for the biological synthesis. The greater the extent of fatigue, the more active these processes are. Fatigue determines the efficiency of rehabilitation. It has been found that after weight-lifting training some types of rehabilitation eg electrostimulation, thermotreatment are most effective during the 2–5 hours after the exercise. On the other hand, rehabilitation types such as massage proved to be most effective immediately after the training. These observations help to add value to different types of methods after exercise.

CONCLUSION

The correlation between compensatory mechanisms are dependant on the type of recovery. The compensatory processes (the supercompensatory phase) is correlated with fatigue ie if the degree of fatigue is great there can be no such phase. Due to improved coaching methods, the period of rehabilitation is shortened after exhausting training and competitions. The effectiveness of different types of rehabilitation depends upon the processes of fatigue. Some types of rehabilitation prove the most efficient not immediately after exhaustion but 2–5 hours after training and competitions.

REFERENCES

Dubrovskiy, V.I., 1991, Reabilitation in sports, M, FIS.
Edington, D., Edgerton, V., 1976, The biology of physical capacity, Hougton Miffein Comp., Boston.
Hollmann, W., Hettinger, T., 1980, Sportmedizin Arbeite - und Trainingsgrundeggen, Stuttgard - New York.
Margaria, R., 1976, Biomechanics and energetics of muscular exercise, Oxford.
Volkov, V.M., Zhillo, Zh., Kostjutchenkov, V.N., Ermakov, V.V., Bakhrakh, I.I., Ganjushkin, A.D., 1994, Agents of the recovery in sports, Smolensk, Smyadynj.

29

ENERGETIC, BIOMECHANICAL, AND ELECTROMYOGRAPHIC CHARACTERISTICS OF ELITE KAYAKERS AND CANOEISTS

V. Timofeev, K. Gorodetsky, A. Sokolov, and S. Shklyaruk

St. Petersburg Research Institute of Physical Culture
Dynamo Ave. 2, 197042 St. Petersburg, Russia

INTRODUCTION

Canoeing and kayaking are general disciplines on the program of the World Championships and Olympic Games. Scientific research of each of these kinds of paddling and the comparison between them have great significance for the progress of sport results and the development of technique and training methods. It is especially interesting to use a multi-disciplinary approach, which allows the construction of the most complete model of each of these separate sports activities. This communication describes an attempt to compare kayaking and canoeing, having analysed at once some aspects of their activity with the help of various research methods.

METHODS

31 males (17 kayakers and 14 canoeists) with a mean age of 21 years (range 18–25) were studied in two series of experiments. All these sportsmen successfully competed in World Championships and Olympic Games during the years 1984–1994. The testing procedure included a time trial over 1000 m distance as fast as possible in kayak-single or canoe-single without opponents.

Measurements and Accounts of Energetic Parameters

The Douglas bag method was used in all oxygen uptake determinations. Gas samples were analyzed with the PGA-O2 M (S.Y.). Ventilation volume was determined by means of a spirometer (Germany).

Peak oxygen uptake determinations were made with back extrapolation methods (VO2be) as previously described (Di Prampero et al, 1976). The testing protocol included

Current Research in Sports Sciences, edited by Rogozkin and Maughan.
Plenum Press, New York, 1996

collection of three samples of expired air during 0.5–1; 1.5–2;2.5–3 min of the recovery period. The correlation between real and estimated values was 0.969 (n=34, p < 0.01). Oxygen intake was calculated according the formula

$$VO2 = t1000/60*(VO2av - 0.3),$$

where VO2—oxygen intake, L; VO2av – average VO2 consumption during a distance race [L/min] – 1; t—time to cover a distance of 1000 m, s.

$$VO2av = 1.668 + 0.578xVO2be,$$

where 1.668; 0.578—coefficients of regression. The correlation between real values and calculated ones was 0.848 (n=34; p < 0.001).

The alactic oxygen debt was determined under the following program:

- the average lactate consumption value of oxygen(VO2La) was taken from current values of oxygen intake in the recovery period (at VO2be < 4.7 L/min it was accepted that VO2La = = 0.8 L/min, and at VO2be > 4.7 L/min VO2La = 1.294 L/min)
- measured values of oxygen consumption in recovery period were identified on a semi-logarithmic system of coordinates and the line of best fit through all three points was drawn. The alactic oxygen debt (AlO Debt, L) was calculated according to the formula :

$$AlO\ Debt= VO2be''/ k'',$$

where VO2be''—consumption of oxygen at the end of a load (at t = 0) after subtraction of VO2La from current values of VO2 in the recovery period, L/min; k''—a constant of speed of payment of oxygen debt, 1/min

$$k'' = \ln 2 / t0.5,$$

where t0.5—time of half-payment of oxygen debt after correction, minutes. The correlation between real values of the fast fraction of the debt, calculated according to Margaria (1976), and that calculated following the technique described above was 0.963 (n=26, p < 0.01).
- Blood samples were taken during the 3rd and the 7th minutes of recovery period to evaluate the maximum blood lactate concentration. Analysis was made according Gutmann and Wahlefeld (1974).
- Total energy cost supply (E, kJ) was estimated from oxygen intake, alactic oxygen debt and lactate according to the method described by Margaria (1976).

Measurements and Accounts of Biomechanical Parameters

The force on the blade of the paddle, the time of underwater and air transfer periods (mechanogram) and boat speed in a cycle of paddling were recorded on a data recorder with a frequency of 100 Cps. The device was inside the boat during the test. For force measurements on the blade of paddle, individual strain gauges were used.

The time of the underwater and air transfer periods was registered with the help of a contact switch fixed on the paddle. The speed of the boat was registered with the help of an electromagnetic hydrometric turbine, fixed on the bow of the boat.

All recordings were made according to the techniques described by Dolnic and Gangenko (1978).

Measurement and Analysis of EMG

EMG was registered with the help of bipolar electrodes, connected to an electromyographic amplifier EMU-10 (USSR) and was recorded on a digital independent data recorder (SOTR-32, KMT, Germany) at a frequency of 1000 Hz. The data recorder was in a motor boat, which followed the sportsmen on a parallel course. In canoeists the electrical activity was recorded in a projection on the skin of arms, trunk and leg; in kayakers recordings were made on a projection on the skin of symmetrical muscles of the trunk and arms. The horizontal acceleration on the boat was registered by means of a tensiometer (Bell & Howell, USA), fixed on the stern of the boat.

Statistics

Average values and mean quadratic deviations of energetic and biomechanic parameters were calculated. Statistical significance was assessed using the Mann-Whitney test.

RESULTS

The results of measurement of biomechanical parameters are shown in table 1.

It is clear that single-kayakers, paddling on both sides, achieved a paddling rate of 88.5 (81–84)/min, whereas canoeists, making strokes only on one side, achieved rates of 50.7 (46–55)/min. The difference is statistically significant ($p < 0.001$). However, the amplitude of the paddling movement in the water of the canoeists was 2.27 (1.90–2.45) m, whereas in kayakers it was 1.70 (1.58–1.77) m. The differences were again statistically significant ($p < 0.01$). Peak force on the blade was accordingly 296.2 (283–327) N and 246.4 (209–280) N respectively; the average force on the blade was not different between kayakers and canoeists ($p>0.05$). The mechanical force on the blade in kayakers was 373.0 (279.1–422.4) W, and in canoeists it was 289.2 (275.4--346.9) W: this difference is statistically significant ($p < 0.001$). Kayakers completed the distance of 1000 m in a time of 251.87 (237.81- -266.0) s, and canoeists took 277.81 (266.00–286.48) s.

These data are contained in Table 2.

Average weight of the kayakers was 88.31 (81–94) kg, and of canoeists was 82.36 (70–93) kg. The distinctions are significant ($p < 0.01$). VO_2be of kayakers was 5.354 (4.5–5.9) L/min and 61.1 (54.1--69.1) ml/kg/min. Absolute and relative values of VO_2be in kayakers appeared to be accordingly 5.045 (4.7–5.4) L/min and 61.4 (57.1–67.1) ml/kg/min. The distinction of absolute values was statistically significant ($p < 0.05$).

- Oxygen intake, calculated taking into consideration the time taken to cover the measured distance and the percentage contribution of aerobic processes in general energy supply was higher in canoeists ($p<0.05$) than in kayakers (19.2 (17.9–20.7) L against 18.5 (15.8–20.0) L and 77 (75–84)% of the total energy cost against 74

Table 1. Biomechanical characteristics of elite kayakers and canoeists

Group N	t1000m s	Rate l/min	Fp N	Fav N	L m	W W	ME %	Est J/l
Kayakers (n=13)	251.87 ± 8.15	88.54 ± 4.67	246.4 ± 22.7	152.9 ± 13.6	1.701 ± 0.289	373.0 ± 38.1	18.4 ± 1.9	1422.4 ± 145.0
Canoeists (n=11)	277.81 ± 6.02	50.73 ± 3.26	296.3 ± 15.1	157.2 ± 9.1	2.268 ± 0.188	299.2 ± 20.6	16.0 ± 1.2	2222.1 ± 122.3
Significance level	<0.001	<0.0001	<0.001	>0.05	<0.00001	<0.001	<0.001	<0.00001

t1000m—time of paddling distance; Fp—peak stroke force on the blade; Fav—average stroke force on the blade; L—length of stroke; W—mechanical power on the blade; ME—net mechanical efficiency; Est—energetic cost of one stroke.

Table 2. Energetic characteristics of elite kayakers and canoeists

Group N	Weight kg	VO2be L/min	VO2 ml/kg/min	VO2 L	%	AlODebt L	La mMol/L	E kJ	E W	E W/kg
Kayakers (n=13)	88.31 ± 4.66	5.354 ± 0.371	61.4 ± 4.7	18.498 ± 1.293	74.1 ± 1.1	3.934 ± 0.457	14.2 ± 3.9	522.122 ± 28.307	2072.3 ± 165.0	23.7 ± 1.4
Canoeists (n=11)	82.36 ± 5.76	5.045 ± 0.258	61.4 ± 3.2	19.183 ± 1.101	77.2 ± 2.2	3.862 ± 0.773	11.7 ± 2.2	520.327 ± 38.165	1877.1 ± 138.0	22.8 ± 1.4
Significance level	<0.05	<0.05	<0.05	<0.05	<0.05	>0.05	<0.05	>0.05	<0.05	>0.05

VO2be—peak oxygen uptake determined with back extrapolation methods; VO2—oxygen intake; AIO Debt—alactate oxygen debt; La—lactate ; E—energy cost

(68–84)% accordingly. Blood lactates in kayakers were 14.2 (5.9–19.3) mmol/L: in canoeists blood lactate reached 11.7 (6.5–14.0) mmol/L ($p < 0.05$). The average values of the contribution of anaerobic glycolysis in kayakers and canoeists were not significantly different ($p>0.05$). Alactic oxygen debt in kayakers was 3.9 (3.4–5.2) L, and in canoeists it was 3.9 (2.8–5.) L ($p>0.05$). Total energy cost in kayakers was 522 (487–559) kJ or 2072 (1839–2352) W or 23.7 (20–28) W/kg. In canoeing these parameters were 520 (447–577) kJ, 1878 (1666–2089) W, and 22.8 (21–26) W/kg respectively. Distinctions of average values were statistically significant only in comparison of absolute values of total power capacity (W) ($p < 0.05$).

• The net efficiency coefficient in kayaking was 18.4 (15–22)%, and in canoeing it was 16.1 (14–17)%, and the energy cost of one stroke in kayakers was 1422 (1200–1680) J, and in canoeists it was 2222 (2024–2346) J ($p < 0.01$, $p < 0.00001$ respectively).

Figure 1 illustrates the bioelectrical activity of the muscle during one stroke in canoe (fragment of real record).

It is clear that the acceleration of boat begins before the moment of contact between blade and water. This fact was marked in 100% of analysed strokes. The peak of acceleration occurs in the first third of the underwater period, at 0.06–0.15 s after contact. The increase of boat acceleration is accompanied by an increase of EMG activity in triceps and quadriceps muscles (100% cases). It is clear that, in the majority of cases, the maximum electrical activity of muscles occurs in the first half of the underwater period (68% of total cases).

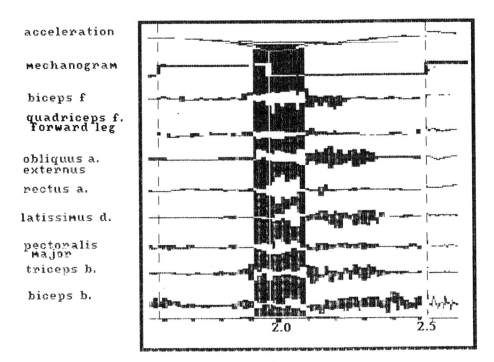

Figure 1. EMG of canoeist at the beginning of 1000 m distance.

In Triceps, the first maximum occurred at the moment of contact between water and blade, and the second peak occurred in the second third of the underwater period (100% of cases). Biceps activity was characterized by the occurrence of the second peak in the last third of the underwater period. In this phase, the sportsman executes steering manoeuvres and removes the paddle from the water. At the same time, the first peak of activity on the electromyogram occurs at the moment of contact between water and blade (18% of cases), or in the first third of the underwater period (54% of cases). It should be noticed that in an interval between two contractions both muscles show marked background activity. Comparison of the myographic activity of various muscular groups involved at various phases of the stroke cycle, shows that muscles of the shoulder on the "working" side and muscles of the legs participate in maintenance of propulsive force and in fulfilment of auxiliary functions. For example, development of a power impulse, preservation of posture, management of the boat and other activities all require muscle activation.

The maxima of activity of a pair of m.biceps, brachii-m and triceps brachii do not generally coincide. M. quadriceps femoris and m. biceps femoris of the forward leg generally contract simultaneously, fixing the position of the forward leg. However, the most unexpected result was observed on analysis of a pair (m. latissimus dorsi and m. pectoralis major) which contract synchronously in 100% of cases. This shows that contraction of this muscle occurs along with extension of the shoulder in the underwater period.

Figure 2 illustrates a typical picture of bioelectric activity of muscles in kayaking (fragment of real record).

It is clear that the greatest activity of the muscles of the left-hand half of body is displayed in the stroke from the same side. It is also possible to reach the same conclusion concerning the muscles of the right half of the body. The activity of the majority of the muscles is displayed in the first half of the underwater period (50–90% of cases). The background activity of latissimus dorsi, pectoralis major, triceps and biceps brachii remains high during the air transfer period. Most frequently it takes place directly before the

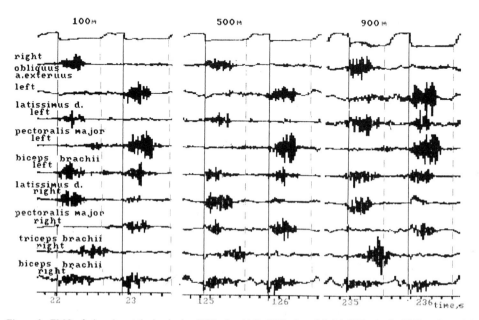

Figure 2. EMG of a kayaker at the beginning (100 m), middle (500 m), and finish (900 m) of a 1000 m time trial.

contact between water and paddle. Latissimus dorsi and m. pectoralis major muscles increase their activity during the underwater period at the opposite side in 60% of cases, but the level is lower than during the underwater period at the same side. In 98% of cases triceps is active in the beginning and at the end of the underwater period. M. biceps brachii displays activity in the first and in the second half of the underwater period. Triceps and biceps are active during the air transfer period.

CONCLUSION

Covering a distance of 1000m in a canoe makes greater demands on the aerobic energy supply than occurs in kayak racing. Kayak racing is characterized by greater activation of glycolysis than occurs in canoe racing.

An important feature distinguishing movements of canoeists, is the greater power production in each stroke, which provides an advantage over the kayaker in parameters of amplitude of the paddling movement and the peak force on the blade. A distinctive feature of the kayaker's movements is the high stroke rate, which gives an advantage over the canoeist in characteristics of mechanical capacity and efficiency, and also speed of movement.

Bioelectric activity of the main muscles in top-class kayakers and canoeists when covering a competitive distance is concentrated in the first half of the underwater period. In kayakers and in canoeists, bioelectric activity of the trunk muscles coincides with the active phase of movement, in accordance with their leading role in providing propulsive effect. M. biceps brachii and m. triceps brachii have a more complex pattern of bioelectric activity, that relates to their participation in management of paddle and boat.

REFERENCES

Di Prampero, P.E., Cortili, G., Magnani, P., Saibene, F., 1976, Energy cost of speed skating and efficiency of work against air resistance, J.Appl.Physiol. 40:584–591.
Dolnik, Y.A., Gangenko, Y.V., 1978, Investigation of parameters of the working activities of women in kayaking on a distance of 500 m.. Rowing and canoeing, Moscow, Physical Culture and Sport :71–77.
Gutmann, J., Wahlefeld, A.,1974, Determination of lactate with LDH and NAD, Methods of Enzymatic analysis, ed H.U.Bergmeyer. Acad.Press 3:1464–1469.
Margaria, R., 1976, Biomechanics and energetics of muscular exercise, Oxford, Clarendon press :146.

30

MYOCARDIAL DYSTROPHY IN ATHLETES

A New Approach to an Old Problem

V. Zemtsovsky

Pediatric Medical Academia
St. Petersburg, Russia

INTRODUCTION

It is recognized that moderate and properly selected physical exercises play an important role in strengthening of health and prophylaxis of diseases of cardiovascular system (CVS) (Blair et al, 1989). On the contrary, excessive physical loads, so typical of modern sports, can result in overstrain of CVS. Today different causes of high risk of sudden death due to physical loads (Jaeger, 1990; Maron, 1993), high frequency of various arrhythmias and idiopathic calcinosis of valve apparatus and under valvular structures in athletes are widely discussed in the literature. These facts are explained in absolutely different ways among which the least attention is given to the role of CVS overstrain and possibility of "pathologic athlete's heart". The concept of "pathologic athlete's heart" and myocardial dystrophy (MD) was proposed by G.Ph.Lang (1936). The term MD in the narrow sense of the word was considered by G.Ph.Lang as all myocardial impairments of non-inflammatory and non-coronary genesis provoked by various extracardiac causes. To indicate the myocardial impairments in athletes, the term MD due to physical overload was proposed later (Dembo, 1991). A.G. Dembo proposed that this diagnosis be applied in all cases of repolarisation disturbances (RD) on the resting ECG. After 30 years since the moment of the first publication of the article about MD in athletes, it has been shown that athletes with RD according to some mean value parameters of CVS function in the rest and under physical loads are different from those in athletes with a normal resting ECG (Dibner, 1986; Dembo, 1991) However, the validity of MD diagnosis only on the basis of the resting ECG analysis is justly called into question by many researchers (Sarkisov, Vtyurin, 1971; Danko et al, 1986). Yet it would be wrong to refuse from the views about MD that allow the combination of a number of the said facts proving a possibility of development of pathologic athlete's heart at hyperfunction caused by physical loads. The foregoing considerations served as the basis for finding new approaches to the problem of MD in athletes which are discussed in the present article.

Current Research in Sports Sciences, edited by Rogozkin and Maughan.
Plenum Press, New York, 1996

RESULTS

Etiology

The main etiologic factor underlying development of MD is apparently excessive physical load. The term excessive is considered by A.G.Dembo (1991) as loads in excess of organism capacities. Along with it one cannot but note that apart from physical over-strain in the development of MD, emotional overstrain also plays a role (Jaeger, 1990). Personal peculiarities of the athlete, first and foremost such features as the level of psychological strain and restlessness, play an important role in the development of emotional stress.

Pathogenesis

Pathogenic mechanisms involved in MD development are varied and as a whole similar to those of heart stress lesions we mentioned above. Since sports activity could be viewed as a series of physical and emotional stresses such a parallel seems to be quite justified. It is well known that both physical and emotional stresses lead to high levels of circulating catecholamines (CA). Apart from the toxic and hypoxic influence of high CA concentration the existence of the most important mechanisms of myocardial lesions triggered by CA surplus has been proved. It is an accumulation of Ca^{++} in cardiomyocytes and activation of lipid peroxidation. Thus the role played by the activation of the sympathoadrenal system in the development of one type of MD which M.Kushakowsky proposed to call "hyperadrenergic" is quite obvious. In the case of extremely intense and exhausting training another way of dystrophic changes is possible. Here we are dealing with a gradual decrease of the athlete's adaptation resources and with "wear" of structures responsible for the adaptation. In that case the level of CA in the myocardium and/or the sensitivity of adrenoreceptors to CA activation turns out to be sharply decreased. Thus, we have all grounds to speak about a "hypoadrenergic" type of MD in athletes. Disturbances of electrolyte balance and especially balance of $K+$ and Ca^{++} ions can play an important role in the pathogenesis of MD. The results of previously mentioned EchoCG studies have confirmed the role of calcium accumulation in myocardium under physical loads, and showed that the so-called idiopathic calcinosis of mitral and aortic rings are observed more often in athletes that in those who take no part in sports. It is clear that in clinical practice it is difficult to define the role of one or another electrolyte in any particular case of MD. Therefore the "electrolyte imbalance" type of MD should be distinguished. One of the most important pathogenic mechanisms of MD is myocardial hypertrophy (MH) which occurs in case of unbalanced increase of myocardial mass. It is known now that the development of MH as a response to hyperfunction is genetically determined and develops faster in hearts having a less powerful system of resynthesis of ATP and Ca^{++} pump. In this case the growth of arterioles and capillaries lags behind the increase of cardiomiocytes. An increase in the diffusion distance and relative hypoxia lead to the fact that even at rest almost all reserve capillaries turn out to be open and coronary reserve is reduced. This alone can be considered as a basis for developing dystrophia. This introduces the concept of a "compensatory-hypertrophic" type of MD. Formation of one or another type among the four pathogenic types of MD mentioned above includes, according to data from the literature, heritable factors and endogenous prostanoids, hypoxic hypoxia and intoxication from chronic infection foci (Dembo, 1991). Along with catecholamines an important part in MD pathogenesis is played by other hormones, and most significant of all are the corticosteroids and ACTH. Other mechanisms can be very important in MD development in athletes. The case in point is the state of cellular membranes and membranous lipids which are primarily responsible for adaptation (Blair et al, 1989). Determination of fluidity characteristics of cellular membranes depends in its turn on the relative content of cholesterol and on the ratio between saturated and non-saturated fatty acids in lipid molecules. An increase in fluidity of membranes provides electrogenic disturbances and accumulation of intercellular Ca^{++}, hypertrophy, dystrophy and disturbances of myocardial contractility.

All the said proves the fact that all pathogenic mechanisms of MD are sophisticated and different. There are good grounds to believe that as a rule formation of MD includes several pathogenic mechanisms, and the role of each can change depending on individual features of the organism, the nature of training and on the stage of the process.

Pathologic Anatomy

Multiple experimental research showed that during physical overstrain a destructive change of myocardial ultrastructure can take place: enlargement of nuclei, swelling of mitochondria, disturbances of their matrix, and destruction of inner membranes. Along with this the focal disintegration of myofibrils, a decrease of ribosome quantity, enlargement of SRP channels, and inner oedema have been found. In some rats exposed to physical loads, researchers have discovered signs of focal MD/clod-like disintegration of cardiomyocytes, loss of protein and fat components, fibroid and necrotic change of myocardium, perivascular haemorrhagic disease. In this case diffusive and morphological changes are of the mosaic type. While on the subject of pathologic anatomy of MD we can hardly pass by in silence a question about the connection of MD with the compensatory hypertrophy which develops as a response to hyperfunction. The experimental research has confirmed that dystrophic changes of myocardium in animals were especially strong when marked MH appeared. Thus the experimental research and data of autopsy of the athletes who died from sudden death confirm that myocardial dystrophy develops as a response to physical overloads and can be both of diffusive and microfocal in nature.

Course

MD can be acute and chronic. On acute physical overload, whether in the untrained person running after a departing train or in the finishing spurt of unprepared athlete, metabolic changes develop quite often and result in the appearance of apparent or hidden signs of a reduction in contractile capacity of the myocardium, changes in the ECG, or in arrythmia. Such changes can be transient or so deep that they can cause electric instability of the myocardium and sudden death. The chronic course of MD can go on for years and have various clinical manifestations.

Clinical Picture

Now we have ample experience of the fact that despite of scarcity of clinical demonstration one can recognize four main clinical versions of MD: asymptomatic, arrhythmic, with disturbance of contractile function, and a mixed one. The asymptomatic version of MD is the most typical for this type of pathology. Since MD is developed gradually as a rule and is connected with progressive "wear" of the adaptive mechanisms, it may not manifest itself clinically for a long time. In the majority of cases the athletes, even with marked RD on ECG or MH on EchoCG, do not complain, although cardiology and giddiness are noticed at the changing of body position. The case history often has indications of a reduction of sporting capacity and slow recovery after training or competition. Sometimes it is possible to detect violations of sporting discipline, participation in competitions while in diseased state, or unusual physical and emotional loads in terms of duration and intensity. Athletes with MD are distinguished in a history case of the catarrhal type more often than healthy athletes. MD does not have a major effect on physical working capacity. Results of clinical examination of athletes are quite poor too, and they do not allow identification of any distinctive signs. Nevertheless a marked sinus bradycardia (less than 45 per minutes) in combination with rhythm rigidity or, on the contrary, an acute marked sinus arrhythmia with marked irregularity of respiratory waves on the rhythmogram provides reason enough to carry out a more thorough examination. The presence of relative tachycardia at rest, arterial hypertension and/or the hyperkinetic type of blood circulation (HrTC) are of equal importance to suspect MD development. On auscultation, athletes with MD show systolic murmurs over apex cordis and basis cordis more often than healthy athletes. There can be two causes constituting the basis for such murmurs:

Change of Heart Geometry in the Process of Adaptation to Loads with Relative Discrepancy between Sizes of Cavities and Openings. It is believed that at MD such a discrepancy becomes important in terms of hemodynamics.

Developmental Anomalies of the Valvular Apparatus: Mitral Valve, Chords Dystopia and Additional Chordas. Such anomalies not only often cause systolic murmurs but also increase the risk of development of dystrophic changes of myocardium during athletic training. The main instrumental methods allowing diagnosis at the asymptomatic stage of MD are ECG, DopplerEchoCG(DEchCG) and functional tests. Taking into account the fact that to carry out the entire complex of functional and pharmacologic tests is rather hard, complicated and expensive, the typical report of research used for MD diagnostic should include tests with physical loads, propranolol and potassium chloride. That in its turn allows us to define the pathogenic type of MD. In the cases when RD do not respond to functional and pharmacological tests, it is necessary to carry out differential diagnosis between compensatory-hypertrophic types of MD and disturbances of electrogenesis not connected with MD. In such cases the DEchoCG is of a great importance. The non-balanced MH on EchoCG can be diagnosed not only on the basis of an increase of absolute parameters of myocardium mass of the left ventricle (MLV) (norm of MLV < 180 g) but also at the relative predominance of myocardium mass over the diastolic volume (DV) of the opening of a MLV (ratio MLV/DV < 0.8) Yet it should be emphasized that absence of absolute and relative EchoCG criteria of MH does not exclude from our viewpoint the presence of pathological hypertrophy which is often of the nest nature. One of the earlier manifestations of MH and developing MD can be disturbance of myocardium distension or an increase in its rigidity. Use of DEchoCG allows us to judge the degree of myocardium distension from the ratio between flows of fast and atrial filling. The arrhythmic version of MD development should be diagnosed when different disturbances of rhythm and conduction at rest and/or at the test with physical loads appear against the background of sports training. More often than not the case is point is the extrasystolic arrhythmias or paroxymal disorders of the cardiac rhythm. Such arrhythmias may not be accompanied by other complaints or signs of circulatory insufficiency for a long time. We think that apart from inheritance, which is considered to play the main role in development of MD, the physical overload has a great effect on myocardium change in left ventricle of the heart and on beginning of arrhythmias. All that allows us to consider MD as one of the manifestations of the arrhythmical version of the clinical course of MD. It is worth noting that G.Ph.Lang (1936) wrote that under physical overloads the dystrophic change appears first of all in the myocardium of the right ventricle. Apart from extrasystolic arrhythmias and paroxymal disorders the syndrome of suppressed sino-atrial node (SSSN) should be ascribed to manifestations of MD. If other signs of SSSN arrhythmia are absent it should be considered as a manifestation of controlled hypoadrenergia of athletes. But quite often SSSN arrhythmias are combined with RD on ECG at rest and accompanied by deterioration of adaptation of athletes to physical loads that gives grounds to suspect MD. At the same time SSSN arrhythmias should not be identified as a sinus sick syndrome (SSS) which is characterized by degenerative changes in the conduction system of the SA zone and of the atrium. Unlike SSS, SSSN arrhythmias are reversible and generally disappear after termination of intense physical training. MD can at times be accompanied by disturbances of the contractile capacity of the heart. The proof of this possibility comes from examination of athletes after exhaustive competitions (Seals et al.,1988). When estimating contractile capacity of the heart in athletes, and generally disturbances are not immediately apparent, the most commonly used criteria of cardiac insufficiency should be applied. Such disturbances can be detected only during CVS examination during or immediately after physical loads. It is absolutely clear that all listed clinical versions of MD course can occur in various combinations, from isolated RD on ECG at rest up to detection of sufficient changes in morphology and function of the heart.

MD Stages

Until recently, the above mentioned approach has predominated in the determination of MD stage and according to this approach the stage of the process had been determined relating to the degree of RD manifestation on ECG at rest. But nowadays disadvantages of this classification have become evident. The main one is a marked discrepancy between degree of display of ECG changes on the one hand, and the clinical manifestation of MD, level of physical efficiency, sign of disturbance in morphology and function of the heart on the other hand. We suggest the use of results of other clinical and instrumental research along with signs of disturbances of myocardium electrical activity at rest. In accordance with the suggested approach the first stage of MD is characterized by different clinical and/or ECG signs of MD detected at rest and disappearing after an exercise test. In the second stage the clinical signs of MD at rest are supplemented by changes during exercise. The first and second stages should not be associated with irreversible morphological changes of the myocardium and conductive system. The third stage is characterized by signs proving irreversible morphological changes of the myocardium (myodystrophic cardiosclerosis, fibrosis and/or calcinosis). To define the stage of MD requires use of the set of diagnostic tools and methods of research including not only collection of medical history, physical research, registration of DEchoCG and ECG at rest but also an exercise stress test controlled by blood pressure, ECG and monitoring of heart rhythm as well as stress DEchoCG. It should be emphasized that MD very often has a focal nature. In such an event the existence of small fibrosis or calcinosis foci does not necessarily have an impact on adaptation to physical loads, does not decrease the pump function of the heart and turns out to be an occasional finding during research. From this it is inferred that signs of decreased contractile capacity of the heart (3rd type of immediate adaptational reaction) can be absent in the third stage of MD, and on the contrary, be the only one and the earliest manifestation of MD detected during exercise, that is, at the second stage of MD. The suggested approach to stage estimation can be applied to all pathogenic types of MD. A question of definition of MD stage of compensatory hypertrophic type deserves a special discussion. As it has been shown by research dedicated to comparison of EchoCG and ECG capacities of ECG diagnostic of MH in athletes are quite restricted. The main reason for these restrictions is the fact that the MH does not reach such values as those observed in the clinical picture. In any case, when the value of MLV does not exceed 200 g the amplitude signs of hypertrophy have quite low sensitivity and specificity. At the same time it is LV with values of about 180–200 g which is a borderline between normal and pathological for athletes. ECG can provide a certain help in diagnosis of pathological hypertrophy. The inclusion of RD in amplitude signs of MH (decrease of amplitude or T inversion and ST depression) should be considered as an illustration of the development of non-balanced hypertrophy which in fact cannot be separated from MD. The important and earlier diagnostic sign of the compensatory-hypertrophic type of MD is the already mentioned increase of myocardium rigidity. Thus the diagnosis of compensatory-hypertrophic type of MD should be based on the combined analysis of ECG and DEchoCG, naturally taking account also of any available clinical data. The importance of timely discerning of this type of MD is hard to overestimate because it is this type which is associated with sudden death of young individuals during sport training. The point is that during MH special conditions are formed which are favourable for desynchronization of electrical process and electrical instability of the myocardium. It is not an accident that the cardiomyopathy is called one of the causes of sudden death of athletes (Zemtsovsky, 1994). We consider it as one of the manifestations of MD.

All the above allows us to define MD as a disease caused by a discrepancy between the volume and/or intensity of physical training and adaptational capacities of the CVS manifested in disturbances of electrical activity, non-balanced hypertrophy, electrical instability and decreases of the contractile capacity of the myocardium. Such understanding of the MD in athletes allows us to expect successful treatment of "pathologic sportive heart" and prophylaxis of sudden death connected with physical overloads.

REFERENCES

Blair, S. N., Kohl, H. W., Poffenbarger, R. S. et al.,1989, Physical Fitness and all-cause mortality a prespective study of healthy men and women, JAMA 262: 2395–2401

Danko, J. I., Cusnetsov, G. I., Loginov, V. G., 1986, Motivation as one of causes repolarisation disturbances in athletes, 3 Russian congress of sports medicine, Sverdlovsk :118.

Dembo, A. G., 1991, Dieseases of cardiovascular system, - In "Dieseases and injures in Athletes". - Edited by A.G.Dembo, Leningrad, Medicine :336.

Dibner, R. D., 1986, About differencial diagnosis of chronic hear overstrain in athletes, Cardiologia :108–111.

Jaeger, 1990, La most subite dans la pratique du sport, Comment en reduire l'incidence? Ann.cardiol.Angeol. 39:565–570

Lang, G. Ph., 1936, Questions of cardiology, M., Medicine : 189.

Maron, B. J., 1993, Hypertrophic cardiomyopathy in athletes, The Physical and Sportsmedicine 21:83–91.

Sarkisov, D. C., Vtyurin, B. V., 1971, Myocardial dystrophia, it reversibility and way of prophylaxis in book Myocardial dystrophya, Leningrad :69–78.

Seals, D. R. et al., 1988, Left ventricular difsunction after prolonged strenuous exercise in healthyy subjects, Am.J.Cardiol. 61:875–879.

Zemtsovsky, E. V., 1994, Myocardial dystrophya in athletes, Cardiology 34:656–74.

31

NEW METHODS FOR THE MEASUREMENT OF SPECIFIC WORKING CAPACITY OF TOP-LEVEL ATHLETES (DYNAMICS OF TRAINING DEVELOPMENT)

F. A. Iordanskaya

Russian Research Institute of Physical Culture
Kazakova St., 18, 103064, Moscow, Russia

INTRODUCTION

Diagnosis and evaluation of specific working capacity of athletes is one of the main tools for monitoring training development and forming future training plans.

Sporting performance is partly determined by physiological adaptations of the human body. Training is the basis of adaptation. Without true knowledge of the adaptation of organs and functional systems of athletes representing different sports the training cannot be effective. The level of functional systems adaptation is defined exactly by testing specific working capacity.

Regarding scientific aspects, the problem of studying specific working capacity includes three basic tasks that should be decided. They are - designing a special test, selection of methods and parameters giving most information and working out criteria of its evaluation.

The main requirements for load testing are frequency, reproduction and limits of exercise, and also compliance with the structure of motor activity.

The complex of methods for studying adaptation to a special test include
methods most adequately reflecting adaptation possibilities of a human body in relation to the above mentioned structure of motor activity.

The aim of the study was to work out tests and programs for diagnostics and evaluation of specific working capacity in football and volleyball players.

METHODS

One hundred football and 43 volleyball players of different age and skill level were studied. Specific working capacity was measured using special tests on both treadmill and cycle ergometer.

Current Research in Sports Sciences, edited by Rogozkin and Maughan.
Plenum Press, New York, 1996

Many methods for evaluating cardio-vascular and vegetative nervous systems, neuromuscular, energy supply and oxygen transportation systems were used.

RESULTS

The Study of Special Working Capacity of Football Players

The range of intensity and volume of movements of football players is wide and most clearly dependent on a player's position in a team. At the same time, the modern attacking style of play requires high level of universal training. To measure specific working level of players, a method of repeated load tests were used; three series of 30m repeated in 5 reps at maximum speed with 3–4 min rest between series. Subsequently, special tests in the form of shuttle running 50 m repeated in 7 reps were used.

Football has changed recently and has become faster with powerful attacking, spectacular playing style and increased requirements for physical and functional state of players. This was clearly displayed at the 1994 World Cup in the United States. Thus, it was necessary to revise the methods for assessing a players functional fitness.

Improvement of the shuttle run test was by shortening and increasing the number of repetitions: 25 m distance with 180 degrees turn, 14 reps with distance speed registration which increased running intensity and players special endurance. Tests were carried out on a playing field after a warm-up.

According to the increase of speed endurance elements in this test we expanded the program of diagnostics of main leading physiological systems.

To determine the most significant methods and parameters of players functional evaluation, the treadmill test was carried out on 64 players of different age (14–15, 16–17, 18–19 and older 20 years) and skill. Factor analysis of the data showed six main factors which cover 65.7% of young and 69.9% of adult qualified players.

Fig.1 presents a diagram of factor structure of functional state of young and trained football players where factor 1 = morphofunctional development, 2 = effectiveness of respiratory system, 3 = functional state of vegetative nervous system, 4 = state of neuromuscular system, 5 = physical capacity with its energy supply, and 6 = playing position and energy abilities with regard to position.

Top-level players have shown significance of physical capacity and its energy supply. Correlation analysis (fig.2) showed that the age of players has a close relationship with playing position and maximum running speed. Playing position correlates with ECG values and motor reaction time (MTR) which becomes longer in midfield players and forwards. Specific working capacity of adult footballers (fig.3) has a significant relationship with values of respiratory, cardio-vascular systems and motor reaction time. From analysis of the experiment the program of diagnostics has been worked out. Since gas exchange analyses are impossible during a field test, the program of examining blood energy supply values (lactate, acid-alkaline state /BE/, etc) was expanded. Running time analysis during the special test permitted us to evaluate both the speed ability and speed endurance of individual players with regard to their position.

Although the starting speed of forwards was better their speed in last sections of the distance decreased, and the speed of defenders increased. It was shown that adaptation changes in the metabolic cost of exercise in goalkeepers and forwards was equivalent to 261.2 ±12.81ms and 261 ±16.26ms respectively. CPK level regulating energy exchange and speed abilities increased from goalkeepers to defenders and midfield players reaching optimal values at rest and after the test in forwards (139.0 ±18.11).

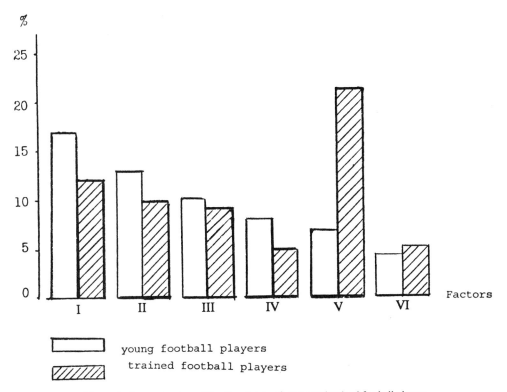

Figure 1. Factor structure of functional state of young and trained football players.

The blood haemoglobin level after the test increased in most athletes. A reduction of haemoglobin after the test indicated that correction using iron containing remedies was necessary. A dynamics study carried out during the competition period showed that an increase in performance level was accompanied by an increase of blood oxygen transportation values, improvement of energy power of CPK and an increase in economy.

Correlation analysis also showed interdependency of different factors which determine the adaptation abilities of players and special performance capacity level (fig.3).

Summing up the comparative analyses of the evaluation of players adaptation abilities with regard to their playing position: the high neuromuscular capacities of goalkeepers while maintaining stability under special load tests should be noted. Forwards have a high metabolic potential and good neuromuscular qualities. Most intense adaptation to exercise was found in defenders.

Modern football, which is characterized by universality of requirements for players of different positions in the process of long-term adaptation has an influence on forming the morphofunctional features of players of different positions bringing these features together.

The Study of Specific Working Capacity of Volleyball Players

The level of development of modern volleyball makes it necessary to take into account many factors which determine a players overall ability: technical and tactical, physi-

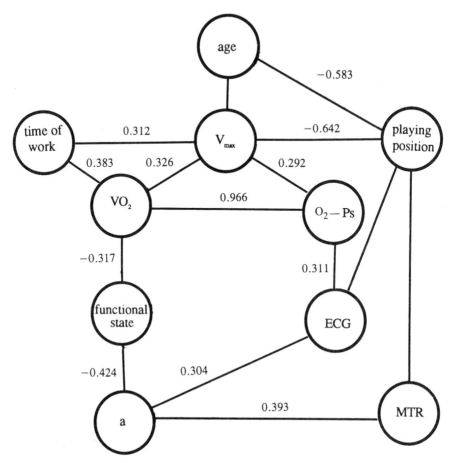

Figure 2. Correlation analysis of some factors of functional state of young football players.

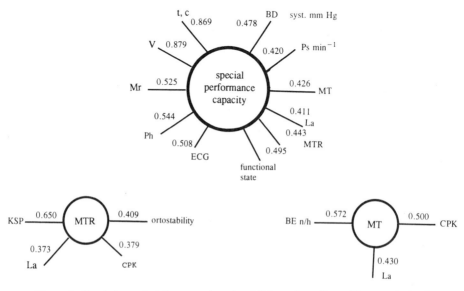

Figure 3. Correlation analysis between adaptation abilities and specific working capacity level.

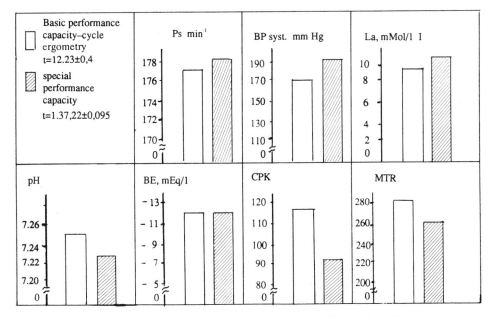

Figure 4. Main parameters of functional and diagnostic testing of basic and specific capacity.

cal, functional, psychological preparation, playing activity, state of health, adaptation abilities.

This study deals with means the control of training development, diagnostics and evaluation of specific working capacity.

Shuttle running was carried out; running 6 m repeated 10 reps (taking into account size and marking of a playing court), touching floor by hand, 180 degree turn, two series with 2 min rest pause.

To select the methods and parameters giving most information (with regard for specifity of motor activity) a test evaluating playing activity was carried out on a student volleyball team. Subsequent correlation analysis found a significant relationship between the main playing parameters (service, reception, attack block, defence) and functional parameters which determine adaptation effectiveness. Thus, selection of parameters giving most information for a complex program of specific working capacity of players was carried out.

To substantiate this a comparative test study of basic working capacity of 16 top-level players (aged 19–20 years) was carried out. To determine basic working capacity, cycle ergometry studies were undertaken: starting power = 630 kg m/min, pedalling speed = 70 revs/min, 210 kg m/min load increase every 2 min till exhaustion.

Fig.4 presents the main parameters of functional and diagnostic testing of specific and basic working capacity. The results of the study and comparative analysis showed that this test allows us to determine the functional abilities of the main physiological systems of the human body.

The quantity of the test, the possibility of its reproduction, regulation of exercise, and the simplicity of testing allowed it to be undertaken in a top-level team during the 5 month competition period of the country championships.

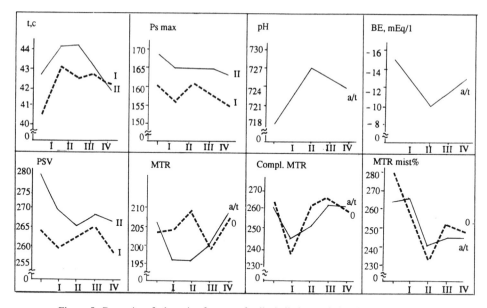

Figure 5. Dynamics of adaptation features of volleyball players during the championship.

During 5 months five tests were carried out: except for the first, all tests were carried out on the third day after closing of the next round of the championship. The volume of training between rounds was as follows: between initial test and 1st round, 67 hours, 1st and 2nd, 49 hours, 2nd and 3rd, 71 hours, and between 3rd and 4th rounds, 65 hours.

Fig. 5 shows certain dynamics of changes during the championship together with functional economy and improvement of neuromuscular reactions, particularly in motor reaction time.

These changes were accompanied by a significant improvement of special speed qualities and jumping endurance.

Individual evaluation showed that in the highest team, the level of functional fitness fell on the 3rd day and coincided with the highest sporting performance of the test of the team in the last round prize-winning places of championship.

CONCLUSION

Thus, tests and diagnostic programmes approved during the competition period in top-level football and volleyball teams showed that an increase in training level was accompanied by an improvement in blood oxygen transport systems, energy of CPK , and an increase in function economy. Both teams performed successfully in the countries championship: the football team became Russian champions and the volleyball team medal winners.

The highest level of functional fitness was displayed by:

- an improvement of blood oxygen transportation function (high level of haemoglobin and its increase after submaximum load);
- maintaining the optimal level of values of ferment exchange which is indicative of good regulation of energy process;

- improved vegetative provision of specific working capacity and development of parasympathic orientation of functions regulation;
- improvement of functional economy.

Diagnosis of poor adaptation in individual players allowed us to recommend means of correction and provided an optimization of the state.

32

FINGER DERMATOGLYPHS AS MARKERS OF THE FUNCTIONAL FEATURES

T. Abramova, T. Nikitina, E. Shafranova, N. Kotchetkova, and G. Secamova

Russian Institute of Physical Culture
11 2-nd Roschinckaya Street, Apartment 115, 113191 Moscow, Russia

INTRODUCTION

The knowledge of genetic indicators marking functional features is the basis of sport selection and early orientation.

One of the simplest and reliable markers using for the genetic potential definition is finger dermatoglyphs (FD). FD are disposed on the arm finger distal phalanges, formed for embryogenes stage, being unchangeable during human ontogenesis, associated with tactile function and characterized by high rate of variety.

The main signs of FD are finger pattern, ridge count and total pattern complexity. The techniques of FD determination is very simple, the procedure takes no longer than 10 minutes for one person.

FD was applied into medicine, legal examination, ethnic anthropology, genetics Olympic sport is the optimal model of external functional human manifestation.

Summarizing all said above it should be reasonable to research prospects of using of FD possibility as a marker of human genetic potential particularly for top athletes.

MATERIAL AND METHOD

The Material of This Study

a. the FD data on more than 600 top athletes in age from 18 to 27 years (participants of Olympic games and World Championships in 1980–1994) in 30 kinds of sports demonstrating such general functional features as short distance, endurance, coordination, games (the most representative kind of sport present in table 1);

b. the data of FD and about 80 functional features on more than 150 top rowers.

The Methods of This Study

a. standard procedure of taking finger-prints, standard determination of the main FD features: patterns (arch - A, loop - L, whirl - W), total ridge count (TRC), total pat-

Current Research in Sports Sciences, edited by Rogozkin and Maughan.
Plenum Press, New York, 1996

Table 1. FD in top athletes of Olympic kinds of sport

N	Kind of sport	n	D10 X	D10 s.d.	TRC X	TRC s.d.	Finger patterns,% A	L	W
1	Speed skating (short distance)	22	9.9	3.93	95.7	45.6	17.2	66.0	16.8
2	Short-track	7	10.3	1.85	98.7	30.0	8.6	80.0	11.4
3	Ski-rasing	17	12.2	3.90	115.5	40.6	6.0	71.0	23.0
4	Biathlon	17	12.6	2.51	130.5	36.5	5.3	63.5	31.2
5	Cycle racing	22	12.6	2.61	139.1	28.3	1.4	70.9	27.7
6	Archery	14	12.7	3.22	139.3	33.4	2.1	62.1	29.3
7	Shoting	11	13.0	3.78	121.2	35.9	5.5	59.1	35.4
8	Cycle track	17	13.0	2.22	128.3	34.1	1.2	67.6	31.2
9	Ski jump	9	13.1	2.80	121.2	30.4	2.0	64.4	33.3
10	Soccer	22	13.4	3.78	142.6	39.5	3.2	62.7	34.1
11	Volleyball	15	13.7	3.35	145.0	41.3	0	62.7	37.3
12	Basketball	18	13.9	3.58	140.4	40.5	0	60.0	40.0
13	Freestyle	14	13.9	3.22	133.8	39.0	2.9	57.9	39.2
14	Free wrestling	24	14.1	2.05	165.4	31.4	0	59.2	40.8
15	Nordic combination	26	14.2	2.97	144.9	39.6	0.4	57.7	41.9
16	Weight lifting	51	14.3	2.80	144.7	58.7	1.3	49.6	49.1
17	Fencing	13	14.4	3.43	130.7	38.5	2.2	53.1	45.4
18	Speed skating (combination)	37	14.4	2.35	147.6	31.6	0.2	52.0	47.8
19	Boxing	38	14.5	2.52	146.4	39.6	1.3	48.2	50.5

tern complexity (D10), pattern percentage (Gladkova T.D., 1966; Cummins H. and Midlo Ch., 1942).

b. usually used tests for determination of maximum strength, max. power, PWC 170, anaerobic threshold, endurance (3000m run), special work capacity, vestibular balance, learning capacity, skill, total body size and body composition

c. the variation, correlation and factor analysis (Urbah V.U., 1964), autoclassification (autoclassification was made in Institute of Control Science of RAS).

RESULTS

FD and Kinds of Sport

The experimental data shows that D10 and TRC in discussed kinds of sport include all population variation. D10 and TRC values classify different group of kind of sport, where functional features appear extremely clear, such as short distances, endurance, games, coordination.

Kind of sport with speed-strength orientation are characterized by low D10 and TRC significances. Coordination kinds of sport are characterized by high D10 and TRC significances. In accordance with the population norm endurance kinds of sport are placed in the intermediate position (table 1).

It is typical for games, that playing function is connected with FD: e.g., basketball defenders have got the maximal meaning of D10, TRC, W, and lack of A, on the contrary, basketball centers have got the minimum meaning of D10, TRC, W, and the maximum meaning of A . The same we can see in each of presented games (table 2).

Table 2. FD in top athlete of soccer, volleyball, and basketball

Kind of sport	n	D10 X	D10 s.d.	TRC X	TRC s.d.	Finger patterns,% A	Finger patterns,% L	Finger patterns,% W
Soccer	22							
forwards	5	9.8	1.29	106.9	42.1	8.0	86.0	6.0
halfbacks	9	14.0	3.03	153.8	47.3	3.3	60.0	36.7
defenders	5	14.2	3.00	147.0	41.8	0	58.0	42.0
goalkeepers	3	16.0	1.76	161.3	23.2	0	40.0	60.0
Volleyball	15							
playmakers	3	11.3	1.18	142.0	24.8	0	86.7	13.3
quards	12	14.3	2.45	147.0	18.9	0	56.7	43.3
Basketball	18							
N 5	3	11.0	1.18	130.3	13.0	0	90.0	10.0
N 4	3	8.0	3.59	57.5	41.0	20.0	80.0	0
N 3	4	12.0	2.37	105.1	26.3	0	80.0	20.0
N 2	4	13.5	3.31	119.5	76.9	0	65.0	35.0
N 1	4	16.5	2.42	157.0	18.4	0	35.0	65.0

Analysis of FD of speed skaters showed that short distance and endurance are characterized by different meanings of D10, TRC and A, L, W on principal: such as short explosive activity corresponds to high percentage of A and L, low percentage of W, the lows meaning of D10, such as endurance corresponds to lack A, decreased percentage of L and increase of W, high meanings of D10. This difference are concerned not only total indices, but also finger pattern distribution (table 3).

FD and Functional Features

Correlation and factor analysis expose the fact of correlation between finger pattern combination and particular functional features. The high level of D10 and TRC correlates with the high level of skill and threshold; the low meanings of D10 and TRCcorrelate with the high level of strength, special work capacity (abs.), but the lowest level of coordination, learning capacity, vestibulary balance.

Using autoclassification analysis we have formed impartial classes, differentiated by combination of FD and functional features (table 4). Results of autoclassification analysis correspond to correlation and factor analysis results, and show more definite quantitative limits of correlation indices.

Table 3. Finger pattern types (%)* in speed skaters

Kind of sport	Pattern type	Finger 1	Finger 2	Finger 3	Finger 4	Finger 5
Speed skating	A	13.2	26.3	18.5	13.2	13.2
(short distance)	L	55.3	52.6	71.0	50.0	76.3
	W	31.5	21.1	10.5	36.8	10.5
Speed skating	A	—	10.0	3.3	—	—
(endurance)	L	43.3	30.0	56.7	46.7	90.0
	W	56.7	60.0	40.0	53.3	10.0

* each figure is mean of (right + left)/2.

Table 4. FD and the manifestation level of functional features

| Class | D10 | TRC | Functional features at min and max levels | |
			Min	Max
I	5.5	28.0	Body size Special work capacity Strength Endurance Coordination	Strength (relative)
II	6.0	48.0	Coordination	Strength
III	11.6	126.0	Strength (relative) Special work capacity Strength (abs.)	Body size
IV	13.1	134.0	Body size Strength (abs.)	Endurance Coordination
V	17.5	163.0	Strength (relative)	Coordination

Discussion

The present research results exposed the fact of great prognostic informativity of FD: according to FD features we can distinguish the group of kinds of sport and also identify the potential of concrete physical features, such as strength, endurance, level of regulation of neuro-muscular coordination.

Such complicated data have not been obtained previously. There were some investigations in children, as a rule only one of physical quality was taken for comparison with FD. No experiments have not been done in top athletes of different kinds of sport (Nikituc, 1988; Sergienko, 1990).

So, we can conclude our data of high-level sportsmen research may be used for express diagnostics of the genetic potential, for early orientation and selection choosing the group of kinds of sport, for choosing and correcting role in sport of children and juniors. It can also be used in professional orientation of adult people.

ACKNOWLEDGMENTS

This investigation has been carried out under the state budget finances.

REFERENCES

Cummins H. and Midlo Ch. 1942. Palmar and plantar dermatoglyphics in primates. Philadelphia.
Genetic markers in anthropogenetic and medicine/ed Nikituk B.A. 1988. Thesis of the 4th USSR symposium.
Gladkova T.D. 1966. Papillary pattern palms and soles in primates and human. 150.
Sergienko L.P. 1990. Genetic and sport. 171.
Urbah V.U. 1964. Biometric methods. Science. 416.

33

MEASUREMENT OF ANAEROBIC PERFORMANCE CHARACTERISTICS OF ATHLETES

Heikki K. Rusko

KIHU - Research Institute for Olympic Sports
Jyväskylä, Finland

ABSTRACT

As reviewed by Vandewalle et al. (Sports Med. 4: 268–289, 1987) none of the previous methods has enabled accurate measurement of all the different determinants of maximal anaerobic performance. Consequently, a new maximal anaerobic running power (MARP) test (Rusko et al., Eur. J. Appl. Physiol. 66: 97–101, 1993) was developed. It consisted of n.20s runs on a treadmill with a 100s recovery between the runs. The speed of the treadmill was increased for each consecutive 20s run until exhaustion. The height of counter-movement jumps (CMJ) and blood lactate concentration (bLa) were measured after each run. Based on bLa vs. running power and CMJ vs. running power-curves submaximal and maximal indices of anaerobic running power were calculated and expressed as the oxygen demand of running. The results give information on 1) maximal anaerobic running power, 2) bLa concentration during submaximal sprinting, 3) peak bLa after exhaustion, 4) the height of CMJ, and 5) the decrease in the height of CMJ relative to the running speed and blood lactate concentration. The 400m runners have attained the maximal anaerobic running power of 120 ml kg^{-1} min^{-1} (200% of VO2max) in the MARP test. The anaerobic contribution to the energy demand of the last completed 20s sprint is about 70%. Repeatability of the test results has been high.

INTRODUCTION

Previous methods for determining the performance capacity of athletes for sport events requiring maximal mechanical and anaerobic power and anaerobic capacity include vertical jump tests (Sargent 1924 cited by Vandewalle et al. 1987; Bosco et al. 1983), staircase running test (Margaria et al. 1966), and cycle ergometer tests (Ayalon et al. 1974; Bar-or 1987, Szögy and Cherebetiu 1974). Exhausting running on a treadmill has also been applied for untrained and trained subjects (Thomson and Garvie 1981; Schnabel

Current Research in Sports Sciences, edited by Rogozkin and Maughan.
Plenum Press, New York, 1996

217

and Kindermann 1983, Medbø et al. 1988). As reviewed by Vandewalle et al. (1987) and Green and Dawson (1993) none of these tests enables accurate measurement of all the different determinants of maximal anaerobic performance. The jumping tests, staircase running test and some bicycle tests have given information on the alactic anaerobic power and on the force and velocity components of power while the all-out and constant-load tests to exhaustion have tried to measure the anaerobic capacity (Vandewalle et al. 1987). The latter tests are questionable because the involvement of aerobic processes increases as the duration of exercise increases, and the mechanical efficiency of maximal exercises is difficult to assess (Di Prampero 1981; Vandewalle et al. 1987).

In our laboratory we have developed a new Maximal Anaerobic Running Power test (MARP-test), which can be used to determine both the energetic and neuromuscular components of maximal anaerobic performance. It was planned that the test should include submaximal exercise intensifies so that the results could be used in the prescription of training for short intense muscular performances in a similar way the results of the so-called aerobic-anaerobic threshold tests are used in endurance sports (e.g. Kindermann et al. 1979). In addition, the modifications of the new test for endurance sports and for rehabilitation purposes have been studied.

THE MAXIMAL ANAEROBIC RUNNING POWER TEST

The new MARP-test has recently been described in detail (Rusko et al. 1993, Paavolainen et al. 1994). Before the MARP-test the subject performs a warm-up including vertical jumps with a preparatory counter-movement. After a short rest period the subject performs three maximal counter-movement jumps (CMJ) on a contact mat or on a force platform. The height of rise of the centre of gravity (CMJ-height) is calculated from the flight time. Fingertip blood samples are taken and resting blood lactate concentration is determined.

The MARP-test consists of n.20s runs on a treadmill with a 100s recovery between the runs, figure 1. A 5s acceleration phase is not included in the running time. The slope

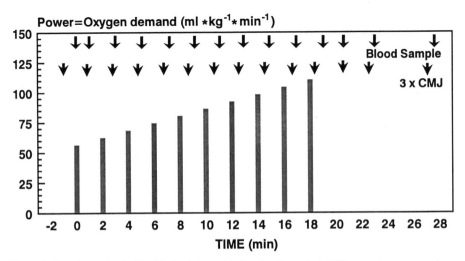

Figure 1. Exercise protocol of the Maximal Anaerobic Running Power test. CMJ = counter-movement jump.

of the treadmill is kept constant (originally 5° but at present we recommend 4°) during the whole procedure. The first 20s run is performed at low velocity (3–5 m.s^{-1}) depending on the performance level of the subject and corresponding approximately to the velocity at exhaustion in the maximal aerobic power test. Thereafter, the velocity is increased by 0.38m.s^{-1} for each consecutive 20s run until exhaustion (corresponding a 6 ml kg^{-1} min^{-1} increase in the VO$_2$-demand of running). Because the mechanical power is difficult to measure during treadmill running the power is expressed as the oxygen demand of running using the formula of the American College of Sports Medicine (ACSM 1986).

$$VO_2 \ (ml \ kg^{-1} \ min^{-1}) = 12 \ v \ (m \ s^{-1}) + 54 \ g \ (frac) \ v \ (m \ s^{-1}) + 3.5$$

where v = the speed of the treadmill and g = the slope of the treadmill.

Blood lactate concentration is analyzed from fingertip blood samples taken 40s after each run and 2.5, 5.0 and 10.0 min after exhaustion. The CMJ-height is calculated from the flight time of the maximal voluntary counter-movement jumps performed 15, 25 and 35s after each run. In addition, three CMJs are performed 2.5, 5.0 and 10.0 min after exhaustion.

The exhaustion in the MARP-test is determined as the point, where the subject can no longer run at the speed of the treadmill. The maximal running power (P$_{max}$) is calculated from the oxygen demand of the last completed 20s run and from the oxygen demand and exhaustion time of the following faster run. If the exhaustion time is 10s the P$_{max}$ is calculated to be 1 ml.kg^{-1}.min^{-1} higher than that of the previous run. Each additional 2s in the exhaustion time of the last run increases the P$_{max}$ by 1 ml.kg^{-1}.min^{-1}. The calculation of the P$_{max}$, as described above, is based on the relation between power and exhaustion time (e.g. Margaria et al. 1964) and on our experience that if an athlete was just able to complete the 20s run at a certain speed he always was able to run at least 7–9s at the next faster speed. Consequently, 10s exhaustion time at the faster speed was chosen to represent a better performance than the power of the preceding 20s run.

Based on blood lactate vs. running power (or velocity) and CMJ vs. running power (or velocity) -curves submaximal indices of anaerobic running power e.g. at 3, 5 and 10 mM blood lactate levels are intrapolated (P$_{3mM}$, P$_{5mM}$ and P$_{10mM}$, respectively) , figure 2. The 3 - 5 mM level describes the onset of blood lactate accumulation and the 10 mM represents the average lactate concentration during maximal anaerobic performance to exhaustion, e.g. during the 400 m sprint on the track (Nummela et al. 1991).

INTERPRETATION OF THE MARP-TEST RESULTS

The results of the test give information on 1) maximal anaerobic running power which has been shown to be related to e.g. 100m, 400m and 800m time on a track (Rusko et al. 1993, Vuorimaa et al. 1991), 2) bLa concentration during submaximal sprinting and running power at fixed submaximal bLa (e.g. 3, 5 and 10mm) which has been suggested to roughly describe the anaerobic sprinting economy (Rusko et al. 1993, Nummela et al. 1994), 3) peak bLa after exhaustion which has been used to describe the anaerobic lactic capacity, 4) the height of CMJ which describes the force-velocity characteristics of the leg muscles (Rusko et al. 1993, Paavolainen et al. 1994), and 5) decrease in the height of CMJ relative to the running velocity or power and to blood lactate concentration which reflects the fatigue in the force production of leg muscles (Rusko et al. 1993, Paavolainen et al. 1994).

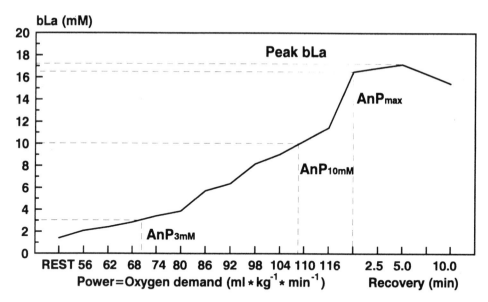

Figure 2. Typical blood lactate vs. sprinting power curve and calculation of the submaximal and maximal indices of anaerobic running power. AnP_{3mM} anaerobic running power at 3 mM blood lactate level, AnP_{10mM} anaerobic running power at 10 mM blood lactate level, AnP_{max} maximal anaerobic running power, PeakBLa = peak blood lactate concentration during recovery after exhaustion.

The high correlation between the 400m sprinting speed on track and P_{max} (r=0.89, p<0.001) as well as the fact that the maximal 20m sprinting speed on track and P_{10mM} correlated with both the P_{max} and the 400m speed would indicate that the new test measures the same determinants of performance as the 400m run on track (Rusko et al. 1993).

The 400m runners have attained the maximal anaerobic running power of 120, endurance athletes 116 and physically active nonathletes 108 ml.kg^{-1}.min^{-1} (VO$_2$ -demand 170–218% of VO$_{2max}$) in the MARP test, figure 3 (Rusko et al. 1993, Paavolainen et al. 1994, Nummela et al. 1995). Maximal blood lactate concentration at exhaustion has been 11–17 mM and the anaerobic contribution to the energy yield of the last completed 20s sprint has been about 70% (Rusko et al. 1993, Paavolainen et al. 1994, Nummela et al. 1995). These results demonstrate that similar maximal values are obtained in the MARP-test and in the 400m run on a track (Nummela et al. 1995).

Power at submaximal blood lactate concentration seems to be an important determinant of maximal anaerobic performance and aerobic power seems to be an important determinant of P_{3mM} and sprinting economy in both endurance and sprint athletes (Rusko et al. 1993, Paavolainen et al. 1994). This may be related to the repetitive nature and long duration of the MARP-test (Paavolainen et al. 1994, Nummela et al. 1995).

The CMJ-height begins to decrease when blood lactate concentration approaches 10 mM level (Rusko et al. 1993, Paavolainen et al. 1994). The decrease is observed in sprint runners while in endurance athletes the CMJ-height may slightly increase during the MARP-test, figure 4.

These results suggest that considerable fatigue takes place in sprinters at the end of the test where fast twitch muscle fibres are probably increasingly recruited. The accumulation of fatigue due to the great number of lumps and the long duration of the test may have influenced more the sprinters who are known to have greater percentage of fast twitch fi-

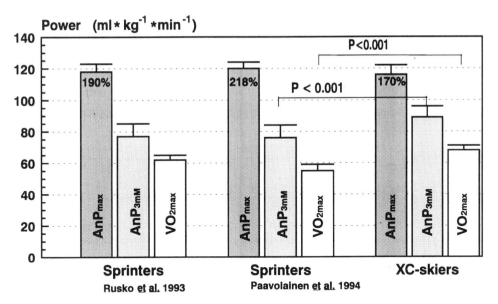

Figure 3. Maximal (AnP$_{max}$) and submaximal (AnP$_{3mM}$) anaerobic running power and maximal aerobic running power (VO$_{2max}$)in three groups of subjects. The numbers in the upper part of the bars indicate the percentage value of AnP$_{max}$ relative to the VO$_{2max}$.

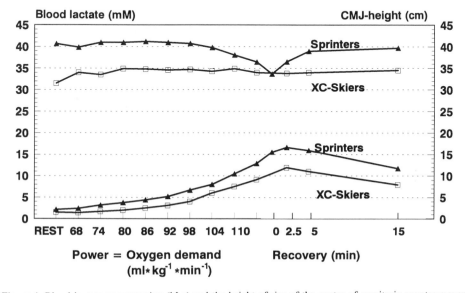

Figure 4. Blood lactate concentration (bLa) and the height of rise of the centre of gravity in counter-movement jump (CMJ~ height) of sprinters and cross-country skiers during the maximal anaerobic running power test. Modified from Paavolainen et al. 1994.

bres in their muscle so that they have been unable to reach a significantly better P_{max} than the endurance athletes (Paavolainen et al. 1994).

The test-retest reliability for the power indices in the MARP test have been as follows: r=0.92 (p<0.001) for P_{max}, r=0.80 (p<0.001) for P_{10mM}, and r=0.67 (p<0.01) for P_{5mM}.

The results of the MARP test have also been compared to the corresponding results of the Wingate test. Although four out of seven of the correlations between the corresponding variables of the MARP and Wingate tests were significant they were not high (r=0.52–0.59) (Nummela et al. 1995).

MODIFICATIONS OF THE MARP-TEST

Noakes (1988) and Green & Patla (1992) have demonstrated that endurance performance of athletes may be limited by either 1) physiological factors related to oxygen transport and utilization or 2) "muscle power", force-velocity characteristics and neuromuscular factors. Therefore, the MARP test versions of one, three, or five 20s sprints at each speed have also been studied. Marathon runners have attained similar maximal anaerobic running power in the different MARP test modifications (1.20s, 3.20s or 5.20s at each velocity) while sprinters and middle-distance runners have attained the highest power when running the 1.20s version (Vuorimaa et al. 1991).

Bicycle ergometer versions of the test have recently been developed and compared. Maximal anaerobic pedalling power was similar in the MARP test and in the two different pedalling tests in which either the pedalling frequency was kept constant and resistance was increased or the resistance was kept constant and pedalling frequency was increased for each consecutive 20s exercise bout (Tossavainen et al. 1994). However, the latter pedalling power test gave greater blood lactate concentration at submaximal exercise intensifies than the former test modification and induced a greater drop in the CMJ-height immediately after exhaustion (Tossavainen et al. 1994). A training study on male sprint runners has shown that the variables of the MARP test reflect the sprint training induced changes in the anaerobic performance capacity (Nummela et al. 1994).

In conclusion, the new test allows the evaluation of several determinants of maximal anaerobic performance including changes in the force-generating capacity of leg muscles and blood lactate concentration relative to the speed of the sprint running. The submaximal indices of anaerobic power may be used in the prescription of training for short intense muscular performance.

REFERENCES

1. American College of Sports Medicine: Guidelines for Graded Exercise Testing and Exercise Prescription, 3rd ed. Philadelphia, Lea and Febiger, 1986, p. 168.
2. Ayalon, A., 0. Inbar, and 0. Bar-Or. Relationships among measurements of explosive strength and anaerobic power. In: International Series on Sports Sciences, vol 1. Biomechanics IV. Nelson, R.C., and C.A. Morehouse (eds.). Baltimore, University Park Press, 1974, pp. 527–537.
3. Bar-Or, 0. The Wingate anaerobic test: An update on methodology, reliability and validity. Sports Med. 4: 381-394, 1987.
4. Bosco, C., P. Luhtanen, and P.V. Komi. A simple method for measurement of mechanical power in jumping. Eur. J. Appl. Physiol. 50: 273–282, 1983.
5. Di Prampero, P.E. Energetics of muscular exercise. Rev. Physiol. Biochem. Pharmacol. 89: 143–222, 1981.

6. Kindermann, W., G. Simon, and J. Keul. The significance of the aerobic-anaerobic transition for the determination of work load intensifies during endurance training. Eur. J. Appl. Physiol. 42: 25–34, 1979.

7. Green, H.J., A.E. Patla. Maximal aerobic power: neuromuscular and metabolic considerations. Med. Sci. Sports Exerc. 24: 38–46, 1992.

8. Margaria, R., P. Aghemo, and E. Rovelli. Measurement of muscular power (anaerobic) in man. J. Appl. Physiol. 21: 1662–1664, 1966.

9. Noakes, T.D. Implications of exercise testing for prediction of athletic performance: A contemporary perspective. Med. Sci. Sports Exerc. 20: 319–330, 1988.

10. Nummela, A., M. Alberts, R.P. Rijntjes, P. Luhtanen, and H.Rusko. Reliability and validity of the maximal anaerobic running power test. Int. J. Sports Med. 1995. (Accepted for publication)

11. Nummela, A., A. Mero, and H. Rusko. The effect of sprint training on the anaerobic performance characteristics in the MARP-test. Med. Sci. Sports Exerc. 26: S636, 1994.

12. Nummela, A., T. Vuorimaa, and H. Rusko. Changes in force production, blood lactate and EMG activity in the 400 m sprint. J. Sports Sci. 10: 217–228, 1992.

13. Paavolainen, L., K. Hdkkinen, A. Nummela, and H. Rusko. Neuromuscular characteristics and fatigue in endurance and sprint athletes during a new anaerobic power test. Eur. J. Appl. Physiol. 69: 119–126, 1994.

14. Rusko, H., A. Nummela, and A. Mero. A new method for the evaluation of anaerobic running power in athletes. Eur. J. Appl. Physiol. 66: 97–101, 1993.

15. Schnabel, A., and W. Kindermann. Assessment of anaerobic capacity in runners. Eur. J. Appl. Physiol. 52: 42–46, 1983.

16. Szögy, A., and G. Cherebetiu. Minutentest auf dem Fahrradergometer zur Bestimmung der anaeroben Kapazitat. Eur. J. Appl. Physiol. 33: 171–176, 1974.

17. Thomson, J.M., and K.J. Garvie. A laboratory method for determination of anaerobic expenditure during sprinting. Can. J. Sports Sci. 6: 21–26, 1981.

18. Tossavainen, M., A. Nummela, L. Paavolainen, A. Mero, and H.K. Rusko. New maximal anaerobic pedalling power test. Med. Sci. Sports Exerc. 26: S852, 1994.

19. Vandewalle, H., G. Pérès, and H. Monod. Standard anaerobic exercise tests. Sports Med. 4: 268–289, 1987.

20. Vuorimaa, T., K. Häkkinen, P. Vähäsäyrinki, and H. Rusko. Comparison of three tests measuring maximal running power in sprinters, middle distance runners and marathon runners. In Book of Abstracts, II IOC World Congress on Sport Sciences. 1991 COOB'92, S.A., Barcelona, 1991, p. 257.

34

PHYSICAL ACTIVITY, PHYSICAL FITNESS, AND HEALTH

Steven N. Blair

Cooper Institute for Aerobics Research
12330 Preston Road, Dallas, Texas 75230

Interest in the healthful effects of physical activity dates back to ancient times, but scientific study of the issue did not receive emphasis until the end of the 19th century. Early studies focused on physiology, and, in fact, three exercise physiologists (August Krogh, A.V. Hill, and Otto Meyerhof) received Nobel Prizes in the early 1920s for their research (Montoye, 1992). In the 1989 McCloy Lecture, Professor Park (1990) provided an excellent review of early scientific studies of exercise. Montoye extended these observations by including more recent reports on the relation of physical inactivity to health and disease.

This report examines recent evidence on the relations between physical activity, physical fitness, and health and draws conclusions about the possible causal nature of these associations. It also evaluates the public health burden of sedentary lifestyles in the United States and provides suggestions and recommendations for increasing participation in physical activity in the population.

EPIDEMIOLOGICAL STUDIES ON ACTIVITY, FITNESS, AND HEALTH

Information about the relation of physical activity to health and disease has grown rapidly over the past few decades. Epidemiological studies on this issue follow the pattern of evaluating the relation of physical activity or fitness to fatal or nonfatal disease end points and, more recently, to functional disability. Various study designs, including retrospective, cross-sectional, and prospective approaches, have been used, and dependent variables such as cardiovascular disease, cancer, diabetes, osteoporosis, obesity, hypertension, stroke, musculoskeletal disability, and all-cause mortality have been examined. More than 100 large, population-based studies on the relation of physical activity or fitness to health have been published in the peer-reviewed literature, most appearing in the past 30 years. Table 1 briefly summarizes this research. Substantial evidence supports the hypothesis that sedentary habits increase the risk of morbidity and mortality from a number

Current Research in Sports Sciences, edited by Rogozkin and Maughan.
Plenum Press, New York, 1996

Table 1. Summary results of studies investigating the relationship of physical activity or physical fitness to selected chronic diseases or conditions, 1963–1993

Disease or condition	Number of studies[a]	Trends across activity or fitness categories and strength of evidence[b]
All-cause mortality	***	↓↓↓
Coronary artery disease	***	↓↓↓
Hypertension	**	↓↓
Obesity	***	↓↓
Stroke	**	↓
Peripheral vascular disease	*	→
Cancer		
Colon	***	↓↓
Rectum	***	→
Stomach	*	→
Breast	*	↓
Prostate	**	↓
Lung	*	↓
Pancreas	*	→
Non-insulin dependent diabetes	*	↓↓
Osteoarthritis	*	→
Osteoporosis	**	↓↓
Functional capability	**	↓↓

Note. From "How much physical activity is good for health?" by S.N. Blair, H.W. Kohl, N.F. Gordon, and R.S. Paffenbarger, 1992, *Annual Review of Public Health*, 13, p. 121. Copyright 1992 by Annual Reviews, Inc., Palo Alto, CA. Adapted by permission.[a] * Few studies, probably less than 5; ** several studies, approximately 5-10; *** many studies, more than 10.[b]No apparent difference in disease rates across activity or fitness categories; some evidence of reduced disease rates across activity or fitness categories; good evidence of reduced disease rates across activity or fitness categories, control of potential confounders, good methods, some evidence of biological mechanisms; excellent evidence of reduced disease rates across activity or fitness categories, good control of potential confounders, excellent methods, extensive evidence of biological mechanisms, relationship is considered causal.

of chronic diseases. The strongest evidence for a causal association is for coronary artery disease, hypertension, colon cancer, obesity, functional capability, and non-insulin-dependent diabetes mellitus. Lower risk of mortality from all-causes and an increase in longevity also accompany a physically active way of life.

It is not possible in this report to exhaustively review all of the epidemiological studies on physical activity, physical fitness, and health. Extensive recent reviews on these topics are available for those interested in more detail (Blair, Kohl, Gordon, & Paffenbarger, 1992; Bouchard, Shephard, & Stephens, in press; Bouchard, Shephard, Stephens, Sutton, & McPherson, 1990). Here I will discuss a few important recent studies on selected topics, and present some recent data my colleagues and I have gathered.

Cardiovascular Disease

Powell, Thompson, Caspersen, and Kendrick (1987) provide an excellent review of physical activity and coronary heart disease. After taking into account varying qualities of study designs, they reported a nearly twofold increased risk for coronary heart disease in sedentary individuals compared with those who are active. One of the principal findings of the review by Powell et al. was that the studies with better designs and methods were nearly unanimous in demonstrating a significant inverse relation between physical activity and coronary heart disease. Using similar methods, I recently extended the review of Powell and colleagues to the studies published between 1987 and 1992 (Blair, in press). These later studies were universally well done, and, as found by Powell et al. for the studies with good methods, all showed an inverse relation between activity or fitness and disease risk. The British Regional Heart Study is a splendid example of the recent generation of epidemiological reports on activity and cardiovascular disease that used a prospective design, careful assessment of physical activity (with a validation of the method), complete follow-up with precise definitions and measurement of end-points, and overall excellent epidemiological and statistical methods (Shaper & Wannamethee, 1991; Wannamethee & Shaper, 1992).

The British Regional Heart Study follows 7,735 men aged 40–59 years at baseline who were drawn from general medical practices in 24 towns in England, Wales, and Scotland. An extensive leisure-time physical activity questionnaire was used to obtain the usual pattern of walking or cycling, recreational pursuits, and vigorous sporting activities. A six-category physical activity scale (ranging from *inactive* to *vigorous*) was based on the habitual activity pattern data. Information on both frequency and intensity of activity was used to establish the categorical scores. Study participants were followed for fatal and nonfatal heart attacks and strokes by established National Health Service registers.

Heart Attack and Physical Activity. There were 488 heart attacks (217 fatal and 271 nonfatal) in the cohort during 8 years of follow-up. Heart attack rates and relative risks across the physical activity categories are presented in Table 2. The death rates indicate a steep inverse gradient across activity levels. These death rates include all men in the study, but when the group was divided into those who did and did not have ischemic heart disease (IHD) at their baseline visit, similar patterns were noted. Also shown in Table 2 are adjusted relative risks across activity categories for the group of men who were initially free of IHD. Relative risks with additional adjustment for other clinical risk factors are included in the paper, and the results are similar to those presented here. The last column of Table 2 shows relative risks of heart attack when all data from vigorous sporting activity were excluded from the physical activity category calculations. Some earlier reports, most notably by Morris, Clayton, Everitt, Semmence, and Burgess (1990), indicate that vigorous activity is necessary for protection against heart disease. This hypothesis is not sup-

Table 2. Physical activity and risk of heart attack in the British Regional Heart Study

Physical Activity Category	Crude heart attack rate[a]/1,000/year (all men)	Relative risk (RR) (95% CI) in men free of heart disease at baseline	
		Adjusted RR[b]	Adjusted RR[b,c]
Inactive	12.9	1.0	1.0
Occasional	9.1	0.8 (0.5–1.2)	0.8 (0.5–1.2)
Light	8.2	0.8 (0.5–1.2)	0.8 (0.5–1.2)
Moderate	5.8	0.4 (0.2–0.8)	0.4 (0.2–0.8)
Moderately vigorous	4.9	0.4 (0.2–0.8)	0.3 (0.2–1.3)
Vigorous	5.8	0.8 (0.4–1.4)	—

Note. CI = confidence interval. Data from Wannamethee & Shaper, 1992.
[a]These rates were virtually identical to age-adjusted rates.
[b]Adjusted for age, body mass index, social class, and smoking status.
[c]All vigorous sporting activity was excluded from calculation of the physical activity score.

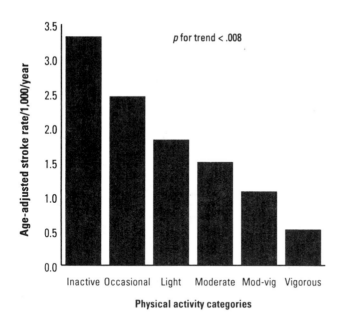

Figure 1. Age-adjusted stroke rate by physical activity categories in the British Regional Heart Study. (Subjects were 7,735 men aged 40–59 years at baseline who were followed for 9.5 years, during which time 128 strokes occurred. Mod-vig = moderately vigorous. Data from Wannmethee and Shaper, 1992.)

ported by the British Regional Heart Study data, and the role of vigorous activity in prevention of heart disease, if any, remains unclear. Additional investigation is needed to address this important question.

The data in Table 2 show an upturn in heart attack risk in the vigorous activity category. This observation also has been reported by Paffenbarger, Hyde, Wing, and Hsieh (1986) for the Harvard Alumni cohort. Whether this trend is truth, chance, or caused by some confounding variable is unclear. In an addendum to their paper, Shaper and Wannamethee's (1991) preliminary analyses indicate that the increase in risk in the vigorous activity group may be related to the influence of hypertension. Our data on fitness and mortality in hypertensive men is consistent with this possibility (Blair, Kohl, Barlow, & Gibbons, 1991).

Stroke. Wannamethee and Shaper (1992) also report on the relation of physical activity to risk of stroke in men in the British Regional Heart Study after 9.5 years of follow-up. During this interval there were 128 fatal and nonfatal strokes in the cohort. The age-adjusted risk of stroke for the six physical activity groups is presented in Figure 1. These data are for all men in the study, but, as with the heart attack data, the results were similar in men with or without heart disease or stroke at baseline. The steep and significant ($p < .008$) inverse gradient illustrated by the data from this valuable study provides the strongest support to date for the hypothesis that activity protects against stroke.

Obesity. Farmers have known for millennia that animals can be fattened more easily when they are penned than when they are allowed to run free. It also is widely believed that physical inactivity promotes the development of obesity in humans and that active individuals are less likely to become overweight. This belief may well be true, but confirmatory data are scarce. Cross-sectional associations between physical activity and body mass or composition are abundant. Both children and adults who spend considerable time in sedentary pursuits, such as television viewing, are heavier than their peers who spend less time in this inactive behaviour. Population-based surveys invariably show that more active and fit persons weigh less than sedentary and unfit persons. What is lacking are well-designed prospective studies, with good assessments of physical activity, in well-characterized populations with sufficient sample size and follow-up to provide adequate statistical power to test the hypotheses that regular physical activity reduces the risk of weight gain over time and that inactivity increases the risk of becoming overweight or obese. No large-scale clinical trials have been conducted to test these hypotheses, nor are they likely in the foreseeable future due to logistical complexity and cost.

National Health and Nutrition Examination Survey-I. Williamson et al. (1993) evaluated the relation of physical activity to weight gain over a 10-year follow-up of 3,515 men and 5,810 women in the National Health and Nutrition Examination Survey-I Epidemiologic Follow-up Study. Baseline physical activity was not related to weight gain in either men or women. Men and women who were sedentary at both baseline and follow-up were much more likely to experience significant weight gain (>13 kg) than participants who were active at both examinations [relative risk of 2.3 (95% confidence interval {CI} = 0.9–5.8) in men and 7.1 (95% CI = 2.2–23.3) in women]. Strengths of this study are a large and representative population and a thorough analysis of the data. The physical activity assessment was relatively crude, and undoubtedly resulted in substantial misclassification, but this should bias the results towards the null hypothesis.

Harvard Alumni Study. Recent data from the Harvard Alumni Study show a weight loss in men who increased their physical activity (Lee & Paffenbarger, 1992). Activity was assessed in 1962 or 1966 and in 1977 by mail-back questionnaires in 11,703 men. Weight was also self-reported. Men who lost >5 kg had increased their physical activity by 1,246 kJ per week, and men who lost >1 but <5 kg increased activity by 701 kJ per week. Men with no change or weight gain had no significant changes in physical activity. Strengths of this study were a more detailed assessment of physical activity than was used in most other reports on this topic, the large sample size, and a lengthy follow-up. As with the other studies described here, the period of activity change coincided with the period of weight change, so it is difficult to know whether activity change or weight change came first.

Aerobics Center Longitudinal Study. We performed a series of preliminary analyses in the Aerobics Center Longitudinal Study population that help overcome some of the weaknesses of earlier reports (Blair, Barlow, & Kohl, 1993). A cohort of 3,736 men (*M* age = 42.2 years) and 530 women (*M* age=42.9 years) completed three voluntary preventive medical examinations. A principal design feature of the study was evaluation of change in fitness from the first to the second examination as a predictor of change in body composition to the third examination. The average interval between the first and last examinations was 6.3 years and the average weight at baseline was 58.4 (SD = 8.8) kg in women and 81.3 (SD = 10.6) kg in men. The men lost an average of 0.1 kg during the study, while the women gained an average of 0.9 kg. Physical fitness was measured at each examination with a maximal exercise test on a treadmill, and this measure was considered an objective marker of habitual physical activity patterns. After multivariate control for the influences of age, baseline height and weight, smoking status, and baseline physical fitness, weight gain was inversely associated with baseline fitness and an increase in fitness in men, and with an increase in fitness in women. Several other analyses in this study support the hypothesis that increases in activity or fitness are inversely associated with weight change.

Summary. Given the principle of energy balance, it is logical that an active way of life might reduce the risk of becoming overweight or obese, and current studies support this hypothesis, but fall short of allowing a definitive conclusion. There is a need for well designed studies in large populations, with multiple assessments of activity and fitness over time, and periodic weight measurements. Extensive data on potential confounding influences, such as diet, also should be obtained.

Physical Activity or Physical Fitness

Death rates in sedentary individuals are approximately twice as high as for active persons (Paffenbarger et al., 1986; Powell et al., 1987). In general, the prospective studies on physical fitness (Blair et al., 1989; Ekelund et al., 1988) show a stronger inverse association with mortality than is seen in the studies on physical activity. In these (and other) studies, the least fit group (typically the bottom quintile or quartile) has a three- or fourfold increased risk for all-cause mortality and a seven- or eightfold increase for cardiovascular disease mortality when compared with the top quartile or quintile. This difference between activity and fitness studies is most likely due to less misclassification in fitness assessment than in activity assessment. Fitness is measured objectively by ergometric methods, and activity is usually measured in population-based studies by self-report. Lower precision in activity assessment leads to greater measurement error, which tends to bias results towards the null. Thus, fitness studies may give a more accurate estimate of the true impact of sedentary living habits on disease risk than do studies on activity. This notion is based on the assumption that increases in physical activity lead to increases in physical fitness, but this has been firmly established by numerous well-controlled exercise training studies over the past several decades. There are alternative explanations, however. Hereditary influences might cause more confounding for the fitness studies than for activity studies, or other unknown confounders also may be factors.

Belgian Factory Workers. Unfortunately, few well-designed studies directly compare the combined effects of physical activity and physical fitness on mortality. One report from Belgium includes data from a 5-year follow-up of 2,363 men who were 40–55 years of age at baseline (Sobolski et al., 1987). Both physical activity and physical fitness were assessed at baseline, and ischemic heart disease risk was compared across levels of activity and fitness. The investigators concluded that physical fitness, but not physical activity, protects against heart disease. This study had many good features, but one serious limitation, I believe, jeopardizes the conclusion. There were only 31 events available for analysis, and this may not have been enough to provide adequate statistical power to test the relation of activity or fitness to disease.

Copenhagen Men. A recent report by Hein, Suadicani, and Gyntelberg (1992) also addressed the issue of the relative importance of activity and fitness in terms of heart disease risk. They followed 4,999 men aged 40–59 years for 17 years. Leisure-time and occupational physical activity was measured by questionnaire, and physical fitness was determined by cycle ergometry. All-cause mortality rates across physical fitness quintiles are presented in Figure 2 for sedentary and active men. There is no association between fitness and mortality in the sedentary men, but a significant inverse trend is evident among

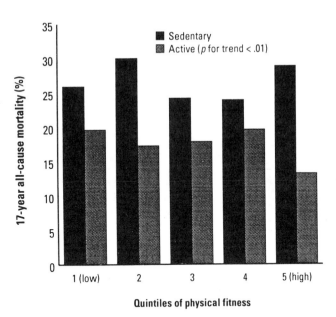

Figure 2. Age-adjusted all-cause mortality for sedentary and active men across quintiles of physical fitness. (Subjects were 4,999 men aged 40–59 years at baseline. Follow-up was for 17 years, with 941 deaths occurring in the cohort. Data from Hein, Suadicani, & Gyntelberg, 1992.)

the active men. It is noteworthy that the active men within each fitness stratum had lower death rates than the inactive men in the same stratum. This difference was greatest in the fittest men.

Overall, this study supports the hypothesis that both activity and fitness provide protection against heart disease and all-cause mortality. It is somewhat puzzling that the benefits of fitness are limited to active men.

Aerobics Center Longitudinal Study. We have performed some preliminary analyses to evaluate the independent associations of activity and fitness on all-cause mortality (Blair & Kohl, 1988). We followed 10,233 men for 8.3 years after a medical examination that included a maximal exercise test to evaluate physical fitness. Physical activity was assessed by a brief questionnaire. There were 240 deaths during 85,050 man-years of follow-up. In contrast to the Copenhagen study, we found an inverse all-cause mortality gradient across fitness quintiles in sedentary, but not in active, men. However, when multivariate models were constructed with all-cause mortality as the dependent variable, and age, physical activity, physical fitness, and other risk factors as independent variables, both activity and fitness were inversely associated with mortality.

Summary. Additional prospective studies with measurement of both activity and fitness with morbidity and mortality follow-up are needed to evaluate more thoroughly the independent influences of these two exposures. Such studies may provide new insights regarding the specific dose of activity needed to reduce risk, and may help quantify the specific costs of a sedentary way of life.

Change in Physical Activity and Mortality

Epidemiological studies on activity or fitness and disease risk are frequent criticized for selection bias. This could be a factor if individuals were sedentary because of some

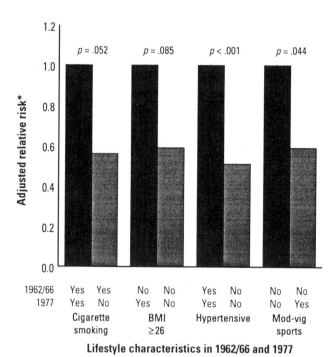

Figure 3. Adjusted relative risks (each relative risk is adjusted for age and all other variables in the figure) for coronary heart disease mortality by changes in lifestyle characteristics. [The black bars represent men who had unfavourable characteristics at baseline (in 1962 or 1966) and at follow-up (in 1977). The gray bars show the adjusted relative risks for men who made favourable changes on the variable of interest between baseline and follow-up. There were 10,269 men in the cohort, with 130 deaths from coronary heart disease. *Adjusted for age and other factors in the figure. BMI = body mass index; Mod-vig = moderately vigorous. Data taken from Paffenbarger et al., 1993.]

underlying and undetected problem such as chronic disease, which also might increase their mortality risk. Although epidemiologists use various techniques to address this and other possible biases, many think that current evidence is insufficient to confirm a causal hypothesis between inactivity and disease. Epidemiological studies with a one-time assessment of activity or fitness with subsequent follow-up cannot predict with certainty what would have happened if sedentary or unfit individuals converted to an active lifestyle. Perhaps they would receive no health benefit. We now have additional data with which to address this important question.

A 1993 paper from the Harvard Alumni study is one of the most important reports from physical activity epidemiology of the past 25 years (Paffenbarger et al., 1993). Paffenbarger et al. report on the relation of changes in physical activity and other lifestyle characteristics to mortality in 10,269 Harvard Alumni. These men completed mail-back questionnaires in 1962 or 1966 and again in 1977; thus changes in lifestyles could be determined over an 11- to 15-year period. These changes were evaluated in relation to mortality during continued follow-up from 1977 to 1985. There were 476 deaths during 90,650 man-years of observation. Men who were initially sedentary, but who started participating in moderately vigorous sports (intensity of 4.5 METs or greater) by 1977, had a 23% lower risk of death than the men who remained inactive. This result was consistent across age groups.

Paffenbarger et al. (1993) also examined the relation between changes in lifestyle characteristics and risk of death from coronary heart disease. Results of these analyses are shown in Figure 3. It is striking that the favourable changes (or maintenance of low risk status) for the four risk indicators appear to have comparable impact on reduction in risk of death from coronary heart disease. Beginning moderately vigorous sports is as important to risk reduction as stopping smoking.

Aerobics Center Longitudinal Study. We evaluated the relation of change in activity to mortality in 10,288 men who received at least two preventive medical examinations at the Cooper Clinic during the interval 1970 to 1989. The average interval between the first and second examination was approximately 5 years. Subjects were followed for mortality from the time of their second examination until date of death or until December 31, 1989 for the survivors. Total follow-up was 52,069 man-years, during which 275 men died. Physical activity was assessed by a brief self-report questionnaire. Participants were given a list of 18 common physical activities, and were asked to indicate their participation in any of the activities during the past month. Men who did not indicate any participation were given a physical activity score of 1. Men who participated in any activity other than walking, jogging, or running were given a score of 2. Walkers, joggers, and runners also provided information on the number of times they participated per week, and the number of miles per occasion. This information was used to calculate miles per week in walking, jogging, or running. Men who covered 1–10 miles per week received an activity score of 3, men with 11–20 miles per week a score of 4, and those with more than 20 miles per week a score of 5. Thus, the physical activity scale was an ordinal variable with a range of 1 to 5. For stratified analyses, men with a score of 1 were considered sedentary, and all others were classified as active. For proportional hazards modelling, the full range of scores was entered into the equation.

There were 3,807 men classified as sedentary at baseline. All men were reclassified on physical activity at the second examination, and change in physical activity was determined. Some men remained sedentary, and there were 46 deaths during 5,999 man-years of follow-up in that group. There were 71 deaths during 17,593 man-years of follow-up in men who converted from sedentary at baseline to active by the second examination. The age-adjusted all-cause death rates were 79.8/10,000 man-years in the persistently inactive category, and 39.3/10,000 man-years for the men who changed from sedentary to active status during the study. The age-adjusted relative risk of death was 0.49 (95% CI = 0.34–0.72) in the men who became active, when compared with those who remained sedentary. These analyses were performed separately in men classified as apparently healthy at baseline, and men who already had evidence of heart disease, stroke, hypertension, cancer, or non-insulin-dependent diabetes mellitus, and the results were essentially identical to the overall analyses. All-cause mortality rates were calculated for the two activity change categories for men in the 20–39 years, 40–49 years, 50–59 years, and 60 years age groups. In every age group, men who started an activity program had lower death rates. The reduction in risk declined across age groups, ranging from 75% in the 20–39 years group to 16% in the 60 years and older men, with reductions of 64% and 37% for men in their 40s and 50s, respectively.

As shown above, starting an activity program reduced risk of dying by 51% in the men who became physically active when compared with those who remained sedentary. For comparison, we also calculated the reduction in risk for other favourable lifestyle changes. Cut-points to designate high risk status were any cigarette smoking, systolic blood pressure 140 mmHg, cholesterol 6.2 mM, and body mass index (weight in kg/height in m^2) 27.0. The percentage reduction in risk of age-adjusted all-cause mortality for changes in risk factors is presented in Figure 4. We view these results as preliminary, and they need further evaluation. However, the results do support the hypothesis that important health benefits can be gained by changing from a sedentary to an active way of life, and that these benefits are comparable to other favourable lifestyle changes.

It is possible that the relation between changing physical activity habits and mortality risk is due to the confounding influences of other variables. This issue was addressed

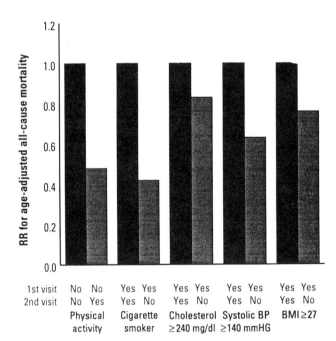

Figure 4. Relative risks (RRs) for age-adjusted all-cause mortality are shown for changes in lifestyle characteristics in 10,288 men with two examinations at the Cooper Clinic during 1970 to 1989. (There were 275 deaths during 52,069 man-years of follow-up. The black bars represent men who were at risk on the variable at both examinations, and the gray bars represent men who made favourable changes in risk factors from the first to the second examination. BP = blood pressure; BMI = body mass index.)

in a multivariate framework with all-cause mortality as the dependent variable, and age, height, weight, health status, family history of cardiovascular disease, physical activity score (full range of 1 to 5), smoking, cholesterol, fasting blood glucose, systolic blood pressure, and changes in risk factors (weight, physical activity, smoking, cholesterol, systolic blood pressure, fasting blood glucose) as independent variables. Change in physical activity was calculated by subtracting the activity score at the first examination from the score at the second examination; we then dummy coded the change score with no change (or a change score of 0) as the reference category and included increases and decreases in the model. Smoking habit change also was dummy coded with continuing smokers as the reference category. Other risk factor change scores were determined by subtracting the value of the first examination from the second, and entering the resulting value as a continuous variable in the analysis. The results of this analysis are presented in Table 3. The principal finding from this analysis for the current topic is that men who increased their physical activity had a significant reduction in risk of dying ($p = .0002$), after adjustment for other variables in the model.

Summary. These data on the beneficial impact of starting an exercise program provide an important and new piece of information about the health value of physical activity. The findings from these two large prospective studies add considerably strengthen the inference that sedentary habits increase the risk of early death and that starting an activity program in middle age can provide significant protection against coronary heart disease.

The Public Health Burden of Sedentary Living Habits

The previous section summarizes the health hazards of physical inactivity. Substantial elevations in mortality risk are seen in sedentary and unfit men and women. If we assume that these associations are causal, we can estimate the impact of inactivity on

Table 3. Proportional hazards model for all-cause mortality, 10,193 men (272 deaths), Aerobics Center Longitudinal Study, 1971–1989

Variable	Coefficient	Standard error	Probability
Age (years)	0.0697	0.0069	0.0001
Baseline activity score	0.0792	0.0647	0.22
Decrease in activity[a]	−0.1454	0.1856	0.43
Increase in activity[a]	−0.5597	0.1493	0.0002
Health status[b]	1.9267	0.2077	0.0001
Baseline systolic BP (mmHg)	−0.0026	0.0047	0.57
Baseline glucose (mM)	−0.0089	0.0116	0.44
Baseline cholesterol (mM)	0.0338	0.0653	0.61
Baseline weight (kg)	0.0250	0.0061	0.0001
Baseline height (m)	−1.1961	1.2058	0.32
Difference in systolic BP (mmHg)	−0.0090	0.0042	0.03
Difference in glucose (mM)	−0.0020	0.0023	0.38
Difference in cholesterol (mM)	0.0026	0.0008	0.0006
Difference in weight (kg)	−0.0030	0.0097	0.76
Started smoking[c]	−1.8412	0.4402	0.0001
Stopped smoking[c]	−1.3907	0.3053	0.0001
Non-smoker[c]	−1.2124	0.1709	0.0001
Family history[d]	−11.9126	1545.0	0.99

Note. BP = blood pressure

[a]Relative to no change in physical activity.

[b]Healthy = 0, Unhealthy = 1.

[c]Relative to smoker at both visits.

[d]History of coronary heart disease in either parent.

national mortality statistics by application of population attributable risk (PAR) calculations. The PAR is an estimate of the number or percentage of deaths in a population that is considered to be caused by (or attributed to) a risk factor The PAR is dependent on the strength of the association between the exposure (or risk factor, in this case physical inactivity or low fitness) and the outcome (here we will use all-cause mortality), and also is influenced by the prevalence of the exposure in the population of interest. The definition of PAR used here is that it represents the percentage of deaths in the U.S. population that is thought to be due to sedentary habits or a low level of physical fitness.

The PAR has been calculated for low levels of activity or fitness in two large prospective studies and also for the U.S. population. These estimates vary due to different definitions for the at-risk population, measurement techniques, and populations. Paffenbarger et al. (1986) estimate that 16.1% of the deaths in the Harvard Alumni Study population could be attributed to sedentary lifestyle. In the Aerobics Center Longitudinal Study, 9.0% of the deaths in men and 15.3% of deaths in women were attributed to low physical fitness (Blair et al., 1989). The PARs for low activity or fitness in these studies were generally comparable to the PARs for other well-established risk factors such as smoking, high cholesterol, or high blood pressure. Hahn, Teutsch, Rothenberg, and Marks (1990) estimated that approximately 250,000 deaths in the U.S. from nine chronic diseases were due to sedentary habits. This number was similar to the estimates for obesity, hypertension, and hypercholesterolemia, but less than the estimate for cigarette smoking, which was 361,000 deaths.

We cannot assume that the PAR provides a realistic projection of the number of deaths that we might be able to prevent with intervention programs. It is not possible to get all smokers to quit or persuade all sedentary individuals to take up vigorous exercise. To reasonably estimate how much we might be able to alter national death rates by promoting physical activity, it is necessary to establish specific definitions for the risk factor, to estimate the population prevalence using those definitions, and to project what percentage of the currently sedentary population might be expected to become more active. In two recent reports researchers have considered these factors and modeled the potential reduction in deaths. Blackburn and Jacobs (1993) used definitions and estimates of physical activity from several surveys they conducted in Minnesota. With these procedures they classified 37% of men as sedentary, 41% as getting some activity, and 22% as vigorously active; corresponding figures for women were 44, 44, and 12%, respectively. They proposed that it might be feasible to shift the population distribution of activity to 25% sedentary, 50% moderately active, and 25% vigorously active for the men; and 30% sedentary, 50% moderately active, and 20% vigorously active for the women. Changes of this magnitude would be expected to reduce deaths by 5% in men and 6% in women.

Powell and Blair (in press) used physical activity categories and population estimates of physical activity described in Healthy People 2000 (U.S. Department of Health and Human Services, Public Health Service, 1991): sedentary (24%), irregularly active (54%), regularly active (10%), and vigorously active (12%). We then modeled the reduction in deaths from coronary heart disease, colon cancer, and diabetes if 50% of each of the first three activity categories shifted upwards to the next category. This calculation gives an estimate of preventing 66,000 deaths or 3% of the total deaths in the U.S. in 1988.

The reduction in deaths predicted by these calculations may seem small. They are, however, conservative, and this approach does not take into account other benefits of an active way of life such as an improved sense of well-being, better stress management, enhanced functional capacity (especially for the elderly population), and the enjoyment of active recreational pursuits. Furthermore, a shift downwards of 3–5% in total deaths is a sizable number of individuals and would be an impressive public health achievement.

RESEARCH, POLICY, AND PROGRAM RECOMMENDATIONS

This is an exciting time to be involved in exercise science. The magnitude of research, policy, and program initiatives in the United States is great, and probably would astonish Dr. McCloy and his contemporaries. In the past 20 to 30 years, there has been an enormous increase in the number of highly qualified investigators in several branches of science, from molecular biology to population-based studies, who are involved in research on exercise, physical activity, and physical fitness. Many public and private organizations are engaged in programs in which physical activity is a key component. Some summary examples of recent activities are presented in Table 4. Members of the American Alliance for Health, Physical Education, Recreation and Dance and others interested in exercise science and sports medicine can take credit for the impressive recent advances in programs and research involving physical activity, and we can look forward to an exciting future.

Table 4. Material related to exercise and health summarized from seven recent conferences or reports on physical activity and health

Conference/report	Purpose	Main recommendations
American Heart Association (AHA) Committee on Exercise and Cardiac Rehabilitation of the Council on Cardiology (Fletcher et al., 1992)	To present a current position statement on exercise and health for all Americans	Named inactivity as a risk factor with the same status as cholesterol, smoking, and hypertension. **Recommendations:** (1) Basic knowledge of anatomic, biochemical, and physiological changes that result from various patterns of physical activity in persons of different ages is needed, and a determination of whether a certain minimal-intensity threshold of physical activity is required for benefit; (2) Evaluation of the biomedical and economic impact of exercise; (3) Inclusion of adequate numbers of the elderly and women in future studies; and (4) Future advancement and investigation should be not only about the benefits of physical activity, but also about methods used to enhance distribution of knowledge to all people.
Statement on Exercise		
American Heart Association Prevention III Conference (Blair et al., 1993)	To evaluate six major health hazards and summarize state of the art activities for reducing each health hazard, identify gaps in research, and give detailed suggestions on professional and public education programs and public policy initiatives	**Research needs:** (1) define how physical activity relates to heart disease in specific ways, as well as how much activity is needed for beneficial effects; (2) specify factors that facilitate adoption and maintenance of activity; and (3) define benefits of physical activity in improving employee well-being, performance, and productivity. **Program and intervention needs:** (1) Physical activity education and intervention programs for health care professionals and the public should present alternatives to the traditional exercise prescription; (2) Life-style interventions in which increased activity is integrated into daily routines is encouraged; and (3) Intervention strategies to increase physical activity should include modified physical environment to enhance physical activity: stimuli that prompt physical activity; and reinforcement.
A conference entitled "Behaviour Change and Compliance: Keys to Improving Cardiovascular Health"		**Policy needs:** (1) Employer recognition of employee needs for time, space, facilities, and other incentives to pursue physical activity objectives; (2) Increased support for mass transit; (3) Insurance coverage for exercise counselling and rehabilitative services; (4) Comprehensive, integrated school fitness programs to promote lifelong fitness; (5) AHA support to all organizations to meet Healthy People 2000 objectives; and (6) Information and materials on physical activity developed for counselling use by medical professionals.
Committee on Health Objectives for the Year 2000 (U.S. DHHS Public Health Service, 1991)	To formulate national health objectives for the year 2000	**Intervention and research needs:** 1. To improve the "health span" or "active life expectancy" of all Americans. 2. To see an increase in all Americans engaging in regular, appropriate physical exercise, such as walking, swimming, or other moderate aerobic activities by the year 2000. 3. To prolong the period of independent living, with particular attention to quality of life. 4. To increase public and professional awareness of the benefits of physical activity. 5. To reach employees with worksite physical activity programs. 6. To improve surveillance and evaluation systems in the area of physical activity and fitness. 7. To achieve access to preventive services for all Americans. 8. To reduce health disparities among Americans.
Compilation of information for the United States Public Health Service		

(continued)

Table 4. (*Continued*)

Conference/report	Purpose	Main recommendations
International Society and Federation of Cardiology (Bijnen et al., 1992) *A position statement for the World Health Organization (WHO)*	To notify the international community that a relation exists between physical inactivity and coronary heart disease	**Recommendations:** 1. Physical inactivity should be considered an important risk factor in coronary heart disease and should be included in prevention policies, particularly in industrialized countries. 2. Because physical inactivity is a modifiable risk factor, industrialized countries should set policy objectives to promote physical activity. Exercise should be enjoyed and a physically active lifestyle should be promoted worldwide. 3. Regular light-to-moderate physical activity, documented in current studies as having significant health benefits, should be encouraged. This may have a positive effect on inactive people, who are most likely to adopt less intense forms of exercise.
National Heart, Lung, and Blood Institute Workshop on Physical Activity and Cardiovascular Health (Haskell et al., 1992; King et al., 1992) *Workshop to review state-of-the-science research on physical activity and cardiovascular health, with special emphasis on children and women*	To review the relation between physical activity and cardio-vascular health and give advice on directions for future research priorities and programs	**Recommendations for research on:** 1. Determinants and methods of changing individual patterns of physical activity, focusing on structure, supervision, frequency, intensity, and units of time. 2. Stages of physical activity behaviour with interventions specifically tailored to each stage of change. 3. Influence of developmental milestones on readiness and ability to be regularly active. 4. Effectiveness of existing theoretical models of behaviour change for explaining and increasing physical activity. 5. Other risk reduction areas and physical activity interventions. 6. Evaluation of methods in physical activity assessments and interventions designed to increase adoption, maintenance, and relapse prevention of physical activity in a variety of settings. 7. Physical environment, social, and biological factors that influence the adoption and maintenance of physical activity and their relation to other risk factors.

Table 4. (*Continued*)

Conference\report	Purpose	Main recommendations
The 1992 International Conference on Physical Activity, Fitness & Health (Bouchard et al., 1993) *An international consensus symposium*	To focus on the relations among physical activity, fitness, and health	**Presents an exhaustive listing of 355 research questions and topics, for example:** (1) Can inevitable biological aging be differentiated from secondary aging processes? (2) What are the most appropriate training prescriptions for the various modes of exercise in order to promote health, fitness, and functional capacities in different populations of men and women? (3) Which healthful dietary habits can be encouraged to improve exercise tolerance in both the general population and athletes? (4) Further research is warranted to elucidate the clinical significance of transient exercise-induced changes in immune status and function, and to determine which variables best predict potential changes in host protection. (5) Is there a significant correlation between fitness gains with prolonged aerobic training and changes in anxiety? (6) What is the optimal multidisciplinary program to promote the long-term control of juvenile diabetes? **Presents a model describing the relations among habitual physical activity, health-related fitness, and health status.** Lifestyle behaviours, physical and social environmental conditions, personal attributes, and genetic characteristics all affect this model. Their interrelationships were explored and defined.
Workshop on Physical Activity and Public Health (Public Health Service, 1993) *Sponsored by CDC and ACSM in cooperation with the President's Council on Physical Fitness and Sports*	To emphasize the important health benefits of moderate physical activity	**Recommendations:** 1. Every American adult should accumulate 30 min or more of moderate intensity physical activity on most days. 2. Because most adult Americans do not presently meet the standard in #1 above, almost all should strive to increase their participation in moderate and/or vigorous physical activity. 3. Regular joint flexibility and muscular strengthening exercises are strongly encouraged. 4. Local, state, and federal public health agencies, recreation boards, professional organizations, school groups, and fitness and sports organizations should work together to provide critical fitness information and promote programs to help Americans become more physically active.

ACKNOWLEDGMENTS

I thank the Research Consortium for the honour of being selected as the 1993 C.H. McCloy lecturer. I appreciate the assistance of several of my colleagues who helped me with the lecture and this manuscript: Dr. Harold W. Kohl for his extensive contributions to our research program and for comments on an earlier draft of this report; Carolyn E. Barlow for managing the mortality surveillance system and for data analysis; Laura Becker and Melba Morrow for preparing slides and figures, proofreading, and manuscript preparation; Joni Bokovoy for assisting with the table development and comments on the manuscript; and Dr. Kenneth H. Cooper for establishing the Aerobics Center Longitudinal Study. This research was supported in part by U.S. Public Health Service research grant AG06945 from the National Institute on Aging, Bethesda, MD.

This article is reprinted with permission from the *Research Quarterly for Exercise and Sport*, vol. 64, no. 4 (December, 1993). The *Research Quarterly for Exercise and Sport* is a publication of the American Alliance for Health, Physical Education, Recreation and Dance, 1900 Association Drive, Reston, VA 22091.

REFERENCES

Bijnen, F.C., Mosterd, W. L., & Caspersen, C. J. (1992). Physical inctivity: A risk factor for coronary heart disease. A position statement for the World Health Organization, governments, heart foundations, societies of cardiology and other health professionals (pp. 2–6). Geneva, Switzerland: International Society for the World Health Organization.

Blackburn, H., & Jacobs, D.R. (1993). Physical activity and the risk of coronary heart disease. In J.P. Boustet (Ed.), *Proceedings of the Vth World Congress on Cardiac Rehabilitation* (pp. 403–418). Hampshire, UK: Intercept, Ltd.

Blair, S.N. (in press). Physical activity, fitness, and coronary heart disease. In C. Bouchard, R.J. Shephard, & T. Stephens (Eds.), *Physical activity, fitness, and health.* Champaign, IL: Human Kinetics.

Blair, S.N., Barlow, C.E., & Kohl, H.W., III. (1993). Physical activity, physical fitness, and risk of weight gain in women and men during a 6.5-year follow-up. Manuscript submitted for publication.

Blair, S.N., & Kohl, H.W. (1988). Physical activity or physical fitness: Which is more important for health? *Medicine and Science in Sports and Exercise, 20* (Supplement), S8.

Blair, S.N., Kohl, H.W., III, Barlow, C.E., & Gibbons, L.W. (1991). Physical fitness and all-cause mortality in hypertensive men. *Annals of Medicine, 23,* 307–312.

Blair, S.N., Kohl, H.W., Gordon, N.F., & Paffenbarger, R.S. (1992). How much physical activity is good for health? *Annual Review of Public Health, 13,* 99–12.

Blair, S.N., Kohl, H.W., III, Paffenbarger, R.S., Jr., Clark, D.G., Cooper, K.H., & Gibbons, L.W. (1989). Physical fitness and all-cause mortality: A prospective study of healthy men and women. *The Journal of the American Medical Association, 262,* 2395–2401.

Blair, S. N., Powell, K. E., Bazarre, T. L., Early, J. L., Epstein, L. H., Green, L. W., Harris, S. S., Haskell, W. L., King, A. C., Koplan, J., Marcus, B., Paffenbarger, Jr., R. S., & Yeager, K. K. Physical inactivity. *Circulation, 88,* 1402–1405.

Bouchard, C., Shephard, R.J., & Stephens, T. (Eds.). (in press). *Physical activity, fitness, and health.* Champaign, IL: Human Kinetics.

Bouchard, C., Shephard, R.J., Stephens, T., Sutton, J., & McPherson, B. (Eds.). (1990). *Exercise, fitness, and health: A consensus of current knowledge.* Champaign, IL: Human Kinetics.

Ekelund, L.-G., Haskell, W.L., Johnson, J.L., Whaley, F.S., Criqui, M.H., & Sheps, D.S. (1988). Physical fitness as a predictor of cardiovascular mortality in asymptomatic North American men. The Lipid Research Clinics Mortality Follow-up Study. *The New England Journal of Medicine, 319,* 1379–1384.

Fletcher, G.F., Blair, S.N., Blumenthal, J., Caspersen, C., Chaitman, B., Epstein, S., Falls, H., Sivarajan Froelicher, E.S., Froelicher, V.F., & Pina, I.L. (1992). Statement on exercise: Benefits and recommendations for physical activity programs for all Americans. *Circulation, 86,* 340–344.

Hahn, R.A., Teutsch, S.M., Rothenberg, R.B., & Marks, J.S. (1990). Excess deaths from nine chronic diseases in the United States, 1986. *Journal of the American Medical Association, 264*, 2654–2659.

Haskell, W. L., Leon, A. S., Caspersen, C. J., Froehlicher, V. H., Hagberg, J. M., Harlan, W., Holloszy, J. O., Regensteiner, J. G., Thompson, P. D., Washburn, R. A., & Wilson, P. W. (1992). Cardiovascular benefits and assessment of physical activity and physical fitness in adults. *Medicine and Science in Sports and Exercise, 24* (Suppl. 6), S221-S236.

Hein, H.O., Suadicani, P., & Gyntelberg, F. (1992). Physical fitness or physical activity as a predictor of ischaemic heart disease? A 17-year follow-up in the Copenhagen Male Study. *Journal of Internal Medicine, 232*, 471–479.

King, A.C., Blair, S. N., Bild, D. E., Dishman, R. K., Dubbert, P. M., Marcus, B. H., Oldridge, N. B., Paffenbarger, Jr., R. S., Powell, K. E., & Yeager, K. K. (1992). Determinants of physical activity and interventions in adults. *Medicine and Science in Sports and Exercise, 24*(Suppl. 6), S237-S248.

Lee, I-M., & Paffenbarger, R.S., Jr. (1992). Change in body weight and longevity. *The Journal of the American Medical Association, 268*, 2045–2049.

Montoye, H.J. (1992). The Raymond Pearl Memorial Lecture, 1991: Health, exercise, and athletics: A millennium of observation - a century of research. *American Journal of Human Biology, 4*, 69–82.

Morris, J.N., Clayton, D.G., Everitt, M.G., Semmence, A.M., & Burgess, E.H. (1990). Exercise in leisure-time: Coronary attack and death rates. *British Heart Journal, 63*, 325–334.

Paffenbarger, R.S., Jr., Hyde, R.T., Wing, A.L., & Hsieh, C-c. (1986). Physical activity, all-cause mortality, and longevity of college alumni. *The New England Journal of Medicine, 314*, 605–613.

Paffenbarger, R.S., Jr., Hyde, R.T., Wing, A.L., Lee, I-M., Jung, D.L., & Kampert, J.B. (1993). The association of changes in physical-activity level and other lifestyle characteristics with mortality among men. *The New England Journal of Medicine, 328*, 538–545.

Park, R.J. (1990). 1989 C.H. McCloy Research Lecture: Health, exercise, and the biomedical impulse, 1870–1914. *Research Quarterly for Exercise and Sport, 61*, 126–140.

Powell, K.E., & Blair, S.N. (in press). The public health burdens of sedentary living habits: theoretical but realistic estimates. *Medicine and Science in Sports and Exercise.*

Powell, KE., Thompson, P.D., Caspersen, C.J., & Kendrick, J.S. (1987). Physical activity and the incidence of coronary heart disease. *Annual Review of Public Health, 8*, 253–287.

Shaper, A.G., & Wannamethee, G. (1991). Physical activity and ischaemic heart disease in middle-aged British men. *British Heart Journal, 66*, 384–394.

Sobolski, J., Kornitzer, M., De Backer, G., Dramaix, M., Abramowicz, M., Degre, S., & Denolin, H. (1987). Protection against ischemic heart disease in the Belgian Physical Fitness Study: Physical fitness rather than physical activity? *American Journal of Epidemiology, 125*, 601–610.

U. S. Centers for Disease Control and Prevention and American College of Sports Medicine. (1993). Workshop on physical activity and public health: Summary statement. Atlanta, Georgia.

U.S. Department of Health and Human Services, Public Health Service. (1991). Healthy People 2000: National health promotion and disease prevention objectives (DHHS publication no. 91–50212). Washington DC: Author.

Wannamethee, G., & Shaper, A.G. (1992). Physical activity and stroke in British middle aged men. *British Medical Journal, 304*, 597–601.

Williamson, D.F., Madans, J., Anda, R.F., Kleinman, J.C., Kahn, H.S., & Byers, T. (1993). Recreational physical activity and ten-year weight change in a US national cohort. *International Journal of Obesity, 17*, 279–286.

35

THE VALUE OF PHYSICAL CULTURE IN A HEALTHY LIFE-STYLE

L. I. Lubysheva

Russian State Academy of Physical Culture
Sirenevyi St.4, Moscow, Russia

In the modern age of social transformation and scientific and technologic progress, the development of the different aspects of human life is changing ever more rapidly. This process had affected the area of humanity's culture, including also its physical culture. The new stage of development physical culture was formed under the influence of society's practical needs and demands to raise a whole generation's physical preparedness to carry out manual labour as the first and the most important condition of human existance. At the same time, as systems of education evolved, physical culture was included in them because of its role in the formation of motor skills and abilities. These last characteristics were considered necessary for realisation of the individual's capacity for physical activity. But its cultural role as an efficacious means of intellectual, moral and aesthetic education remained not more than a paper declaration. These tendencies were the root of many of the flaws in the system of human physical culture. All the state and social interests in physical activity were gradually removed into the area of preparation of elite sports performers. Progressively greater financial support attracted the best specialists who were directed into this area. This policy was a strong impediment to the development of physical education, sports for children and youth, and sports for all. That is the basis of the social need for an all-embracing use of physical culture's opportunities to develop a new healthy life style, as well as for a new approach among the young generation in relation to the value of physical culture in he development of vitality and creative activity.

According to our classification, the potential value of physical culture can be divided conventionally into its components which are represented by intellectual, motor, technological, mobilizational and intentional components. The content of the intellectual valuables constitutes the knowledge about methods and means of human physical potential development, about the regularities of its natural and stimulated evolution, rules of physical training, fundamentals of biomechanics, morphology and physiology of motorsystems, history of sports and philosophy of physical activity and healthy life style. The valuables of motor character are represented by the best patterns of motor activity which are achieved as a result of physical education and sporting preparedness; personal achievements in man's physical education and sporting activity, and his or her real physi-

Current Research in Sports Sciences, edited by Rogozkin and Maughan.
Plenum Press, New York, 1996

243

cal potential. The technological valuables of physical culture are formed by skills of using methods and means to develop the individual's physical potential and to assimilate other valuables of physical culture. The physical culture value of intentional character is of a great social importance for society and personality. Unfortunately, they are not well understood at the present time. These values reflect the degree of public mental maturity, the prestige accorded to physical culture in a given society, its popularity in different groups of the population, and, most importantly, the intention and readiness of people to work towards a permanent development and improvement of their physical potential. When analyzing the effectiveness of the assimilation of these valuables, we can evaluate in real terms the needs of society and every individual in respect of physical culture and assess the true state of affairs in this important area of culture.

We relate the social-psychological purposes of people to the same group of values. The last of these are determined by the character, structure and direction of needs, motivation and value orientations to physical activity and to participation in sports. The mobilisational values of physical culture relate to the particular importance of preparedness for vigorous activity among the socially active young generation. The capacity for strict organization of one's time, internal discipline and accuracy, the speed of situational evaluations and of decision taking, the persistence while reaching for an objective, the ability to take victory or failure easily, and finally simply to be in control of of a difficult situation. In life it is often necessary to possess an optimum balance of physical and spiritual manifestations, but also to possess the ability to function at close to the maximum level when confronted with any one of a number of challenges (disease, injury, accidental situation, natural calamity, socially determined sudden and long-term loading of psychological, mental and physical character, etc). The probability of the occurence of such critical situations is not so little that it can be ignored. That is why the organism's ability to mobilise its resources in similar conditions must also be formed by means of training as well as other physical, psychological and functional means. There is no doubt of the fact that physical and sporting activity facilitates such development because of its mediative influence on the creation of the reserves of functional and psychological capacity, and due to physical and mental rehearsal of future responses to unexpected external influences, demanding rapid mobilisation of the body's functional reserves.

We relate the formation of the new system of ideas about physical culture's values to the further humanisation and democratisation of society. We consider a human being as the measure of all thing she is or can be. All is good that is good for a person, and all is bad that is bad for him or her. Such an approach involves the refusal to accept a utilitarian approach to physical culture's values and a radical reconsideration of physical education's purposeswith regard to the direction of the development of its humanistic and cultural functions. Let us consider the main scientific approaches to the assimilation of physical culture's values. For the effective realization of the intellectual component of physical education ideas it is necessary to include in this process the potential of the physical cultural information that we possess at the present time using the existing informational structures (television, radio, papers, journals, books etc). At the same time we are absolutely convinced that radical transformations in an individual's participation in physical activity can take place only with the availability of information and education support regarding the adoption of a healthy life style. The system of information and education assistance must organise all the means of such influences on man and society, using professional pedagogical and psychological assistance (3,6,7).

The peculiarity of our approach to the motor values of physical culture lies in its use of motor activity and of the development of skills to apply it to the self-organisation of a healthy life style. The effectiveness of the assimilation of motor values is closely connected with the individual's choice of forms and means in the organistaion of his or her physical activity. Freedom of choice on the basis of maximum information about one's physical potential and taking into consideration the personal needs and wishes must be a principal determinant of the effectiveness of the assimilation of motor values (1, 4).

Regarding the problem of the assimilation of technological values, it is necessary to underline the opportunities for creative transfer of progressive methods and means, developed in the elite sports area, into the practice of physical education. Here we mean the specific transfer of elite sports technologies adapted to the aims and conditions of sport for all and physical education. The main point of this transfer of imformation is the assimilation of training philosophy in the interests of physical improvement (2). The methodical arsenal of physical education and sports for all must be essentially enriched by adoption the well established technologies of physical training developed for elite sport and the techniques of movement teaching usedd in the training of elite athletes. The essence of this borrowing does not, of course, presuppose the mechanical transfer of sport training loading volumes and intensity into the practice of physical education and sport for all. The main idea is to use training technologies and methods of movement improvement which were created by elite athletes and coaches after much trial and error, with many mistakes and sacrifices (2).

As to the perspectives of the assimilation of intentional values, it is necessary to note that they exert a strong influence upon the area of physical culture. The strength of these values determines the financial, economic and industrial activity of different state, social and commercial structures which can create the conditions for effective physical activity of the population for development of physical culture and sport. At the same time, society's efforts to ensure the existence of the appropriate economic, legal, ethical, moral, informational and other conditions for the development of physical culture can serve as a criterion of civilization quality of that society (8, 9).

The perspectives of the assimilation of mobilisation values can be widened by efforts to influence the mass media with regard to the establishment of appropriate mental orientations in society and in individuals with respect to health improvement. This aim can be achieved by means of healthy lifestyle propaganda, by the explanation of the protective effects of exercise against disease, by the creation of a suitable sports image among business people with regard to a particular style of health, strength, endurance and physical. Contemporary society has established concepts of the ideal personal abilities and individuality development priorities. In this connection the assimilation of the whole complex of physical culture's values becomes the strongest stimulus for the personality's self-realization. It is possible on the basis of new opportunities for self-organization through the attaintment of higher levels of functional, physical and psychological conditions as a priority objective of physical activity.

REFERENCES

1. Balsevich, V. K., Karpeev, A. G., Martin, E. E., 1981, Hereditary and environmental determination of biomechanical characteristics in human motion ontogenesis, VIII-th International Congress of Biomechanics, Nagoya :275.
2. Balsevich, V. K., 1994, Conversion of high technologies of training: Topical aspect of Adapting popular sport and physical education, International Journal of Theory and Practice of Physical Culture :18–20.
3. Biddle S. 1981. The "why" of health related fitness. Bulletin of physical education:17, 28–31

4. Bouchard, C., 1988, Gene-environment interacxtion in human activity in early and modern population, Champaigh: Human Kinetics Books :56–66.
5. Evans, J. & Clarke, G., 1988, Changing the face of physical education, Evans (Ed.) Teachers, Teaching and Control in Physical Education, Lewess: Falmer Press :125–143.
6. Feltz, D. L. & Landers, D. M., 1983, The effects of mental practice on motor skill learning and performance: a meta analysis, Journal of Sport Psychology 5:25–57.
7. Morgan, W. P., 1969, Physical fitness and emotional health: A review, American corrective Therapy Yournal :124–127.
8. Tucker, L. A., 1989, Physical attractiveness, somatotype, and the male personality: a dynamic interactionnal perspective, Journal of Clinical Psychology 40:1226–1234.

36

PHYSICAL ACTIVITY LEVELS OF SCHOOL-AGED CHILDREN IN ST. PETERSBURG

A. G. Komkov, G. L. Antipov, and E. G. Gurinovich

Research Institute of Physical Culture
Dynamo Ave. 2, 197042, St. Petersburg, Russia

INTRODUCTION

The problems of bringing up representatives of the new generation and helping them to become independent in their future life are the main ones in the process of social evolution. An important place is occupied by investigations of the process of health improvement for school-aged children (in its social and pedagogical aspects), analysis of the interrelations between physical fitness and physical activity level, and also the identification of the behaviour characteristics connected with life style. Physical activity is the most significant life style element (Vinogradov, 1990) and forms the natural basis for health reserves gain (Amosov & Bendet, 1994). Human physical activity consists of one's systematic and motive-oriented action directed to the achievement of physical perfection or improvement (Balsevich & Zaporoqanov, 1987).

The investigation of the influence of social factors upon physical activity levels among children (Wold, 1989) and analysis of pedagogical and psychological problems of their rationale for participation in physical activity (Baranovski & Bouchard, 1992; Pate & Dowda, 1993) is of great interest. Analysis of children's attitude to participation in physical education and their self-estimation of their physical status permits us to identify the most significant elements which characterize their physical activity level and the factors which contribute to it. The present investigation was concerned with the dependence of physical activity upon motive-oriented behaviour and analysis of problems concerning self-control, health promotion, development of physical fitness, and the perfecting of psychological activity.

METHODS

A sociological questionnaire was used as the basic investigation method. It was carried within the frames of participation in the International project on "Health Behaviour in

Current Research in Sports Sciences, edited by Rogozkin and Maughan.
Plenum Press, New York, 1996

247

Table 1. Demographic characteristics of school-aged children, who
participated in the questionnaire survey (n = 4059)

Age	Boys	Girls	Total
11	699	686	1385
13	633	686	1319
15	631	724	1355
Total:	1963	2096	4059

School-aged Children" (HBSC), fulfilled under the auspices of the World Health Organisation (WHO) and the WHO European regional bureau. The procedure and organization of international research in school-aged children is regulated by a special research protocol for the 1993–94 study (Health Behaviour in school-aged children, 1993).

The standard version of questionnaire consists of the following sections: the core questions (selected demographic and behavioural questions relevant to major health problems, psychosocial aspects of health and psychosomatic complaints), special foci (the health related aspects of school as a work environment, psychosocial health, injury-related behaviours, social inequality).

24 countries are involved in the investigation process. They each added to their questionnaire a set of questions connected with national needs, their own interests and the profile of the organization. The questionnaire that was used for investigation of physical activity of Saint-Petersburg school-aged children, consisted of 82 questions (164 items). There should be italicized a part relating to self-estimation of health status, physical fitness, psychological status, social behaviour, and achievements in studies.

School-aged children were asked about family structure, father's and mothers's occupation, smoking habits, use of alcohol, eating habits, quantity of peer relations, school-related support, and current level of physical activity. The part of the questionnaire concerning physical activity contained 13 questions (25 items) directed to an analysis of the types and kinds of physical activity, participation in sports activity, attitude to school lessons of physical education, independent physical exercises, and active recreation.

A pilot investigation was carried on in May 1993, in 3 schools of Saint-Petersburg. More than 4500 questionnaires were collected, and after cleaning the data base contained 4059 questionnaires.

For creating the data base an IBM PC, operation system MSDOS 3.5, Norton Commander and relational database Paradox release 3.5 (Borland International, Company: VEK) were used. Data base technology consists of coding answers, clearing and recording. When the table was complete (by using the Paradox menu "Report") has an ASCII file (approximately 1 MB) was created for the Research Center for Health Promotion (University of Bergen). The computing capacity of Paradox is highly limited, therefore

Table 2. Participation in sport activity (%)

	Boys			Girls		
Regularity of phys. exerc.	11	13	15	11	13	15
Never	54.5	55.8	59.4	66.0	71.9	75.0
Less than once a week	8.3	8.1	5.4	9.5	7.0	5.4
Every week	26.6	28.1	23.0	18.7	17.5	16.4
Every day	10.6	8.1	12.2	5.8	3.6	3.2

Table 3. Participation in play activity (%)

Regularity of phys. exerc.	Boys			Girls		
	11	13	15	11	13	15
Never	28.6	25.9	21.4	37.9	33.5	37.7
Less than once a week	26.0	24.6	23.6	26.1	34.3	32.5
Every week	29.5	35.2	38.5	25.9	25.1	23.8
Every day	15.9	14.2	16.5	10.1	7.1	6.1

our own program package was developed. It contains five programs: computing averages and standard deviations; construction of distributions; construction of joint distributions; computing correlation matrix; factor analysis.

RESULTS

Secondary schools aged 11–15 years took part in the investigation (Table 1).

Analysis of the data reveals the real level of physical activity of school-aged children, that includes a measure of the regularity of training sessions under the leadership of a coach, independent physical exercises and active recreation. The investigation showed a tendency towards a decreasing involvement of school-aged children in sports activity when they became older. 10% of boys and 4% of girls are involved in everyday sports activities (Table 2). At the age of 15, 59% of boys and 75% of girls do not go in for sports at all. It is obvious that girls are less active in sports than boys.

15% of boys and 7% of girls take part in some form of independent physical exercise (Table 3). There are no significant changes in play activity accompanied by aging. Among children aged 15 years, 38% of boys and 23% of girls independently take part in sports exercises every week.

12% of boys and 11% of girls carry on some form of out-door activities every day (Table 4). Involvement of children in regular out-door walking practically does not change with aging. 37% of boys and 41% of girls among those surveyed go in for recreational activities more seldom than once a week.

Physical fitness is estimated by school children to be relatively high (Table 5). More than half of school children are satisfied by their physical abilities; they become more critical with aging. Only 8% of boys and 33% of girls among 15 year olds consider their physical fitness to be excellent.

A similar pattern of results is observed in self-estimation of health status (Table 6). The boys consider themselves to be more healthy than girls do. At age 15, 28% of boys and 14% of girls believe that they have good health status.

Table 4. Participation in recreational activity (%)

Regularity of phys. exerc.	Boys			Girls		
	11	13	15	11	13	15
Never	23.0	26.1	24.6	25.7	21.4	24.0
Less than once a week	33.6	33.2	32.2	31.8	36.2	41.7
Every week	28.5	28.4	29.2	28.6	31.6	25.7
Every day	14.9	12.3	9.0	14.0	10.8	8.6

Table 5. Self-esteem of physical fitness (%)

Fitness	Boys			Girls		
	11	13	15	11	13	15
Excellent	16.3	10.7	8.7	8.5	7.0	3.0
Low	42.3	44.4	43.9	38.0	33.7	31.1
Middle	37.6	38.9	38.7	46.9	49.3	54.6
High	3.8	6.0	8.7	6.6	10.1	11.3

Table 6. Self-esteem of health state (%)

	Boys			Girls		
Health status	11	13	15	11	13	15
Excellent	38.1	34.3	28.5	22.9	20.0	14.4
Practically enough	50.1	54.5	59.0	50.3	58.3	59.6
Practically not enough	11.8	11.2	12.5	26.8	21.7	26.0

Table 7. Attitude to PT lessons at school (%)

	Boys			Girls		
Attitude to lessons	11	13	15	11	13	15
Excellent	43.4	20.9	18.4	27.6	13.7	11.3
Average	40.2	48.5	46.4	42.7	42.3	40.9
Indifferent	12.2	21.8	22.8	18.2	27.0	27.9
Negative	4.3	8.8	12.4	11.6	17.0	19.9

Table 8. Necessity of physical activity (%)

	Boys			Girls		
Necessity estimation	11	13	15	11	13	15
Certainly. yes	42.9	39.0	34.1	22.7	19.2	18.2
Likely, no	44.1	46.6	51.0	47.5	50.3	49.9
Likely, yes	10.3	11.4	13.3	22.7	26.4	28.0
Certainly, no	2.7	3.0	1.6	7.0	4.1	3.9

Table 9. Number of hours spent by school children in physical exercise (%)

	Boys			Girls		
Number of hours	11	13	15	11	13	15
4 h and more	13.5	17.3	26.8	7.0	9.2	10.9
2–3 h	41.9	44.5	41.2	35.7	37.6	33.7
1 h and less	30.5	25.6	20.3	34.1	28.0	24.9
Not a bit	14.1	12.6	11.7	23.2	25.2	30.5

Table 10. General level of school-aged children physical activity (%)

Number of hours	Boys			Girls		
	11	13	15	11	13	15
High	21.9	21.1	20.7	16.1	14.3	12.0
Middle	24.2	23.8	22.1	23.8	22.4	21.1
Low	53.9	55.7	57.2	60.1	63.3	66.6

We have also acquired data concerning children's opinion about their body image. Among the school children who completed the questionnaire, 19% of boys and 12% of girls consider their weight to be lower than the normal. More than one third of girls aged 15 consider themselves to be overweight. Only 9% of boys and 10% of girls express an opinion that their height is less than normal. At the age of 15, 41% of boys note that their height is more than average: probably their self-esteem is too high.

Although the results of school children's self-estimation of their physical status is highly subjective, these results reveal sex and age related peculiarities and also analyze their interrelation with the level of physical activity.

Ideas of children about their appearance and attractiveness of their body image change with aging. At the age of 15 only 7% of boys and 6% of girls consider that they look perfect. This demonstrates that boys and girls are equally exacting in their assessment of their appearance. The record of the level of physical activity permits an evaluation of children's attitude to school PT lessons (Table 7). 28% of boys and 19% of girls like PT lessons very much. Unfortunately at the age of 15 children's interest in school PT lessons decreases.

Attitude to future participation in sports activities serves as an important characteristic for evaluation of physical activity level in school-aged children. This changes with age and differs in boys and girls (Table 8). Among children of aged 15, 34% of boys and 18% of girls underlined that they would certainly participate in some form of sports activity in future.

A measure of the place of physical activity in the lives of these school children is indicated by the amount of time they spent in physical exercise per week (Table 9). 19% of boys and 9% of girls devote more than 4 hours to sports each week. At the age of 15, 31% of boys and 55% of girls lack the necessary physical loads. 23% of boys and 11% of girls spend 4 hours per week or more in physical exercise. Among the 15 year old girls 36% do not take part in any physical exercise.

- These results provide data about the general level of physical activity among Saint-Petersburg school-aged children (Table 10). It was found that this characteristic is higher for boys than for girls of all three age groups. At the age of 15, the number of children with a high level of physical activity is lower than that of younger ones (20% of boys and 12% of girls).

DISCUSSION

Notwithstanding the fact that research workers have clearly identified the major role played by physical activity in health promotion, the level of physical activity in different demographic groups of Russia is far from its optimal state at the present time.

School years form the most sensitive period for the development of a physically active life style which can be transferred later to adult life. That is why this research focuses on the investigation of determinants of school-aged children's physical activity.

Investigations in school-aged children connected with different aspects of their self-estimation of physical status, are widely used at schools and in research. At the same time, analysis of children's and young people's attitude to physical activity, physical development and health promotion is of evident interest.

The use of sociological methods in the research process appears to be one of the ways in which the effectiveness of contemporary physical education system for children can be assessed. The standard cycle of sociological investigations permits results obtained in this investigation to be compared with those that will be obtained in a follow-up study. This approach is most effective in international and inter-regional observations aimed at the identification of general and specific features of social processes and events.

The method that has been developed for the estimation of physical activity level in Saint-Petersburg school-aged children as an element of life style is regarded as a suitable instrument for diagnosis of problems in health promotion of children by means of physical activities; it also permits an investigation of the influence of social factors and economic conditions upon the identified process.

CONCLUSION

The results of this questionnaire survey represent generalized data about the involvement of Saint-Petersburg school children in regular physical activity. A steady tendency of decreasing levels of children's physical activity with aging is apparent. It is obvious that the physical activity involvement of boys is higher than that of girls at comparable ages.

The survey has produced qualitative and quantitive data on the characteristics of general physical activity level, and on children's opinions about their health, physical fitness and physical status. Information about the attitude of school-aged children to PT lessons at school and their wish to go in for sports in future was also collected.

The criteria which have been identified above allow us to estimate the correspondence of children's physical activity level with accepted norms and also their current physical condition. It is doubtless the case that programs of physical education must provide effective methods and means for the realization of the whole pedagogical process; this requires a unified approach to education, upbringing and progress.

In this case we talk about working out and using scientifically based methods of pedagogical diagnostics, which are required to optimize the process of education and to provide a correct estimation of results to the benefit of society as a whole. Investigation of the level of physical activity among school-aged children, and introducing new technologies for receiving and using complex scientific information requires the establishment of new regional programs in physical education for children. The standardizing of this method is connected with the use of pedagogical monitoring that includes continued use of sociological questionnaire surveys and objective estimation of physical status of school-aged children based on established testing procedures.

ACKNOWLEDGMENTS

This work was supported by the St.Petersburg town-hall's Committee of Education.

We thank T.Bannikova, M.Zinchenko, D.Ivanova, E.Kirillova, I.Sharobaiko, G.Varlachev for their participation in this study.

REFERENCES

Amosov, N. M., Bendet, L. A., 1994, Physical activity and heart, Kiev.

Balsevich, V. K., Zaporoqanov, V. A., 1987, Human physical activity, Kiev.

Baranovski, T., Bouchard, C., et al, 1992, Assesment, prevalence and cardiovascular benefits of physical activity and fitness in youth, Med. Sci. J. in Sport and Exerc. 24(6):237–247.

Health behaviour in school-aged children: a WHO cross National survey, 1993, Ed. Wold, B., Aaro, L., Smith, C., Bergen.

Pate, R., Dowda, M., 1993, Physical activity in South Caroline Youth, J. of South Caroline Med. Ass. 8:371–376.

Vinogradov, P. A., 1990, Physical activity and health life style, Moscow.

Wold, B., 1989, Physical activity and life style in socialisation perspective, Thesis, University of Bergen.

37

THE EFFECTS OF FITNESS AND QUALITY OF HEALTH ON RATING OF PERCEIVED EXERTION

O. M. Evdokimova

Research Institute of Physical Culture
Dynamo Ave. 2, 197042, St. Petersburg, Russia

INTRODUCTION

It has become obvious in recent decades that technical progress influences the increase of psychosocial stress factors and as a consequence results in a deterioration in the health state and the appearance of sub-clinical changes in a large part of the population (Aro, Hasan,1987; Vein et al., 1988; Bundzen, 1992). In relation to this, there arises a problem in making the correct physical exercise prescription even in healthy subjects, especially if they exercise without supervision and self-control remains the only method of training control (Bundzen, Dibner, 1994).

The rate of perceived exertion (RPE) is recognised as one of the important methods of self-control during physical training (Borg, 1982). However, there is a lack of data about the connection between the RPE measured during exercise and the psychophysical state. There is also a lack of knowledge as to how to use the RPE scale evaluations for psychological well-being in the physical training process.

The purpose of present study was therefore to investigate some peculiarities of the RPE in relation to the level of fitness and quality of health.

SUBJECTS AND METHODS

38 healthy untrained men and 25 healthy athletes between the ages of 18–20 were examined. All studies were carried out between the hours of 8–10 in the morning. A number of different methods of evaluation were used in present study. Firstly individual and personal characteristics were estimated according to the following questionnaires:

1. Strelau questionnaire, giving us the main properties of the central nervous system (Vyatkin, 1978);
2. Spielberger scale estimating levels of state anxiety and the trait anxiety (Hanin, 1976);

Current Research in Sports Sciences, edited by Rogozkin and Maughan.
Plenum Press, New York, 1996

3. Eysenck method measuring levels of neuroticism, extra- and introversion (Mar-ishuk et al, 1984).

After that subjects went through a cycle ergometer test with four progressively in-creasing steps and periods of rest. Each exercise period lasted for 4 minutes, and rest inter-vals for 3 minutes. The power output at each stage was 25, 35, 50 and 75% of the individual predicted maximal oxygen uptake (Pirogova et al, 1986).

Evaluation of the cardiovascular system responses was carried out by electrocardi-ography, impedance cardiography and by measuring blood pressure. Saliva cortisol level at each step was used as an indicator of the hypothalamic-pituitary system response to physical exercise and was measured by the radioimmunological method (Chard, 1978). The rate of perceived exertion was estimated by means of 15-point Borg scale (Borg, 1982). Some pre-clinical disturbances of neuro-psychic status were diagnosed with the help of an automated screening system ("OFIS"; Bundzen, 1992). Statistical calculations were accomplished by correlation and factor analysis methods.

RESULTS

It was established that the athletes had significantly lower estimations (approxi-mately 1 point of RPE scale) compared to the untrained subjects. The linear regression analysis established that RPE was related to exercise power output (Table 1).

In untrained subjects RPE was connected with: the neuroticism level at the 1-st and 2-nd stages of loading; with stroke index at the 3-rd stage, and with properties of the cen-tral nervous system at the 4-th stage. In athletes RPE was connected with the state anxiety level only at the 1-st stage, whereas RPE-stroke volume connections presented at all stages of exercise intensities.

Our two groups did not demonstrate significant differences in their individual and personal characteristics, but cardiovascular system responses to physical exercises were different in athletes and in untrained subjects as a result of the higher level of fitness in athletes. Thus, heart rate in athletes was significantly less both at rest and at all stages of

Table 1. RPE relationships in untrained subjects and in athletes

Stage of loading	Untrained subjects n=38	Athletes n=25
1. 20% VO2max	neuroticism r = 0.393 (p<0.01)	state anxiety r = 0.510 (p<0.01) stroke volume r = −0.445 (p<0.05)
2. 35% VO2max	neuroticism r = 0.356 (p<0.01)	stroke volume r = −0.575 (p<0.01)
3. 50% VO2max	neuroticism r = 0.306 (p<0.05) stroke index r = −0.321 (p<0.05)	stroke volume r = −0.458 (p<0.05)
4. 75% VO2max	strength of the nervous processes concerning excitation r = −0.316 (p<0.05) strength of the nervous processes concerning inhibition r = −0.323 (p<0.05) lability of the nervous processes r = −0.313 (p<0.05)	stroke volume r = −0.411 (p<0.05)

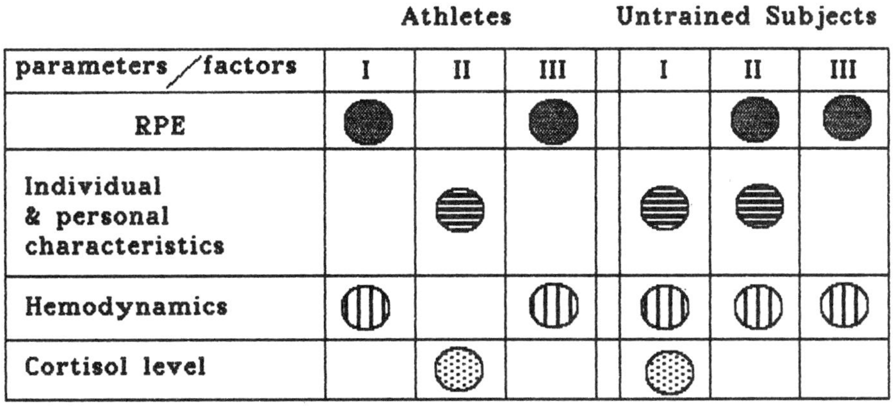

	Athletes			Untrained Subjects		
parameters / factors	I	II	III	I	II	III
RPE	●		●		●	●
Individual & personal characteristics		⊜		⊜	⊜	
Hemodynamics	⦀		⦀	⦀	⦀	⦀
Cortisol level		⊛		⊛		

Figure 1. Summarized data of the factor analysis. There were 3 main factors in the group of untrained subjects and 3 main factors in the group of athletes. The first factor in untrained subjects consisted of significant factor loads concerning individual and personal characteristics, hemodynamic parameters and saliva cortisol level. RPE was placed into the second factor, including individual and personal characteristics and hemodynamic parameters, as well as into the third factor, including only hemodynamic parameters. In athletes the first and third factors were similar and included RPE and hemodynamic parameters. The second factor consisted of factor loads of the cortisol level and individual-personal characteristics.

the cycle ergometer test. Stroke volume in untrained subjects increased initially and decreased from the second to the fourth stage. In athletes stroke volume increased during all stages of the ergometer test. Stroke index dynamics were characterized by similar changes in both groups of subjects.

Hypothalamic-pituitary system responses were also different in the two groups. In untrained subjects the saliva cortisol level gradually increased by the third stage from 5.78 ± 0.66 ng/ml to 7.60 ± 1.03 ng/ml ($p < 0.05$), whereas by the fourth stage it had already decreased to 7.07 ± 0.88 ng/ml ($p > 0.05$). In athletes, who had significantly higher cortisol levels at rest, (8.65 ± 0.72 ng/ml; $p < 0.05$) this level was maintained during the first three stages of the test. At the fourth stage, there was a tendency for an increase. We did not find any correlation between RPE and cortisol level in either group.

Results of the factor analysis are presented in Figure 1.

Computer testing with the help of a system "OFIS" revealed pre-clinical disturbances of the neuro-psychic state in 22 subjects (35%) both in athletes and in untrained subjects. According to the answers to questions of the "OFIS" system all subjects with pre-clinical disturbances of the neuro-psychic state were divided into three groups (Figure 2), each of which had peculiarities of RPE.

RPE in the first group (anxious syndrome) is characterized by an overestimation on the first and second stages compared to healthy subjects. RPE in the second group (neuroasthenic syndrome) is characterized by an overestimation of RPE at all power outputs. RPE in the third group (hypochondriacal syndrome) is characterized by great instability at all loads.

DISCUSSION

As follows from linear regression analysis there are significant relationships between RPE and some individual and personal characteristics, mainly in untrained subjects,

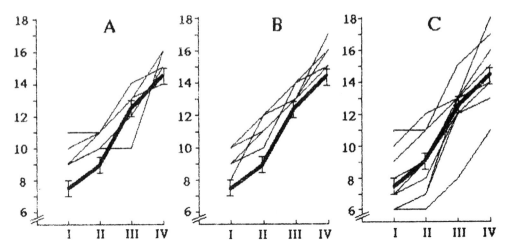

Figure 2. Rate of perceived exertion (RPE) in subjects with pre-clinical disturbances of neuropsychical status and in comparatively healthy subjects.

as well as between RPE and cardiovascular system responses to physical effort mainly in athletes. On the base of the correlation and factor analysis data we suggest that RPE in untrained subjects is connected with both individual - personal characteristics and character of the cardiovascular system responses to physical exercises, followed by the hypothalamic-pituitary system responses. The prevalent influence of these parameters depends on power output of the exercise load at which they are measured. RPE in athletes is mainly related to the hemodynamic responses, whereas the hypothalamic-pituitary system response to physical exercises is connected with the individual and personal characteristics. We consider that not only fitness influences RPE, but quality of health is of great importance in perception of physical effort. It turned out that one third of the subjects had pre-clinical disturbances of the neuro-psychic state. Our data showed that RPE in subjects with pre-clinical disturbances of the neuro-psychic state has different alterations in relation to specificity of the pre-clinical syndrome. The patterns which emerged gave us the possibility of suggesting that the intensity of physical exercises in subjects with pre-clinical disturbances of the neuro-psychic state should be decreased, except for subjects with the anxious syndrome when intensity of training programs may be maintained at a sub-maximal level. Our suggestion does not contradict the generally accepted point of view (Berger et al, 1993). We suppose also that subjective intensity of exercise recommended for these persons should correspond to 12–13 points on the Borg scale. This pleasurable exercise will result in decreased anxiety and tension, improvement in the stress coping process and will promote the positive motivation to physical training (Moses et al, 1989; Long, 1993; Steptoe et al, 1993; Kulikov et al, 1994).

There is convincing evidence from a review of the literature that both aerobic and anaerobic exercises promote psychological benefit (Berger et al, 1993; Weyerer, Kupfer, 1994). It allows us to suggest a combination of aerobic training with the special exercises in relation to the specificity of the pre-clinical syndrome. In the event of anxious syndromes, it is very important to use a prolonged warm-up, aerobic gymnastics, and muscular relaxation. Breathing gymnastics and coordination exercises should be used in the event of a neuroasthenic syndrome along with endurance training. We recommend aerobic conditioning in combination with sport games, strength training and balance exercises in the event of hypochondriacal syndrome.

CONCLUSION

The stable relationship between RPE and hemodynamic parameters occurs as a result of systematic aerobic sport training. It is necessary to emphasize that long-term aerobic exercises lead to a weakening of the relationship between RPE and individual-personal characteristics. We suggest that even minimal worsening of health state, i.e. decrease of the quality of health (such as pre-clinical disturbances of the neuro-psychic state) is accompanied by alterations of RPE during exercise. Exercise prescriptions for these persons should be corrected in relation to the specificity of the pre-clinical syndrome, with the consideration of the individual RPE. Self-control in these cases must be supplemented by objective methods (pulse-control).

REFERENCES

Aro, S., Hasan, J., 1987, Occupational class, psychsocial stress and morbidity. Ann. Clin. Res. 19(2):62–68.

Berger, B. G., Owen, D. R., Man, F., 1993, A brief review of literature and examination of acute mood benefits of exercise in Chechoslovakian and United States swimmers. Int. J. Sport Psychol. 24:130–150.

Borg, G., 1982, Psychophysical bases of perceived exertion, Medicine and Science in Sports and Exercise 5(14):377–381.

Bundzen, P., 1992, Theoretical foundations for the elaboration of diagnostic consulting system "OFIS" and its use in health promotion by means of mass physical culture, Sports and Health., St-Petersburg :5–13.

Bundzen, P., Dibner, R., 1994, Health and sport for mass: Problems and ways of their decision, Theory and Practice of Physical Culture 5(6):6–12.

Chard, T., 1978, An introduction to radioimmuno-assay and related techniques, Elsevier/North-Holland Biomedical Press.

Hanin, Ju. L., 1976, Short Guidelines for application of Spielberger scale, Leningrad.

Kulikov, V. P., Eremushkin, G. G., Aksenov, A. V., Melnikov, S. A., Osypova, I. V., 1994, Effectivness of physical training in free-choice regime in healthy subjects and in patients with myocardial infarction, Cardiology 34:29–31.

Long, B. C., 1993, Aerobic conditioning (Jogging) and stress inoculation interventions, An exploratory study of coping, Int. J. Sport Psychol. 24:94–109.

Marishuk, V. L., Bludov, U. M., Plahtienko, V. A., Serova, L. K., 1984, Methods of psychodiagnostics in sport. Moscow, Education.

Moses, J., Steptoe, A., Mathews, A., Edwards, S., 1989, The effects of exercise training on mental well-being in the normal population: a controlled trial. J. of Psychosomatic Research 33:47–61.

Pirogova, E. A., Ivashenko, L. Ya., Srapko, N. P., 1986, Influence of physical exercises on capacity work and health in human. Kiev, Health.

Steptole, A., Moses, J., Edwards, S., Mathews, A., 1993, Exercise and responsivity to mental stress: discrepancies between the subjective and physiological effects of aerobic training. Int. J. Sport Psychol. 24:110–129.

Vein, A. M., Dyukova, G. M., Stupa, M. V., 1988, Psychosocial factors and illness, Soviet Medicine 3:46–51.

Vyatkin, B. A., 1978, The temperament role in sport activity, Moscow, Physical Culture and Sport.

Weyerer, S., Kupfer, B., 1994, Physical exercise and psychological health, Sports Med. 2(17):108–116.

38

INFLUENCE OF ECOLOGICAL CONDITIONS ON THE PHYSICAL ACTIVITY AND PHYSICAL STATUS OF SCHOOLCHILDREN

K. F. Kozlova

Teachers Training Institute
286018 Vinnitsa, Ukraine

INTRODUCTION

The modern understanding of physical health is based on the complex biosocial characteristics of the individual, which ensure his active and effective functioning in the social environment. The most socially important goal of the process of physical health is not only high absolute indices of all physiological and psychological systems, but also a high level of adaptability of these systems (Malina, 1989; Moravec,1989). Numerous investigations have produced results which chow quite convincingly that the broader the range of adaptive abilities of the individual, the more quickly and more completely he reacts in conditions of changing environment, the more effectively he copes with physical demands, and the better he is able to overcome a range of different illnesses (Masironi and Denolin, 1985; Zwiren, 1988). Among the many questions which are of current scientific interest, few demand such constant attention from the public and arouse so many arguments as the question of the effect of radiation upon man and his environment. As a result of the accident at the Chernobyl electric power station a great number of radionuclides were released into the atmosphere. The accumulation of such radionuclides in the human organism can induce the development of a range of disturbances in the function of the different organs and body systems. Children most of all are susceptible to the effects of these radionuclides. However, the biological influence of small doses of radiation on children, in particular on their physical development has not been comprehensively investigated. In order to study the possible influence of small radiation doses on the physical state of children, we have conducted some investigations in the Vinnitsa and Zhitomir regions, where the level of radioactive pollution reached 1.5–8.0 c/km^2.

METHODS

Children aged 7–17 years took part in this investigation. We studied the principal indices of physical and motor development of both boys and girls within this age range

Current Research in Sports Sciences, edited by Rogozkin and Maughan.
Plenum Press, New York, 1996

(Gettman,1988; Ball, 1992). Observations of age dynamics were carried on during the period 1990–1993. The findings regarding children living in zones with small doses of radiation were compared with analogous findings about children made in 1985 (before the Chernobyl accident).

RESULTS

The results characterising the physical development of children show that the there are some distinct differences in the indices measured on children in 1990 and in 1985 (Table 1).

The height of boys aged 7, 8, 11, 15, 16 and 17 years and of girls aged 7, 11, 12, 13, 14 and 15 years who are living in the zone of radioactive pollution exceeds the mean height index of schoolchildren in 1985 ($p<0.05$). Characterising the age dynamics of girls' growth we see that at the age of 9, 10, 11 and 17 the height index is lower than that in 1985. As for boys, lower height indices in comparison with 1985 are characteristic of the ages of 10, 13, 14, 15, 16 and 17. Comparing the age-specific height indices of children aged 7–17 it should be noted that it is generally agreed that exposure to small levels of ionising radiation for a short period has a stimulating effect on the child's organism. However, this tendency is of short duration (2–3 years), and after that begins an irreversible process of sharp reduction in growth and development of the child's organism. This is corroborated by our sequential observations on the same children over a period of 3 years (1990–1993). The absolute indices the speed of height accretion among the boys aged 10 was reduced from 5.7 cm to 3.5 cm, and among the girls from 4.4 cm to 3.2 cm. The annual accretion rates of body length by boys aged from 7 to 10 was reduced from 4.8% to 2.6%, and in girls from 3.7% to 2.4%. Measurements of the dynamics of body mass gain of schoolchildren aged from 7 to 17 years showed that boys of 7, 13, 15 years of age and girls of 7, 12, 13 and 14 years living in places which were exposed to small doses of radiation demonstrate no consistent differences in body mass compared with the average indices at this age in 1985. In the other age groups, schoolchildren of 1985 show an advantage over children who were living in conditions of heightened radiation. The average annual accretion rates of body mass for boys from 7 to 10 years in absolute indices were: aged from 7 to 8, 4.3 kg (20%), from 8 to 9, 2.8 kg (11%), from 9 to 10, 3.1 kg (11%). The cor-

Table 1. Height of the children living in a radiation-polluted zone (cm)

Age	Boys		Girls	
	1990	1985	1990	1985
7	123.0 ± 1.4	118.7 ± 0.5	125.1 ± 1.1	117.3 ± 0.6
8	127.1 ± 1.0	123.6 ± 0.5	127.8 ± 1.0	128.5 ± 0.5
9	133.1 ± 1.1	133.4 ± 0.5	132.3 ± 1.1	133.0 ± 0.5
10	132.7 ± 3.5	139.1 ± 0.5	135.5 ± 1.8	139.2 ± 0.5
11	144.1 ± 1.0	139.8 ± 0.5	141.3 ± 1.0	144.4 ± 0.5
12	149.1 ± 1.4	149.6 ± 0.5	150.9 ± 1.3	146.5 ± 0.6
13	152.8 ± 1.4	153.6 ± 0.6	157.4 ± 1.4	154.8 ± 0.5
14	159.5 ± 1.4	160.3 ± 0.6	161.4 ± 0.9	156.5 ± 0.6
15	163.5 ± 1.2	165.5 ± 0.5	163.1 ± 1.0	161.1 ± 0.5
16	170.2 ± 2.1	172.8 ± 0.6	163.2 ± 1.1	164.2 ± 0.6
17	173.0 ± 2.6	175.0 ± 0.6	162.5 ± 0.8	165.2 ± 0.6

responding figures for girls were: from 7 to 8, 5.3 kg (26%), from 8 to 9, 2.8 kg (11%), from 9 to 10, 3.0 kg (10.7%). These findings are evidence of retardation of very substantial reductions in the growth rate of pupils exposed to radiation.

The qualitative analysis of indices of physical development of children aged from 7 to 17 years, who lived in places exposed to the influence of small doses of radiation showed that in most cases these children have average or below the average level of physical development. As a result of these investigations, we have identified the indices which characterize the level of motor development of children living in conditions of heightened radiations in comparison to the same characteristics of the level of children's physical preparation which were measured before the Chernobyl accident. All physical examinations of children were made according to the tests which are commonly used in the practice of physical training. The unique environmental conditions imposed by the radiation demand new approaches to methods of physical training. The results obtained demonstrate differences in the level of development of the main physical characteristics of schoolchildren in comparison with measurements made in 1985. We have established that there are no consistent changes in the results for a 30 m running test. Boys and girls appear to respond differently. Boys living in the zone of heightened radiation have better results at the age of 7, 9 and 11 years; these distinctions are reliable ($p<0.05$). In the other age groups, performance in the 30 m run is lower than that of boys in 1985; again, these distinctions are statistically significant ($p<0.05$). Girls have higher mean indices only at the age of 7, 8, 12 and 13; statistically significant distinctions were established only at the age of 7 ($p<0.05$). In the other age groups, the indices of 1985 are higher than those measured in 1990. The qualitative characteristic of development of these schoolchildren's speed ability was compared with the population norms that have been established: children with the mean level of development of speed abilities accounted for 57–62% of the group, values higher than mean were recorded by 9–17%, and low levels by 5–9% of the children.

The dynamometry index was used for estimation of muscular strength: this reflects the level of overall development of strength. An increase in this index with age was observed. The absolute result for boys increased from 11.0 kg at the age 7 to 48.0 kg at 17; for girls the corresponding change was from 10.2 kg to 27.5 kg. According to the dynamometry index, the boys of 1990 have an advantage in terms of muscle strength over the indices of 1985 in the following age groups: 7, 8, 10 and 15 years and the girls at all ages from 12 to 17 years. It seems possible that increased usage of stimulators during the training may account for these differences, which are somewhat surprising in view of the pattern of change in body weight that was observed. 36.1–47.6% of these schoolchildren were close to the mean level of development of strength abilities, and high level were present in 25.60%; low levels were absent in this population group.

The dynamics of development of speed and strength abilities were determined by the length of a standing jump. The results of the investigations carried out showed that the level of development of jump abilities by boys and girls is rising; the results of boys are consistently higher than those of girls. Girls of 1990 have statistically significant distinctions at ages 7, 8, 9, 12, 13 and 15 and boys at ages 7, 8, 11, 12, 13 and 17. It should be noted that the average annual rates of increase of jump length are not equal in the different age groups. Compared with population norms, mean levels of development of speed and strength abilities were present in 34–50% of these children; lower than mean values were achieved by up to 30%; higher than mean by 10–15%, and low values by 8–12%. The same pattern of results was observed in other indices of motor development.

CONCLUSION

On the grounds of the present findings, the following conclusion may be made: the physical and motor development of children exposed to the influence of small doses of ionising radiation is affected in different ways, depending on age and other characteristics. The present data can serve as the starting-point for the establishment of physical standards for children living in areas of increased radiation levels.

REFERENCES

Ball, T. E, 1992, The relative contribution of strength and physical to running and jumping performance of boys 7–11, J.Sports.Med.Phys.Fit. 32:364–71.

Gettman, L. R., 1988, Fitness testinng, In: Resource manual for guidelines for Exercise testing and prescription, Blair, S. N., Pointer, P. (Eds) Philadelphia: Lea and Febiger: 161–170.

Malina, R. M., 1989, Growth and maturation: normal variation and effect of training, In: Perspectives in Exercise Science and Sports Medicine, v.2. Youth, Exercise and Sport, Gisolfi, C. V. and Lamb, D. R. (Eds). Indianapolis, Benchmark Press: 223–265.

Masironi, R., Denolin, H., 1985, Physical activity in disease prevention and treatment, Padova, Piccin Butterworths: 117–147.

Moravec, R., 1989, The influence of performed motor activities on physical development and motor efficiency of 7–18 years old, Teor. praxe tel.vych. 37: 596–606.

Zwiren, L. D., 1988, Exercise prescription for children, In: Resource manual for Guidelines for Exercise Testing and Prescription, Blair, S. N., Painter, P. (Eds) Philadelphiia, Lea and Febiger: 309–314.

39

A SCREENING METHOD FOR FITNESS ASSESSMENT

U. N. Utenko, D. N. Gavrilov, and D. A. Ivanova

Research Institute of Physical Culture
Dynamo Ave. 2, 197042, St. Petersburg, Russia

INTRODUCTION

In recent years the health and physical status of the Russian population, in particular that of those living in St. Petersburg, has deteriorated considerably under the influence of various social, economic and environmental factors. There is little hope for improvement in some of these harmful factors in the near future. Under the prevalent social and economic circumstances, the only commonly accessible means to improve and maintain the health of the general populations appears to be an increased level of fitness training (Bouchard et al, 1988; Gettman, 1989). To make health-related fitness training efficient, it is essential to have a screening method, which would classify those who practice or would like to practice fitness exercises according to their physical abilities, and provide them with comprehensive and suitable recommendations on the type, quantity and intensity of these exercises (ACSM Position Stand, 1988; Blair, 1994). To meet the special requirements of mass population screening, such methods must meet the following requirements:

- To conform to modern scientific knowledge in the field of health-related fitness
- To be absolutely safe and accessible to the vast majority of the subjects
- To be simple and not require any special skills or preliminary training on the part of the subjects
- To provide sufficient capacity for mass population use
- To be standardized and feasible

There are currently available various fitness screening methods, some of which are intended for special categories of subjects, for example for soldiers or students, and are not feasible for others. Some are too complicated, and demand too great an effort from the subjects, or require special equipment and facilities. Most of them do not fully meet the above mentioned requirements of mass screening. Therefore the object of our study was to develop such a method and to prove its efficiency for the improvement and maintenance

Current Research in Sports Sciences, edited by Rogozkin and Maughan.
Plenum Press, New York, 1996

265

of health-related fitness in adult men and women. The study was conducted over a period of 3 years (1991–93) and was aimed at the solution of two main problems:

- To compose a test battery which would allow the acquisition of all the necessary data on each subject's physical status and abilities with the smallest number of test and measuring and in the shortest time
- To provide the subjects with recommendations on the exercise practice, based on general rules and at the same time individualized accordingly to the subject's health and abilities (Pate, 1988; Blair et al, 1989).

METHODS

The subjects were 2592 persons of both sexes, aged from 18 to 54 years, without serious health disturbances and mostly of sedentary lifestyle. The health status was assessed with a modified QPRL questionnaire. The habitual level of physical activity was estimated with 5 questions concerning the amount of everyday movements (Shephard, 1993). After a series of experiments with successively reducing test-batteries, the following set of tests and measurements was established: hand-grip for muscular strength; 1-min sit-ups with crossed hands for muscular endurance; 3-min sit-test (Gavrilov, 1992) for aerobic endurance; sit-and-reach for flexibility; height, weight and vital capacity were measured for evaluation of body development. Three derivative indices were calculated: body mass index, vital index and strength index. On the base of these data, physical status level was assessed.

The mark of each fitness item was adjusted for the subject's sex and age. Recommendations were divided into three sets. The first one establishes the stage of conditioning in accordance with current physical status level: an initial stage for low level, an improvement stage for average level, and a maintenance stage for high level. Each stage implies a certain combination of intensity, duration and frequency of exercise sessions. The next set of recommendations concerns the methods of raising the level of particular fitness items: aerobic endurance, muscular strength and endurance, flexibility, body composition and

Table 1. Evaluation of physical status

Name	Sex	Age	Occupation
		↓	
		Questionnaire "HEALTH"	
		↓	
		Questionnaire "ACTIVITY"	
		↓	
		Evaluation of body development	
Body mass index		Vital index	Strength index
		↓	
		Fitness level evaluation	
Aerobic endurance		Muscular endurance	Flexibility
		↓	
		Physical status level evaluation	
Low		Average	High

Table 2. Individual recommendations for health-related fitness training

1. Common recommendations

Level of ph.status	Stages	Exercise intensity (% HRmax)	Exercise duration (min)	Frequency (times per week)
Low	Initial	50–60	20–30	4–5
Average	Improvement	60–70	30–45	3–4
High	Maintenance	70–80	45–60	2–3

2. Special recommendations for development of fitness factors
Aerobic endurance
Muscular strength & endurance
Flexibility
Body composition
Bearing correction

3. Supplementary recommendations for practice of various types of health-related fitness activity
Walking
Jogging
Bicycling
Swimming
Rowing
Stretching
Weight-lifting
Skiing cross country
Gymnastics
Seasoning

bearing correction. The third set of recommendations is intended to give advice for health-related practice of the most common types of exercise, such as walking, jogging, cycling, weight-lifting, stretching etc.

In order to reduce time consumption during testing and exercise prescription, an automatic system (COFR) was elaborated and produced. The system includes: computer, device for anthropometric measurment, chair for fitness testing (sit-test, sit-ups, sit-and-reach, heart rate gauge) and a platform with altered height for various methods of step-testing.

The CORF system was used for mass population studies and permitted a 4-fold reduction in time consumption and 2-fold reduction of necessary skilled staff. The COFR system was presented at two international exhibitions in St.Petersburg and in Moscow, and was patented as an invention (Utenko et al, 1994). At present the automatic system is installed in one of the city's fitness centers and accessible to everyone who wants to check themselves and to improve their fitness and health.

Table 3. General scheme of the automatic fitness screening system (COFR)

Questionnaire "HEALTH"	Identity data	Questionnaire "ACTIVITY"
Anthropometric device	Computer printer	Fitness-testing device

Evaluation of individual's physical status recommendations:
1. Common
2. Special
3. Supplementary

Table 4. Mean results of testing and measuring (M ± SE)

	Males (n=978)		Females (n=1614)	
Exponents	M	SE	M	SE
Body height (cm)	174.3	3.2	163.8	2.0
Body weight (kg)	72.6	3.1	64.4	1.9
Body mass index (BMI)	24.1	0.6	26.8	0.8
Vital capacity (ml)	3478.0	61.0	2665.0	55.1
Vital index (ml/kg)	47.9	2.7	41.3	2.6
Hand grip (kg)	34.6	6.1	24.8	5.4
Strength index (conv.units)	47.6	2.7	38.5	3.0
Aerobic endurance (points)	7.3	0.5	6.6	0.8
Muscular endurance (repeat)	18.8	4.0	15.6	4.0
Flexibility (cm)	5.4	1.2	8.2	0.8

RESULTS

The mean results show a rather average level of body development and fitness in men as well as in women (Table 4).

More detailed analysis of the data shown in Table 5 confirms that about half of males and females correspond to the average level of body development and fitness and about 20 to 30% are either above or below the average level.

All the participants were provided with individualized recommendations on the type, intensity, duration and frequency of exercise sessions as well as with questionnaire-diaries to mark the completion of the recommendations. At the last stage of the study, 262 volunteers passed through 12 weeks of conditioning training to verify the efficiency of their recommendations. At the end of the period, they were reexamined and the data were compared to the initial values (Table 6).

As can be seen in Table 6, the mean scores in all components was increased by 0.4 to 1.1 points. Overall, 87.2% of the participants improved all or some of fitness items and 61% raised their level of physical status at least at one degree. As for those who carried out from 50 to 75% of recommendations they raised their mean score by about 0.2 point. The subjects who fulfilled less then 50% of recommendations were not taken into account. The method described above, in its field version, was used successfully to evaluate and to

Table 5. General data on the subjects

		Level of body development			Level of physical fitness		
Category (N)	Mean age	Below average	Average	Above average	Below average	Average	Above average
Males (978)	31.5	307	412	259	163	452	363
Females (1614)	27.4	624	653	337	324	868	422
Total (2592)	29.4	931	1065	596	487	1320	785

Table 6. Comparison between scores of body development and physical fitness after 12 weeks of fitness training (in points)

Study's stages	Body development (n=262)			Physical fitness (n=262)		
	Body mass index	Vital index	Strength index	Aerobic endurance	Muscular endurance	Flexibility
Initial examination	3.6	3.4	3.2	2.8	3.2	3.3
Final examination	4.1	3.8	3.6	3.9	4.2	4.0
Difference	0.5	0.4	0.2	1.1	1.0	0.7

raise the fitness level of the volunteers engaged in the service of the Good Will Games - 94 in St.Petersburg.

ACKNOWLEDGMENTS

This work was patented as an invention by the Russian Patentional Committee under N2015682.Bull. N13.15.07.94.

REFERENCES

ACSM Position Stand, 1988, The Recommended Quantity of Exercise for Developing and Maintaining Cardiorespiratory and Muscular Fitness in Healthy Adults.

Blair, S. N. et al., 1989, Resource Manual for Guidelines for Exercise Testing and Prescription, ACSM, Philadelphia.

Blair, S. N., 1994, Physical Activity: an Important Contributor to Health and Function, Dullas.

Bouchard, C. et al., 1988, Assessment of Physical Activity, Fitness and Health in Exercise, Fitness and Health, A Consensus of Current Knowledge, Toronto.

Gavrilov, D. N. et al., 1992, Method of assessing of cardiorespiratory endurance: Description of Invention N 1729485 Bull N 16,30.04.92, Moscow.

Gettman, L. R., 1989, Fitness Testing in Resource Manual for Guidelines for Exercise Testing and Prescription, Philadelphia.

Pate, R. R., 1988, The Evolving Definition of Physical Fitness, GUEET.

Shephard, R. J., 1993, Physical Activity and Fitness Quote, Toronto.

Utenko, V. N. et al.,1994, Equipment for evaluation of Health-Related Fitness: Description of Invention N2015682, Bull.N13, 15.07.94, Moscow.

40

FACTORS OF UNFITNESS (FITLESS FACTORS)

L. A. Kalinkin, E. V. Kuzmichova, V. D. Kriazev, and A. L. Kalinkin

Russian Research Institute of Physical Education
Kazakova St.,18, 103064, Moscow, Russia

INTRODUCTION

There is general agreement about the factors involved in determining the physical state, or fitness, of the human organism. For the sportsman competing at high level it means mobilization of all the body's resources to achieve the extreme limit of possibilities. At the other extreme, two or three times per week of jogging provides enough fitness for the average man. It is necessary therefore to apply special consideration to different groups of professionals, eg surgeons, ballet dancers, musicians, etc. Even within each of these special groups, it is possible to distinguish different ways to achieve the desirable level of fitness. For example, track athletes must use training methods which are very different from those of the weightlifter. This is reflected in the values of VO2max (Maximal Oxygen Consumption) which are 60–80 ml/min/kg for the middle distance or long distance runners and 30–50 ml/min/kg for the weightlifters. A variety of clinical conditions, such as injuries, active infection, heart attack, brain haemorrhage and so on can acutely change the possibility to demonstrate fitness but in fact the fitness level is changed only insignificantly. Many cases are known where sportsmen have demonstrated results at the highest level after a brief medical assistance (a pain killer drug etc.). Many professionals participate successfully in competition in spite of fever or other illnesses.

However there are long term factors which finally change the fitness level of the organism and its readiness for performance of daily occupational tasks. That is because a man's ability to carry out his professional activity depends on an appropriate level of physical fitness, and for this reason any factors that result in a long term decrease in fitness decrease the individual's ability to continue work.

These long term negative action factors may be termed "unfitness" factors. In this category are included also some casual circumstances which do not permit the organism to achieve a championship level of fitness for a given specialization in spite of a good physical body condition.

It is convenient to introduce the following classification of unfitness factors:

- Environmental factors.
- Physiological and pathophysiological factors.
- Psychological unfitness factors.

Current Research in Sports Sciences, edited by Rogozkin and Maughan.
Plenum Press, New York, 1996

271

ENVIRONMENTAL FACTORS

Long term unfavourable environmental or ecological factors can produce a negative influence on physical evolution of organism and body fitness. This is true especially for children. Our research demonstrates that the physical conditions of children 13–14 years old directly depends on the region of city (Moscow) where they live. The standard level of training according to the official physical education programme took place in all these areas. In the regions of poor air quality, pollution by industry and heavy traffic (dust, high percentage CO_2, NO_2, formaldehyde, gasoline etc.) was common. The combined pollution concentration (on scale of Russian standards) could exceed the permitted medical norms by 1.5–10 times. Teenagers and juveniles in bad ecological regions had a level of physical fitness that was on average 10–13% lower compared with good regions. In more seriously affected regions such as Novomoscovsk with a high level of radio-active background (a legacy of the Chernobyl disaster) the reduction in fitness levels was much more obvious, especially for 7-year old children who were born in the year of the Chernobyl disaster. Our research was carried out according to the methods of the European Tests of Physical Fitness (Strasbourg 1993).It was consistently and reliably found that there was a reduction in speed and strength capacity for girls and in endurance fitness for boys (Kalinkin et al, 1993). Our data showed a close relationship between pollution levels and traffic density: pollution was highest in sports complexes (stadia, training grounds etc.) which were surrounded by multiple lanes of traffic. The total pollution index exceeded the permitted norms by 1.3–2.5 times: CO_2 concentration was 3–4 times higher than the permitted level. These examples demonstrate the need for careful architectural design and construction of specific locations for sport facilities in and around major cities, especially in those regions of ecological disasters and in the industrial centres of the Extreme North, Siberia and the Tropics (Kalinkin, 1986). A combination of ecological factors (as air pollution) which one cannot avoid in ordinary life, with endo-ecological ones (such as pesticides) can obviously result in a reduction in physical working capacity and in other signs of impaired health. Our research carried out at the Syktyvkar timber industrial complex demonstrated that a reduction in a variety of health indexes was correlated with serious impairment of neurohumoral regulation in workers at this site.

PHYSIOLOGICAL AND PATHOPHYSIOLOGICAL FACTORS

In this category, we can include:

- factors of gravitational origin (weightlessness etc.);
- reduced physical activity;
- syndrome of "night hypoxia" ;
- bacterial infection of the intestine;
- syndrome of chronic fatigue;
- syndrome of abolition of biostimulators and drugs;
- hyper- or hypo-thermia;
- physiological overloading;
- virus illness of unknown nature.

It is necessary to pay special attention to the syndrome of "night hypoxia" which has a direct influence on physical working capacity. Our research in this subject has shown that long term "night hypoxia" results in a reduction in physical activity which in turn results in a decrease in labour (working) capacity. A special treatment of snoring has been introduced which utilises an apparatus for breathing under increased barometric pressure: this treatment increased PWC 170 (Physical Working Capacity at a heart rate of 170 bpm). People with high blood pressure who also suffer from the syndrome of "night hypoxia" were also treated by means of a period of intensive physical training (one hour of bicycle exercise per day) and experienced good results: during a night when hypoxic symptoms were recorded, the blood pressure remained within the normal limits.

PSYCHOLOGICAL FACTORS

During an examination of the population living in the region affected by fallout from Chernobyl, a phenomenon was observed which has been referred to as "radiation kinesphobia"—a fear of moving through radioactive areas. This results firstly in an increasing but then less noticeable reduction of the capacity for work. After 5 years living in the radioactive area, the speed-endurance characteristics of children aged 5–7 years fell by 2–14% and muscular strength decreased by up to 20%. The explanation of this mental phenomenon lies in the fact that the human body does not possess any sensors for detecting the presence of ionising radiation. The same reaction of man is typical of that to any invisible and insensible variations in the environment. A wide range of different chronic neural and psychic overstrain phenomena are also examples of psychological unfitness factors.

The influence of unfitness factors can usually be compensated by a programme of increased physical training with simultaneous reductions in the factors which induce unfitness. However, this approach is not always possible. Sometimes any effect of treatment with exercise, limitation of psychological factors, or a change of nutritional state can be ineffective if exposure to the adverse ecological conditions continues. Similarly it is very difficult to improve the health of the individual who suffers from radiation kinesphobia, if you can not convince that individual to leave the affected region.

Nevertheless, some methods are under development which have the potential to solve the problem at least to some extent. In particular, a coach can alter training regimens and reduce the physical load in the badly affected areas in proportion to the level of pollution. In our experiments it was shown that a special programme of exercises for cultivation of speed and strength reduced the negative influence of air pollution on the health of children. An endurance training programme was supposed to be carried out in special halls and sport-health-games complexes with a new type of training equipment, including training equipment of psychophysical conjugation STAC (Kuzmichova, 1993).

REFERENCES

Kalinkin, L. A., 1986, Physiological basis of repose regulation, Proceedings of ...conf., Tashkent :229–263.
Kalinkin, L. A. et al., 1993, Ecological problems of physical education, Proceedings of RRIPE-60y (Russian Research Institute of Physical Education) anniversary, Moscow :243–257.
Kuzmichova, E. V., 1993, Methods of estimation of social norms for technical basis of Physical Education, Teoria i practica physicheskoi kultury 9:31–32.

41

PHYSICAL FITNESS IN POLISH CHILDREN AND ADOLESCENTS

R. Przeweda

Academy of Physical Education
Marymoncka St.34, 01813 Warszawa, Poland

INTRODUCTION

In the system of physical and health education in Poland we often evaluate the level of physical fitness of children and youth (Przeweda, Trzesnowski, 1983; Charzewski, Przeweda, 1992) We treat physical fitness as a positive gauge of health - in accordance with the definition of the World Health Organisation which says that health is not lack of disease, but the general condition of the body and the degree of man's adjustment to the biological and social conditions of his life. For more than 60 years population studies of physical fitness have been conducted in Poland, embracing large numbers of young people chosen at random from the whole population. A carefully chosen random sample guarantees that the material gathered this way is a reliable representation of all young people (Przeweda, 1992, 1993).

METHODS

In 1979 we selected by lots about 2,000 schools of various types in Poland in villages and in cities, and in each school we picked at random about 120 students aged 7 to 19. We tested these students, measuring the selected somatic characteristics and applying motor tests, among others the international Physical Fitness Test, which consists of: the 50 m dash, the standing long jump, the 600, 800 or 1000 m run, the hand grip, the flexed-arm hang, the 4 x 10m shuttle run, sit ups and the trunk bend. Ten years later, in 1989, we repeated the same tests in the same schools. In both cases we obtained sample material embracing more than 200,000 tested persons in each series. The material is especially valuable in that it covers the decade 1979–1989, when significant social, economic and also ecological changes were taking place in Poland. The purpose of the tests was, among other things, to look for answer to the questions:

Current Research in Sports Sciences, edited by Rogozkin and Maughan.
Plenum Press, New York, 1996

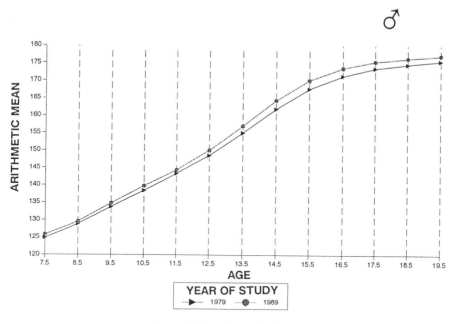

Figure 1. The body height of boys.

- Does acceleration and the secular trend still exist in our young people's growth observed in former years?
- Are there changes of young people's physical proficiency accompanying the changes in growth in subsequent young generations?
- How do the living conditions and growing conditions of the young generation influence growth and physical fitness?

RESULTS

In Poland, as in most countries, a secular trend in young people's growth is observed, which means that in successive generations children are taller than their parents. I have presented the last stage of the secular trend, observed in the decade 1979–1989, in graphs.

As can be seen in Figures 1 and 2, children and young people of 1989 are taller than their counterparts of 1979. These differences are statistically significant.

In the oldest age groups the differences in body height in favour of the year 1989 are on the average 1.8 cm for boys and 2.0 cm for girls. One can thus speak of a continuing occurrence of the secular trend in the direction of a greater height of the population in Poland. The body mass has changed in a similar way. These changes are distinct for boys, while for girls the increase of the average body weight has turned out to be small in the period 1979–1989. Because of this there appeared differences in the ponderal index, which is calculated by dividing the body height by the cube root of the body weight. A higher ponderal index signifies greater slenderness in body build.

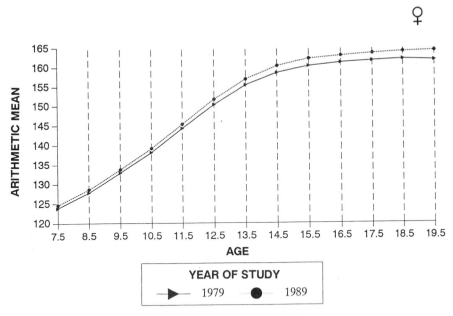

Figure 2. The body height of girls.

For boys (Fig.3) in the decade examined the ponderal index did not change much, while for girls (Fig.4) it is clearly higher.

Polish girls are becoming taller and more slender, which on the one hand is the result of population changes in body build, and on the other hand is an expression of the intention of today's girls to maintain a slim figure, in accordance with the world fashion. On the basis of the findings.of physical fitness tests, we looked for an answer to the question whether generation changes in somatic development entail similar changes in motor development. In my country the existing opinion is that along with the greater body height of young people their physical fitness is declining. The results of the 60 m dash for boys and girls have improved in subsequent studies over the span of 20 years. Similar changes are seen in tests of long jump for boys and girls and the results of other tests, with the exception of strength tests. If we compare the arithmetic average of the results of all children and youth from 1989 with the results of their counterparts from the same schools from 10 years earlier, we can see that today's youth has better results in the following tests: in the 40, 50, 60 and 100 m run; in the standing and running long jump; in the 1 kg and 2 kg pitched ball; in the front trunk bend; in the 4x10 m shuttle run; in the agility run between flags; in the number of sit-ups. All these differences have a high degree of statistical significance. In three tests today's youth is worse than 10 years ago, namely in the hand grip, the bar hang (girls) and pull-ups (boys) and in the softball throw. No statistically significant differences have been determined between the year 1988 and 1979 in the results of the 600 m, 800 m and 1000 m run. In general it can be said that Polish young people are taller than before and better in motor activities that require speed, agility and jumping ability, and worse in static strength rests. Showing the arithmetic average results for all young people does not render the true picture of physical fitness and generation changes in motor skills. The entire truth is revealed when we single out detailed elements from the

♂

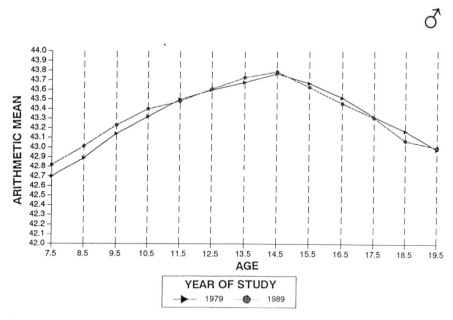

Figure 3. The ponderal index of boys.

♀

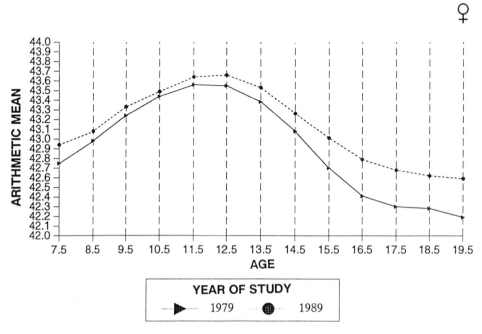

Figure 4. The ponderal index of girls in 1989 compared with 10 years earlier.

Table 1. The results of the 60 m dash for boys and girls

Age	1966 $\bar{x}$	1979 n	1979 $\bar{x}$	1989 n	1989 $\bar{x}$	Difference 1979-1989
Boys						
10.5	11.53	9769	11.09	8325	11.00	0.09
11.5	11.22	9964	10.77	9594	10.70	0.07
12.5	10.85	10457	10.43	10165	10.30	0.13
13.5	10.55	10402	10.08	9948	9.90	0.18
14.5	10.17	10464	9.68	10071	9.50	0.18
15.5	9.52	10665	9.26	10579	9.10	0.16
16.5	9.11	11073	8.96	10744	8.80	0.16
17.5	8.86	10025	8.77	9031	8.70	0.07
18.5	8.73	5696	8.64	5603	8.50	0.14
19.5	—	1826	8.55	1586	8.50	0.05
Girls						
10.5	12.18	9070	11.62	7719	11.40	0.22
11.5	11.80	9442	11.23	9160	11.00	0.23
12.5	11.49	9846	10.86	9777	10.70	0.16
13.5	11.34	10081	10.64	9771	10.40	0.24
14.5	11.06	10246	10.52	9646	10.30	0.22
15.5	10.92	10822	10.47	10448	10.30	0.17
16.5	10.77	11217	10.46	10993	10.30	0.16
17.5	10.70	9722	10.46	9287	10.30	0.16
18.5	10.68	5880	10.46	5079	10.30	0.16
19.5	—	1850	10.49	916	10.30	0.19

general picture. For example, the generation changes in growth and physical fitness of youth differ in accordance with the living conditions. We defined affiliation to a social group by the level of education of the father. As can be seen in Figure 5 for 8-year old boys (before puberty), 14-year old boys (at the start of puberty) and 17-year old boys (after puberty) the average body height is higher the higher educational status of the father. This interdependence has been consistently appearing both in the tests of 1979 and 1989. Also the level of general physical fitness, measured by the sum of 8 tests of the International Physical Fitness Test, points to a dependence both on the level of education of both parents (father and mother) and on the character of the place of residence. The higher the educational status of the parents, the higher the level of physical fitness of the children. This is illustrated in Figure 6, where the results of 8 proficiency tests have been standardized in accordance with the method for the arithmetic average 0 and dispersion 1. This means (in Figure 6 and further) that 0 level is equal to the arithmetic average, all that is above line 0 is higher than the arithmetic average, and below 0 indicates a result which is worse than arithmetic averages. As can be seen from Figure 6, the relation between the physical fitness level of children and the educational level of their parents is more distinct in villages than in cities. In the decade of 1979–1989, under discussion here, there was improvement of the general physical fitness of young people in the case of children whose parents had a higher level of education, i.e. belonged to the upper strata of Polish society. In families where the parents had lower education, we observed a worsening of fitness of children in the years 1979–1989. Similarly, children living in villages had lower general physical fitness between 1979 and 1989, although the arithmetic average of the entire population has indicated an improvement of the general level of physical fitness of Polish children and young people.

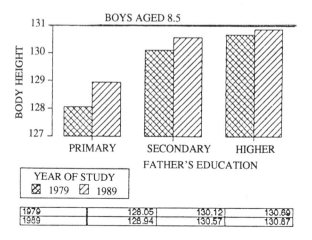

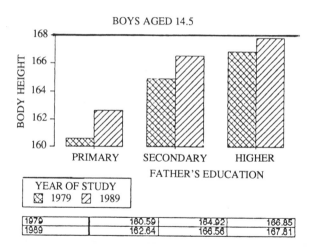

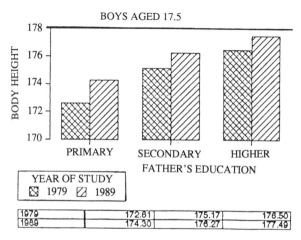

Figure 5. Body height for boys depending on education of the father.

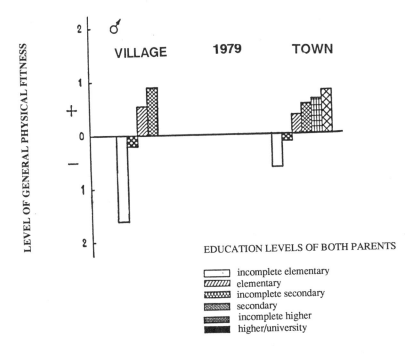

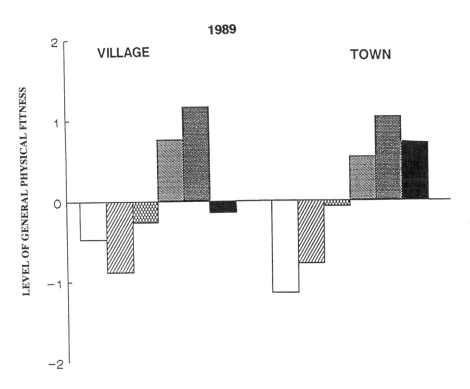

Figure 6. General physical fitness of boys depending on educational level of both parents.

CONCLUSION

The selected test results presented here give rise to numerous conclusions. The conclusions which point to the need of increasing concern in Physical education for children from lower social strata and children living in villages have been communicated to the Polish educational authorities, responsible for the education of young generations.

REFERENCES

Charzewski, J., Przeweda, R., 1992, Certain social determinants of growth and physical fitness of Polish children: Scientific Yearbook of AWF, Warsaw: 121–151.

Przeweda R., Trzensniowski, R., 1983, Changes in physical fitness of Polish school youth (aged 7–9 years), In.:Research in School Physical Education. (Risto Telama et al. Proceedings of the International Symposium/Jyvaskyla: 274–281.   .

Przeweda, R., 1992, Social stratification and physical fitness, Proceedings 6th ICHPER-Europe Congress "Physical Activity for Better Life Style", Prague: 375–378.

Przeweda, R., 1993, Is secular trend also present in the physical fitness of youth? Preceedings The 1993 FIEP World Congress "Physical Activity in the Lifecycle", Wingate Institute, Israel :182–190.

42

INDIVIDUAL APPROACH IN EXERCISE FOR HEALTH

V. D. Sonkin, V. V. Zaytseva, O. V. Tiunova, M. V. Burchik, and
D. A. Phylchenkov

Russian Research Institute of Physical Culture
Elizavetinsky, 10, 107005 Moscow, Russia

INTRODUCTION

Specialists in physical education admit the necessity for health-promoting exercise. There are many studies devoted to the scientific investigation of it's feasibility. Most workers are still influenced by mean-population standards, and training methods take account only of patients age and sex characteristics [1, 2, 3]. This situation is caused by the present undeveloped theoretical foundations of individual approaches in health-care exercise [4]. The optimal degree of detailed necessary and sufficient to develop effective training exercise programs taking account of patients' individual peculiarities is not yet known.

Studies by anthropologists of natural variability in human populations over the past 100 years revealed the main types of human morpho-functional constitution. Krechmer's and Sheldon's classic works proved the existence of a close correlation between psychic and physical organization in humans [5, 6]. There are many studies concerning the peculiarities of adaptational processes and the progress of disease in humans of different constitutional type [7, 8, 9]. We believe that an individual approach in physical education as well as in medicine and other spheres of human studies should take account of a wide complex of inherited properties which form a persons morpho-functional constitution.

Every person is unique but no one pedagog is capable of considering all the individual's peculiarities; some are inherited and unchangeable and some can result from the person's life experience. Usually these properties are studied jointly. This is misleading because a mixed picture prevents the observation of useful changes resulting from adaptational processes. Moreover, attempts to change inherited human properties by means of physical exercise are not only hopeless but may even be harmful to health due to the disturbance of natural balance of the adaptative systems. However, we should take account of hereditary peculiarities while developing methods of pedagogic influence aimed to increase the organism's adaptive reserves and therefore improve health. Consequently, the primary problem of an individual approach is revealing a set of "basic" personal proper-

Current Research in Sports Sciences, edited by Rogozkin and Maughan.
Plenum Press, New York, 1996

ties, i.e. identification of constitutional type, and only after this can we solve the second problem—to choose optimal (and probably different for every constitutional type) ways of physical conditioning and health improvement.

Hypothesis

Before beginning the study we supposed that properties characterizing psychophysical human conditions could be of two kinds:

- conservative, hardly changed, probably - inherited;
- labile, adaptive, which can be changed in wide range.

Properties of the first group can only be used for constitutional type identification. Besides, we supposed that the difference between persons of various constitutional types with respect to basic properties should lead to differences in motor abilities' structure and in the dynamics of adaptive reorganization under standard training influence. This should be clarified in order to develop adequate regimens of health-caring physical activity.

Aim of the Study

To reveal a group of "basic," conservative properties from the number of somatometry and motoric indices and to investigate the difference in structure of physical fitness and dynamics of the adaptational processes in humans of different constitutional type under influence of regulated physical training.

PURPOSES AND METHODS OF INVESTIGATION

Two experiments, differing in problems, methods and subjects were undertaken. The first, was devoted to the study of populational diversity in measured indices of organism's physical state. We used questionnaires to gather data. These are self-testing and were published (in 1991) by one popular Russian newspaper and we have 3888 responses from Russian central region inhabitants aged from 18 to 70. They included: (1) sex; (2) age; (3) body weight; (4) body height; (5, 6) systolic and diastolic blood pressure; (7, 8) heart rate at rest and during 3-d minute of recreation after standard physical exercise; (9) height of standing vertical jump; (10) simple motor reaction time; (11) spinal column flexibility (standard sitting test); (12) maximal number of dynamic exercises during a standard time period for muscles of upper and lower extremities and trunk.

Variability of the property value was determined by coefficient of variance (CV) equal to the ratio of the standard deviation (SD) to the mean of the group. Descriptive statistical analysis was made using STATGRAF software.

The second experiment was with the participation of juniors—students of 1–2-course of Moscow Institute of Telecommunication (260 subjects of 18–20 years) and 66 16–17-year old pupils of Moscow secondary schools. The investigation's aim was to evaluate differences in the dynamics of adaptational processes in representatives of different constitutional types as well as to reveal the most conservative and labile characteristics. Subjects were divided in three somatotype groups based on the expert evaluation of qualified anthropologists (members of Research Institute of Anthropology of Moscow State University). Each group was tested before and after the pedagogical experiment described below.

Testing included measurement of anthropometric and neurodynamic indices, heart rate at rest and during exercise and motor tests. Analysis included data only from those subjects who performed all the test complex.

    a. Anthropometry. To determine somatotype the following standard characteristics were measured in all subjects: body weight, body height, chest, waist and shoulder (exerted and relaxed), thigh and calf circumferences, subscapular, triceps and abdominal skinfold thickness.

    b. Neurodynamic characteristics were measured using a serial device PFK-01 with automated result registration. The subjects, who were isolated from disturbing noise and light, performed following tests:

- Simple senso-motoric reaction: the time period from light stimulus on terminal display until the moment of button pressing, as well as number of mistakes made was recorded.
- Complex senso-motoric reaction—(1) the time of recognition of colour (1 from 6) of light signal on terminal display (latent time period); (2) time between the beginning the action until it's end (finger touch off from a button and pressing another determined button); (3) total time of complex senso-motoric reaction; (4) number of mistakes.
- Tepping-test - total number of local finger movements during 1 minute and dynamics by 15-s intervals.

    c. Measuring of heart rate at rest and during exercise was made using a sport-tester POLAR-2000. Blood pressure (systolic and diastolic) was measured at rest using the Korotkov method.

Motor testing was performed in a sports hall or stadium (running 1000 m and 3000 m), in the morning, under the supervision of a physician. Only healthy subjects with no chronic diseases were allowed to participate. Motor testing was carried out during physical education lessons in the student's group. No more than 3 tests were made in one day, tests of a similar kind (for instance, running tests) were widely separated.

Statistical analysis was performed using calculation of mean and standard deviation as well as variance. Student's two-tailed t-test for paired samples was used to analyze training changes.

RESULTS AND DISCUSSION

First of all we tried to evaluate the divergence of typologic variants in human populations. That is why we made statistical analyses of mass testing results from Russia's central region population. The results are presented in Table 1.

All the data can be divided in two groups by the value of CV. Much less variable (cv = 0.033...–0.188 in women and cv = 0.035...–0.191 in men) are the characteristics placed in first seven lines of the table. These are body dimension characteristics and functional indices of the cardio-vascular system.

The second group of characteristics demonstrates much more variability (from 0.28 to 1.26 in women and from 0.24 to 2.62 in men). Among them there are ordinary characteristics of physical fitness: vertical jump height; simple motor reaction time; spine flexibility; upper extremity muscle strength; trunk muscle strength. These properties are known to be easily trained.

Table 1. Variability (CV) of physiometric and motor test results in men and women of different age

Variables	Women in various age ranges (n)					Men in various age ranges (n)				
	18–29 (237)	30–39 (437)	40–49 (537)	50–59 (477)	60–69 (294)	19–29 (247)	30–39 (541)	40–49 (411)	50–59 (253)	60–69 (454)
Body weight	0.159	0.188	0.175	0.155	0.169	0.139	0.147	0.160	0.154	0.143
Body height	0.033	0.032	0.035	0.035	0.034	0.037	0.038	0.038	0.036	0.035
Ectomorphy index	0.155	0.175	0.166	0.144	0.142	0.134	0.132	0.129	0.134	0.123
Syst. blood pressure	0.091	0.101	0.101	0.084	0.071	0.029	0.070	0.086	0.082	0.074
Diastolic blood pressure	0.122	0.122	0.122	0.112	0.103	0.106	0.084	0.104	0.094	0.087
Resting heart rate	0.120	0.125	0.118	0.119	0.121	0.124	0.127	0.136	0.121	0.129
Recovery heart rate	0.159	0.166	0.173	0.162	0.158	0.191	0.164	0.175	0.152	0.163
Vertical jump	0.322	0.268	0.307	0.351	0.349	0.299	0.270	0.290	0.295	0.387
Reaction time	0.513	0.488	0.492	0.516	0.539	0.473	0.477	0.463	0.488	0.542
Spine flexibility	1.104	1.086	1.137	1.125	1.264	1.327	1.930	1.957	1.914	2.618
Arm muscle strength	0.460	0.440	0.409	0.473	0.435	0.375	0.434	0.429	0.427	0.460
Trunk muscle strength	0.353	0.312	0.328	0.340	0.369	0.290	0.291	0.309	0.289	0.316

Next we analysed the dynamics of these physical state characteristics under the influence of physical training. Table 2 includes test results demonstrated by young men of mesomorph (muscular) somatotype before (in September) and after (in May) physical training during a school year. The training was directed predominantly at the development of total endurance and muscle strength. The set of tests included some neuro-motoric characteristics which were obtained using special psycho-physiologic devices (time of simple motor reaction, time of complex senso-motoric reaction).

As can be seen from table 2, some characteristics should be considered conservative: vertical jump height, speed of single movement, flexibility, blood pressure as well as most

Table 2. Dynamics of psycho-physical state characteristics under physical training in young men

Variable	Before training	After training	Shift %	Significance level, p
HRrest, 1/min	70 ± 3.6	64 ± 3.0	9.3	<0.01
BPsystolic, mm HG	119 ± 7.2	120 ± 6.5	1.0*	>0.05
BPdiastolic, mm Hg	68 ± 6.9	69 ± 6.7	1.4*	>0.05
100 m run, s	14.0 ± 1.2	13.6 ± 1.8	3.0*	>0.05
1000 m run, s	228 ± 42	190 ± 46	10.9	<0.05
3000 m run, s	861 ± 110	818 ± 94	6.0*	>0.01
Vertical jump, cm	38.6 ± 5.9	40.1 ± 4.2	5.2*	>0.05
Pushups during 30 s, times	31.8 ± 4.8	38.3 ± 5.1	20.4	<0.05
Legs up during 20 s, times	10.3 ± 1.9	14.0 ± 2.1	35.9	<0.05
Flexibility, cm	4.0 ± 1.9	4.2 ± 2.1	5.0*	>0.05
Motor reaction speed, cm	25.8 ± 2.6	23.8 ± 3.3	8.4*	>0.05
Simple motor reaction time, ms	322 ± 36.5	307 ± 34.6	4.8*	>0.05
Latent time (LT) of complex senso-motoric reaction, ms	3 21 ± 45.0	313 ± 43.1	2.8*	>0.05
Motoric time (MT) of complex senso-motoric reaction, ms	4 11 ± 20.4	371 ± 18.6	10.0	<0.01
Ratio MT/LT	1.28	1.18	8.0*	>0.05
Tepping-test, times	326 ± 29.5	336 ± 27.6	3.0*	>0.05

*Conservative characteristics; other can be considered as labile

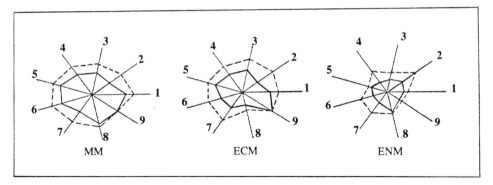

Figure 1. Structure of motor fitness in juniors of three somatotype groups before (solid line) and after (dotted line) a year of physical training : 1—speed of 100 m running; 2—speed of 1000 m running; 3—long jump; 4—pushups during 30 s; 5—pull-up number; 6—trunk strength; 7—number of sittings during 30 s (from lying position); 8—spinal column flexibility; 9—reaction speed.

neuro-motoric indices. At the same time, strength characteristics are labile - similar to the previous experiment result.

Therefore, not only physical development characteristics (i.e. somatotype) can be included in the organism conservative signs but also neurodynamic characteristics which determine the motor abilities of a person. At the same time many characteristics of motor fitness are predominantly labile. This permits us to use them for evaluation of pedagogic influence effectively.

Usually motor test results are improved by training, but their relationship, reflecting the structure of motor fitness, is conservative enough and can be considered as a type of constitutional peculiarities' manifestation.

Fig.1 contains sun-ray diagrams showing the structure of motor fitness (by results of standard test set) in juniors of three different somatotypes (ectomorph (ECM), mesomorph (MM) and endomorph (ENM)) before (A) and after (B) pedagogic experiment. During the experiment (of school-year duration) these three groups were trained by different programs aimed to compensate the lag in motor abilities inherent to their constitutional type. For instance, ECM show the least muscle strength so their experimental program consisted of up to 40% of their time being devoted to strength development; ENM have significantly low endurance so their program consisted of predominant cyclic loads; MM demonstrate in whole the best motor abilities compared with other types, so they developed all the motor components evenly.

Comparing these three diagrams we can see that the structure of motor fitness in three somatotypes is quite different. Probably, it is connected with nervous regulation peculiarities of muscle activity and fiber composition which is known to be scarcely changed during the adaptation process [10].

We believe it to be important that in spite of increased absolute motor fitness by training the individuals structure remains almost unchanged. Consequently, the difference in motoric characteristic development between constitutional types depends predominantly on hereditary characteristics while fitness level is secondary.

This fact is the basis of another useful effect. When humans of different constitutional types are trained in the same manner the results are quite different. This was proved by us in two experimental series (of a school-year duration) involving the participation of

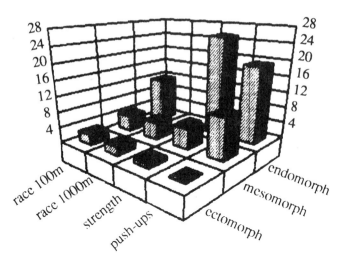

Figure 2. Bar diagram of test results (% increase) in young men of three somatotype groups (MM, ECM, and ENM) after similar physical training.

young men of 18–22 and 15–17 years of age. Some of the experimental results—bar plot—are shown in Fig. 2.

Column height shows % increase in motor test results in young men of three somatotype groups during a year of training on the same program. The increases in different tests are not similar and the greatest development can often be seen in physical ability in a given predominant constitutional type. For instance, endomorph persons are capable of much less endurance than other people, but the increase in this ability for them is the least. On the contrary, their muscle strength is the best developed, and it's increase in endomorphs is greatest.

CONCLUSION

The results of our study confirm our research hypothesis. As tables 1 and 2 show, some characteristics of physical development and neuromotorics are less variable in a population and as a rule are least changed in any physical training program in spite of it's emphasis. Evidently there are "basic" characteristics which form a common human constitutional type. Other characteristics reflect adaptive reorganization and can be changed according to physical training direction. At the same time the structure of motorics appeared to be strong enough.

We suppose that the difference between constitutional types consist not only in body proportion but in skeletal muscle composition, and consequently, in organization of energy and vegetative supplying of muscle activity. If this is true then we can understand easily why the same training influence leads to different results in humans of different constitutional type. But this fact has one more consequence: every type of morpho-functional constitution needs its own regimen of health-caring physical exercise adequate for his morpho-functional peculiarities. Therefore, the problem of individualization of loads and training regimens in health-caring physical culture cannot be settled considering sex and age only but should be solved through an understanding of human constitutional peculiarities, adaptation potential and individual-typologic properties which determine health state and motor abilities.

ACKNOWLEDGMENTS

The authors thank N. S. Smirnova, T. A. Dunaevskaya and N. B. Selverova for their participation in data collection and analysis; G. M. Maslova, I. A. Kornienko and T. V. Panasyuk for their help in discussion of results; D. A. Aisakov—for technical assistance. This research was supported, in part, by Grant of J.Soros' International Foundation "Culture Initiative," reg.N 220 in the programm of Innovations in Humanitary Education in Russia.

REFERENCES

1. Pollock , M. L.,.Wilmore, J. H, Fox, S. M., 1978,Health and fitness through physical activity, N. Y.: John Wiley and Sons Inc.
2. Khruschev, S. V., and Kruglyi, M. M., 1982, Book for Young Athlete's Coach. Moscow: Physkultura i Sport.
3. Cooper, K,.H., 1988, The Aerobics Program for Total Well-Being, Moscow: Physkultura i Sport, 2-nd ed.
4. Matveev, A. P., and Nepopalov, V. N., 1994, Physical culture: physical exercise or human development, Theoriya i Praktika Physicheskoy Kultury (Rus.). 9: 2–6.
5. Krechmer, R., 1930, Body constitutions and character, Moscow: Gosizdat.
6. Sheldon, W. H., and Stevens, S. S, 1942, The varieties of temperament, N. Y.: Harpers.
7. Tanner, J. M., 1977, Growth and Human Constitution, In: Human biology, An Introduction to Human Evolution, Variation, Growth and Ecology, 2-nd ed. Oxford: University Press: 366–471.
8. Khrisanfova, E. N., 1990, Human Constitution and Biochemical Individuality, Moscow: Moscow State University.
9. Kharytonova, L. G., 1991, Types of Adaptation in Sports, Omsk: OGIFK.
10. Hamel, P., Simoneau, J. A., Lortie G., et al., 1986, Heredity and muscle adaptation to endurance training Med.Sci.Sports Exerc., 18, 6: 690–696.

43

COMPARISON OF TESTING METHODS FOR THE ASSESSMENT OF AEROBIC ENDURANCE

D. N. Gavrilov and T. V. Zabalueva

Research Institute of Physical Culture
Dynamo Ave. 2, 197042 St. Petersburg, Russia

INTRODUCTION

In spite of the fact that there are a number of different tests for evaluation of aerobic endurance (Cooper, 1968; Blair, et al., 1989) efforts continue to identify new ones, more especially those that might be simple and practicable. The 1-mile walking test (Kline, 1987) and 2-km walking test (Oja, 1991) have been developed in recent years. Not all existing endurance tests correspond to the requirements of mass population studies as to their simplicity, safety and feasibility. During the elaboration of the automatic screening "COFR" system for evaluation of fitness level (Utenko et al., 1994) for population studies we worked out two new 3-min tests: a modified step-test (40 cm for males and 35 cm for females) and sit-test (ups and downs on a chair). Both tests are based on natural habitual movements and correspond to the requirements of mass studies (Gavrilov, 1992). They were tested during 3 years on more then 2000 subjects of various ages, sex and fitness level. The object of the actual study was the further evaluation of the feasibility of a field version of the tests for mass assessment of aerobic endurance.

SUBJECTS AND METHODS

The subjects were 76 students (42 males and 34 females aged from 18 to 25) of the Pedagogical University in St.Petersburg. Their general data are shown in Table 1 in comparison with Finnish subjects (Oja, 1991). There are no essential differences between the Russian and Finnish subjects.

At the first stage of the study all the subjects were tested with 6 tests: a 5-min step-test (Astrand & Rhyming, 1954), the Harvard step-test, a 3-min modified step-test, a 3-min sit-test and a 2-km walking test and 2-km running. At the second stage 20 subjects were tested with the sit-test and 2-km Finnish walking test; 18 subjects were tested with the cycle ergometry and sit-tests. The heart rate was measured before, during and after exercise. The data from 36 subjects, were statistically analyzed.

Current Research in Sports Sciences, edited by Rogozkin and Maughan.
Plenum Press, New York, 1996

Table 1. Subject's anthropometric data (M±SE)

Exponents	University students		Finnish subjects	
	M	SE	M	SE
Males	(n = 42)		(n = 7)	
Age (years)	19.1	0.2	21.9	0.7
Height (cm)	180.3	1.2	182.0	0.6
Weight (kg)	73.5	1.1	75.2	0.5
BMI	22.9	3.6	22.5	0.7
Vital capacity (l)	4.9	0.1	—	—
Females	(n = 34)		(n = 9)	
Age (years)	19.2	0.9	23.1	0.4
Height (cm)	167.7	7.6	169.1	2.2
Weight (kg)	55.6	2.9	59.4	2.3
BMI	20.2	5.4	20.8	1.0
Vital capacity (l)	3.4	0.1	—	—

RESULTS

The physiological variables of the subjects are shown in Table 2.

Heart rate at the end of the load, the working increase and the relative power (in Watts) in males and females differ according to the power of the load. An authentic correlation is established in males between the level of aerobic endurance in the Finnish 2-km walking test and the 3-min sit-test ($r=0.739$); between the 2-km running and the 2-km walking tests (0.542); between the Harvard step test and the sit-test (0.539); the Astrand step-test and the modified step-test (0.625). There are correlations of 0.3–0.4 between results of the other tests. In females the closest correlations are between the modified step-

Table 2. Physiological data of the subjects (M±SE)

Exponents	Modified step-test	Astrand step-test	Sit- test	Finnish 2-km walking	2-km running	Harvard step-test
Males (n=34)						
HR load	123.0	120.5	128.0	136.2	142.2	147.1
	3.4	2.4	4.4	2.6	3.5	2.8
HR increase	52.8	50.1	63.2	58.6	71.6	81.4
	1.6	1.5	1.8	2.0	2.2	2.4
Sum working HR	179.0	252.0	174.0	896.0	574.0	307.0
Power of load (Wt/kg)	2.5	2.0	2.5	1.7–2.5	3.0	3.3
Relative power	1.5	1.2	1.5	1.2	—	1.9
	0.04	0.03	0.04	0.03	—	0.06
Females (n=22)						
HR load	130.6	137.4	130.7	136.5	155.0	—
	4.3	4.3	6.3	6.2	8.1	
HR increase	65.9	60.7	60.5	60.4	78.9	—
	1.5	1.4	1.8	1.9	2.1	
Sum working HR	197.0	304.0	182.0	1003.0	634.0	—
Power of load (Wt/kg)	2.0	1.7	2.0	1.5–2.0	2.5	—
Relative power	0.8	0.7	0.8	0.7	—	—
	0.02	0.01	0.02	0.02	—	—

test and the sit-test (0.789), the 2-km walking test and the sit-test (0.738), the 2-km running test and the 2-km walking test (0.594), the Astrand step test and the 2-km walking test (0.542). The other are about 0.3–0.4. The questionnaire proved that the vast majority of the subjects prefer the sit-test, the modified step-test and the 2-km walking test.

DISCUSSION

The choice of the test for population studies ought to take into consideration the physiological cost of the load. The 6 tests used in our study are placed according to their power zones: the Astrand step test (1.75–2.0 W/kg) is in the moderate zone; the 2-km walking test (1.5–2.5 W/kg), sit-test and modified step-test (2.0–2.5 W/kg) are between moderate and high zones; the Harvard step test and the 2-km running test (3.0–3.3 W/kg) are situated in the submaximal zone. Comparison of HR values, working increase of HR, load's power and summary HR shows essential differences in the physiological cost of every test. For population studies is very important the time consumption of each test should not be too great. Comparison of time consumption for each of the tests demonstrates that the sit-test and modified step-test consume half as much time as the Harvard step-test and the 2-km running test, 4 times less than the 2-km walking test and 6 times less than cycle ergometry (Table 3).

Comparison of working volume (in W) shows the advantage of the sit-test and the modified step-test as regards the Astrand steptest in 1.5 times, the Harvard step test in 2.5 times, the 2-km running test in 3.5 times and the 2-km walking test in 5.5 times.

CONCLUSION

The tests ordinarily used for evaluation of aerobic endurance do not fully meet the requirements of mass population studies.

As a rule they correspond to submaximal and maximal power zones and do not take into account the individual's age, sex and fitness level. Also they are rather time consuming.

Our 3-min sit-test and modified step-test, based on the habitual natural movements, corresponding to the moderate and high power zones, are simple and sure for all categories of subjects. Comparison with the criterion tests reveals their feasibility for mass population studies.

Table 3. Time consumption for testing one subject and necessary staff

Exponents	Time consumption (min)	Load power (Wt/kg)	Power zones (level)	Necessary staff (number)
2-km walking test	15–20	1.7–2.5	from moderate to great	2 and more
Harvard step-test	10	2.8–3.3	submaximal	1
2-km running test	7–10	3 and more	submaximal	2 and more
Astrand step-test	6–7	1.75–2.0	moderate	1
Modified step-test	4–5	2.0–2.7	from moderate to great	—
Sit-test	4–5	2.0–2.5	from moderate to great	1
Cycle ergometry	15–30		from moderate to maximal	1–2

ACKNOWLEDGMENTS

This work was patented as an invention by the Russian Patentional Committee under N1729485//Bull. N 16. 30.04.92.

REFERENCES

Astrand, P. O., Ruming, I., 1954, A nomogram for calculation of aerobic capacity (physical fitness) from pulse rate during submaximal work, J.of Appl.Physiol. 7:218–221.

Blair, S. N. et al., 1989, Surrogate Measures of Physical Activity and Physical Fitness. J.Epidem. 199(6):1145–1156.

Cooper, K. H., 1968, A means of assessing maximal oxygen intake. Journal of the American Medical Association 203:135–138.

Gavrilov, D. N. et al., 1992, Method of testing of cardiorespiratory endurance: Description of Invention N 1729485, Bull N 16,30.04.92, Moscow.

Kline, C. M. et al., 1987, Prediction of VO2 max from a one-mile track walk, Medicine and Science in Sports and Exercise 19:253–259.

Oja, P., Lankkanen, R. et al., 1991, A 2-km Walking Test for Assessing the Cardiorespiratory Fitness of Healthy Adults, International Journal of Sports Medicine 12:356–362.

Utenko, V. N., Gavrilov, D. N. et al., 1994, Equipment for evaluation of Health-related Fitness: Description of Invention N 2015682, Bull. N 13, 15.07.94, Moscow.

44

EFFECT OF EXERCISE TRAINING AND ACUTE EXERCISE ON ESSENTIAL HYPERTENSIVES

R. D. Dibner, M. M. Shubin, N. Taylor-Tolbert, D. R. Dengel, S. D. McCole, M. D. Brown, and J. M. Hagberg

Preventive Cardiology
Suite 1212, Kaufmann Bldg, 200 Lothrop Street
UPMC, Pittsburgh, Pennsylvania 15213

INTRODUCTION

Hypertension is a major health problem in the United States and Russia, as well as around the world. Essential hypertension, defined as a blood pressure(BP)>140/90 mmHg, is present in approximately 20% of adults in industrialized societies and these prevalence rates rise sharply with age (10). Hypertension is a major risk factor for cardiovascular disease, particularly coronary artery disease (13). Numerous studies indicate that endurance exercise training lowers BP in individuals with mild essential hypertension (BP 140–180/90–105mmHg) with the reduction averaging approximately 10 mmHg for both systolic and diastolic pressure (7,14).However, little is known about this response in women (6).

It has also been shown that casual BP is reduced following an acute bout of exercise in hypertensive individuals (3,7,14,16). The magnitude of this response is generally greater in hypertensive than in normotensive individuals. In addition, this BP-lowering effect of acute exercise has been observed in response to 40–70% of VO$_2$max exercise intensities. To determine if the BP-lowering effect of acute exercise is clinically significant, at least three important questions must be answered. First, is the BP lowering of sufficient magnitude to be considered clinically significant? Second, is the duration of this response sufficient to lower average daily BP? Third, does this response occur under conditions of normal activities of daily living. No experimentally controlled investigation has determined whether BP remains reduced from control levels for 24 hours after exercise.

METHODS

Exercise Training Study

Thirty-six middle-aged (45–55 years) premenopausal women volunteered to take part in this study. Twenty-two women had essential hypertension (average BP 156/94

Current Research in Sports Sciences, edited by Rogozkin and Maughan.
Plenum Press, New York, 1996

295

mmHg) and 14 women had normal BP (average BP 123/77 mmHg). Subjects in both groups were sedentary, defined as not currently performing regular endurance exercise. Hypertensive subjects were included in the study if their systolic and diastolic BP were both in the ranges of 140 to 180 and 90 to 105 mmHg, respectively, and they had a history of essential hypertension. All subjects underwent a medical history and physical examination by a physician to rule out secondary causes of hypertension and to determine if subjects had any contraindications for exercise training (1). Subjects were excluded if they had evidence of coronary, or other organic, heart disease. Subjects underwent one yr of endurance exercise training consisting of jogging 50 min 3 times/wk. For the first month of the program normotensive subjects exercised at heart rates up to 135 b/min (65% heart rate reserve) while hypertensive subjects exercised at heart rates up to 110 b/min (40% heart rate reserve). Thereafter, and for the remainder of the 12 month training program, both groups exercised at heart rates up to 135 b/min (65% heart rate reserve).

All subjects underwent a cardiovascular examination prior to and after 6 and 12 months of exercise training. This evaluation consisted of casual BP determinations, an echocardiographic study, and a cycle ergometer test. Casual BP was measured by auscultation 3 times in a sitting position after 5 min of rest on a single day. M-mode echocardiograms were recorded at rest using the parasternal view in a 30° left lateral supine position. Standard American Society of Echocardiography conventions were used to measure cardiac dimensions. Exercise testing on a cycle ergometer was performed to evaluate work capacity, which was expressed as the sum of the work accomplished during this test in terms of kgm. The initial work rate for this test was 300 kgm/min and it was increased by 150 kgm/min every 3 min. The test was terminated when heart rate reached 130–150 b/min or ST-segment on the ECG was depressed > 1mV.

The significance of changes resulting from endurance exercise training was assessed using repeated measures analyses of variance with the testing of appropriate subhypotheses.

Acute Exercise Study

The subjects in this study were 12 sedentary men with an average age and weight of 59 ± 6 years and 93 ± 16 kg, respectively. Their casual BP was measured weekly for 4 wks using a Hawksley randomized zero sphygmomanometer. Their casual systolic and diastolic BP averaged 152 ± 9 and 96 ± 7 mmHg, respectively, during this month of screening. Subjects underwent a progressive maximal treadmill exercise according to the Bruce protocol to screen for the presence of coronary heart disease (1). Subjects with > 2 mV ST-segment depression during or following this test were excluded from further participation in the study. Subjects also underwent a second maximal treadmill exercise test (5) to assess their VO_2max, which averaged 26 ± 4ml/kg/min.

Subjects then underwent two 24 hour ambulatory BP monitoring sessions using a Space Labs Model 90219 Ambulatory BP Monitor. Both recordings were initiated at the same time of the morning on 2 occasions separated by at least 7 days. One of these days was a control day, which was not preceded by exercise, and the other was the experimental day, which was preceded by 45 min of treadmill walking at 70% of VO_2max. These treatments were administered in random order to the different subjects. On the first of these days subjects recorded their activity in a diary at the time of each BP determination. They were given a copy of their diary and asked to repeat these activities on the second day.

During the 24 hr ambulatory BP recordings, BP was measured every 15–20 min during the waking hrs and every 30 min when the subject was asleep. The data were down loaded to a computer for analysis. Data were automatically edited by the system if they were artifacts or if they were outside of normal physiological ranges. Values were manually deleted from the data set if they differed from the prior or following values by > 15 mmHg and were not explained by differences in physical activity levels. BP data were averaged for each hr of the recording and these hourly averages were then also averaged to determine 24 hr average BP values. Statistical analyses compared the difference in BP between the control and experimental day for a specific hr, or the 24 hr average values, to the null hypothesis of no difference.

The exercise trial preceding the 24 hr ambulatory BP recording on the experimental day consisted of three 15 min treadmill exercise sessions with 5 min of rest between sessions. Treadmill speed and grade were adjusted in the initial 5–10 min of the first session to ensure that subjects were exercising at the desired intensity. Heart rate, BP and VO_2 were determined at the end of each 15 min session.

All data are expressed as mean ±SE. A p<0.05 was considered statistically significant.

RESULTS

Exercise Training Study

Initial work capacity was not different between the hypertensive and normotensive women (1575 ± 64 vs. 1510 ± 69 kgm). However, work capacity increased significantly in the hypertensive group (p<0.05), to 2850 ± 81 kgm, only after 12 months of endurance exercise training. Their change in work capacity after 6 months of exercise training was not significant. On the contrary, normotensive women increased their work capacity with 6 months of exercise training (to 3310 ± 89 kgm, p< 0.001) and increased in further with 12 months of training to 4820 ± 72 kgm (p<0.002). After 12 months of exercise training, the absolute amount of work performed during this test was nearly 70% greater in the normotensive compared to the hypertensive women (4820 ± 72 vs. 2850 ± 81 kgm).

The hemodynamic adaptations elicited with exercise training also had a different time course in the two groups. The hypertensive group showed no significant changes in systolic or diastolic BP after 6 months of exercise training, whereas both were reduced significantly after 12 months of exercise training (Table 1).

The hypertensive group also had significant reductions in heart rate and total peripheral resistance and significant increases in stroke volume and ejection fraction after 12 months, but not after 6 months, of exercise training (Table 1). Since their cardiac output was unchanged, the entire reduction in BP in the hypertensive women with exercise training was the result of reduced total peripheral resistance.

In the normotensive women both systolic and diastolic BP were reduced after 6 months of exercise training and did not change further after 12 months of training (Table 2).

Heart rate, stroke volume, ejection fraction, and total peripheral resistance also changed significantly in the normotensive group after 6 months of exercise training and did not change further with 12 months of exercise training (Table 2). The entire reduction in BP in the normotensive women with exercise training was again the result of a decrease in total peripheral resistance as cardiac output was, if anything, increased as a result of training.

Table 1. Changes in hemodynamic parameters at rest with exercise training
in hypertensive women

| | Baseline | Exercise training | |
		6 months	12 months
HR (b/min)	74 ± 1	73 ± 2	68 ± 1 *
Systolic BP (mmHg)	156 ± 6	155 ± 6	132 ± 2 *
Diastolic BP (mmHg)	94 ± 2	90 ± 2	82 ± 1 *
Mean BP (mmHg)	115 ± 2	112 ± 3	99 ± 2 *
Stroke volume (ml)	69 ± 3	73 ± 2	78 ± 2 *
Ejection fraction (%)	63 ± 1	65 ± 1	69 ± 1 *
Cardiac output (l/min)	5.1 ± 0.1	5.3 ± 0.2	5.3 ± 0.1
TPR (dynes · sec E-1 · cm E-5)	1797 ± 185	1676 ± 159	1493 ± 210 *

Values are means ± SE. HR, heart rate; BP, blood pressure; TPR, total peripheral resistance. *p<0.05 compared to
the baseline value.

Acute Exercise Study

There were no significant differences between baseline BP or heart rate on the control and experimental days in this component of the study. During the exercise sessions, the subjects' VO_2 averaged 1.8 ± 0.1 l/min which was equal to 73 ± 2% of their VO_2max. Their heart rate and systolic and diastolic BP during the exercise session averaged 136 ± 3 b/min, 175 ± 6 mmHg, and 92 ± 3mmHg, respectively. Systolic BP was reduced (p<0.05) during hr 1 (by 14 ± 4 mmHg), 2(18 ± 3 mmHg), 3(11 ± 4 mmHg), 4(8 ± 3 mmHg), and 5(16 ± 6 mmHg) following exercise (Fig 1).

Systolic BP also tended to be lower throughout the night on the exercise day, but the difference was significant at only one time point (Fig 1). The 24-hr average systolic BP was 6 ± 2 mmHg lower(p<0.05) on the day preceded by exercise. Diastolic BP also tended to be reduced during the first 5 hrs following exercise, but the reductions were only significant during the second (by 8 ± 2 mmHg) and fifth (11 ± 3 mmHg) hours following exercise (Fig 1). Mean BP showed the same pattern of changes as diastolic BP (Fig 1). Average 24-hr diastolic (-2 ± 3 mmHg) and mean (-3 ± 3 mmHg) BP did not change significantly on the recording preceded by exercise.

Table 2. Changes in hemodynamic parameters at rest with exercise training
in normotensive women

| | Baseline | Exercise training | |
		6 months	12 months
HR (b/min)	72 ± 1	64 ± 2	60 ± 1 *
Systolic BP (mmHg)	123 ± 3	114 ± 2	112 ± 2 *
Diastolic BP (mmHg)	77 ± 2	71 ± 2	69 ± 1 *
Mean BP (mmHg)	92 ± 3	85 ± 2	83 ± 2 *
Stroke volume (ml)	70 ± 2	88 ± 2	89 ± 2 *
Ejection fraction (%)	67 ± 1	75 ± 1	75 ± 1 *
Cardiac output (l/min)	5.0 ± 0.2	5.6 ± 0.2	5.4 ± 0.1 *
TPR (dynes · sec E-1 · cm E-5)	1465 ± 175	1219 ± 156	1248 ± 146 *

Values are means ± SE. HR, heart rate; BP, blood pressure; TPR, total peripheral resistance. *p<0.05 compared to
the baseline value.

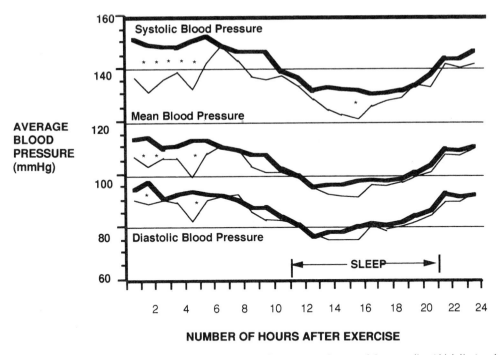

Figure 1. Average hourly systolic, mean, and diastolic blood pressure on the control day recording (thick line) and on the recording preceded by exercise (thin line). The asterisks indicate individual hour time points where the difference in blood pressure between the two trials was significantly different at $p<0.05$.

DISCUSSION

Though numerous studies have assessed the effects of exercise training on the BP of individuals with essential hypertension, less than half of these studies have included female hypertensives as subjects and to date only 3 studies have reported separate data for female hypertensives (12,15,17).Thus, the present data provide important confirmatory evidence documenting the BP-lowering effect of endurance exercise training in middle-aged women with essential hypertension. In fact, as opposed to the conclusion that 75% of studies in hypertensives report significant reductions in BP with exercise training (2,6,9), all 4 studies in women to date have reported significant BP reductions with exercise training. In the present study the reductions in systolic and diastolic BP elicited with 12 months of exercise training averaged 23 and 12 mmHg, respectively. These reductions are substantially greater than the 10.5 and 8.6 mmHg average reductions in systolic and diastolic BP, respectively, we reported in a recent review of all previous exercise training studies in essential hypertensives (6). However, the reductions are similar to those reported in previous exercise training studies in hypertensive women which averaged 19 and 14 mmHg for systolic and diastolic BP, respectively (6). Thus, these data confirm that female hypertensives also elicit substantial reductions in BP with exercise training. In fact, it appears that women may elicit greater and more consistent reductions in BP than are evident in male hypertensives.

The reduction in BP in female hypertensives following exercise training was the result of decreases in their total peripheral resistance as their cardiac output did not change with training. These results are consistent with a number of other studies indicating that a reduction in vascular resistance, and not cardiac output, is responsible for the training-induced reduction in BP in hypertensive individuals (e.g. 9). On the other hand, some previous studies have reported that cardiac output may be reduced as a result of exercise training in persons with essential hypertension (e.g. 9). It is interesting that the normotensive females in the present study also decreased their BP with exercise training and that this was due to a decrease in total peripheral resistance as their cardiac output actually increased somewhat.

Substantial differences in the training-induced changes in work capacity and hemodynamics were evident between the normotensive and hypertensive women in the present study. The normotensive women had significant increases in maximal work capacity, stroke volume, and ejection fraction and significant decreases in heart rate, BP, and total peripheral resistance after only 6 months of training. The hypertensive women elicited these same beneficial cardiovascular adaptations, but they were not evident until after 12 months of training. The slower time course of the hypertensives' cardiovascular adaptations may be due to their decreased exercise training intensity during their initial month of training. However, we have previously noted that 9 months of exercise training also failed to elicit an increase in submaximal exercise stroke volume in older hypertensives (8), a response that is clearly elicited within this time frame in normotensives. Thus, the long-term adaptation to endurance exercise training in hypertensives may develop more slowly than in normotensives.

A number of studies have shown that BP is also reduced after a single bout of exercise in sedentary hypertensive individuals (3,7,14,16). However, most of these studies have assessed this response for only 1–3hrs after exercise and in a very controlled laboratory setting. Thus, it is unclear if BP is reduced for a more clinically significant timeframe such as 12–24 hrs. Also it is not known if this response is evident when subjects leave the laboratory setting and go about their usual daily lifestyle. One previous study has addressed these points utilizing ambulatory BP monitoring in hypertensive individuals (14). However, they only compared the BP after acute exercise to that measured immediately prior to exercise. Since BP exhibits diurnal variations and marked variability during a person's daily activities, such a study design is not adequate to assess the duration and magnitude of this response. In the present study the BP-lowering effect of acute exercise was assessed by comparing the subjects' BP at the same time of day on 2 days—one preceded and one not preceded by 45 min of submaximal exercise.

Systolic BP was reduced for 5 hrs following this single bout of exercise and the average reductions ranged from 8–18 mmHg. Systolic BP also tended to be somewhat lower during sleep on the exercise day. With respect to clinical significance, these hourly reductions in systolic BP resulted in a decrease in average 24 hr systolic BP of 6 ± 2 mmHg. Such a decrease is not as substantial as the reduction that can be elicited with antihypertensive medications (10). However, it is similar to the reduction in systolic BP evident in a number of clinical trials of the efficacy of antihypertensive medications that have shown 10–30% reductions in cardiovascular mortality and morbidity (4). The reductions in diastolic and mean BP tended to be smaller and less consistent following exercise than those for systolic BP which is similar to previous findings in hypertensive individuals studied under more controlled laboratory conditions (7,11).

In summary, the data from the exercise training portion of this study provide evidence confirming that hypertensive women elicit the same, and perhaps even greater and more consistent, reductions in both systolic and diastolic BP as hypertensive men with exercise training. The data from the acute exercise component of this study provide evidence that systolic BP is reduced substantially for up to 5 hrs following a single bout of exercise in older sedentary hypertensive men. Systolic BP is also somewhat, though not significantly, reduced during the hrs of sleep. Diastolic BP is also reduced somewhat during the first 5 hrs following exercise, but the reductions are not as consistent or as substantial as with systolic BP. The overall result of these changes is that the average systolic BP was reduced significantly for the 24 hrs following exercise, while average 24 hr mean and diastolic BP were not.

ACKNOWLEDGMENTS

The research conducted at the University of Maryland was supported by a grant from the Maryland Affiliate of the American Heart Association and from a Biomedical Research Support Grant from the University of Maryland.

REFERENCES

1. American College of Sports Medicine, 1991, Guidelines for exercise testing and exercise prescription (4th Ed),Lea And Febiger: Philadelphia.
2. American College of Sports Medicine. Position Stand: Physical Activity, Physical Fitness, and Hypertension.,1993, Med Sci Sports Exerc 25:i-x.
3. Clirous, J.,Kauome, N., Nadeau, A. et al., 1992, After effects of exercise on regional and systemic hemodynamics in hypertension, Hypertension 19:183–191.
4. Collins, R., Peto, R., MacMahon, S., Hebert, P., et al.,1990, Blood pressure, stroke, and coronary heart desease. Part 2. Short-term reductions in blood pressure—overview of randomized trials in their epidemiological context, Lancet 335:827–838.
5. Dengel, D. R., Pratley, R. E., Hagberg, J. M., Goldberg, A. P., 1994, Impaired insulin sensitivity and maximal responsiveness in older hypertensive men, Hypertension 23:320–324.
6. Hagberg, J. M., Brown, M. D., Does exercise training play a role in the treatment of essential hypertension? In press: J. Cardiovasc. Risk.
7. Hagberg, J. M., Montain, S. J., Martin, W. H., 1987, Blood pressure and hemodynamic responses after exercise in older hypertensives, J. Appl. Physiol. 63:27–276.
8. Hagberg, J. M., Montain, S. J., Martin, W. H., Ehsani, A. A., 1989, Effect of exercise training on 60–69 yr old persons with essential hypertension. Am. J. Cardiol. 64:348–353.
9. Hagberg, J. M., 1988, Exercise, fitness, and hypertension. In: Exercise,Fitness and Health. Ed: C. Bouchard, R. J. Shephard, T. Stephens, B. D. McPherson. Human Kinetics Press: Champaign, IL.:455–466.
10. Kaplan, N., 1994, Clinical hypertension. 6th ed. Baltimore: Williams and Wilkins.
11. Kaufmann, F. L., Hughson, R. L., Schaman, J. P., 1987, Effect of exercise on recovery blood pressure in normotensive and hypertensive subjects, Med. Sci. Sports Exercise 19:17–20.
12. Koga, M., Ideishi, M., Matsusaki, M., Tashiro, E., Kinoshita, A., Ikeda, M., Tanaka, H., Shindo, M., Arakawa, K., 1992, Mild exercise decreases plasma endogenous digitalis-like substance in hypertensive individuals, Hypertension 19 (Suppl. II):II231-II236.
13. Kuller, L. N., Hylley, S. B., Cohen, J. D., et al., 1986, Unexpected effects of treating hypertension in men with EKG abnormalities: a critical analysis. Circulation 73:114–123.
14. Pescatello, L. S., Pargo, A. E., Leoch, C. N., et al., 1991, Short-term effect of dynamic exercise on arterial blood pressure, Circulation 83:1557–1561.
15. Roman, O., Camuzzi, A. L., Villalon, E., Klenner, C., 1981, Physical training program in arterial hypertension, Cardiology 67:230–243.
16. Somers, V. K., Conway, J., Coats, A. et al., 1991, Post exercise hypotension is not sustained in normal and hypertensive humans, Hypertension 18:211–215.
17. Tanabe, Y., Sasaki, J., Urata, H., Kiyonaga, A., Tanaka, H., Shindo, M., Arakawa, K., 1988, Effect of mild aerobic exercise on lipid and apolipoprotein levels in patients with essential hypertension, Jpn. Heart J. 29:199–206.

45

SELF-RESISTANCE GYMNASTICS AND PHYSICAL REHABILITATION OF WHEELCHAIR-DISABLED IN RUSSIA

V. S. Dmitriev, A. I. Osadchih, and E. V. Ozolina

Russian Research Institute of Physical Culture
Kazakova St., 18, 103064, Moscow, Russia.

INTRODUCTION

In the context of democratic processes in Russia the idea of social and physical reha-bilitation of the handicapped has become more popular in society. A new concept of physical rehabilitation of disabled individuals has been developed [4]. Unfortunately the disabled population is increasing in Russia due to widespread use of automobiles, social instability, local wars etc. The overall proportion of spinal cord injured individuals and amputees in the community increases from year to year. In spite of the fact that there is much international scientific and clinical data demonstrating the vital importance of mus-cular activity and sport for rehabilitation of the disabled [7,11] very few institutions in Russia deal with the problem of fitness in special populations. Only a few years ago the first Russian research laboratory was organized in the All-Russia Research Institute of Physical Culture to conduct investigations in such areas as fitness, exercise, sport, and health promotion for disabled. At the moment the laboratory makes investigations in two main directions: a) physiological response to exercise (cardiorespiratory, endocrine and metabolic), and adaptation to muscular activity; b) elaboration of technical devices for physical rehabilitation of the disabled. Experimental data on acute cardiorespiratory, en-docrine and metabolic responses to physical exercise in spinal cord injured patients ob-tained in our laboratory [9] gave preliminary information to start with a detailed description of the specific Russian disabled population. We believe these results to be the basis for the development of a system of norms for fitness tests for Russian handicapped which proved to be quite different in their level of habitual activity, dietary pattern, life-style and mentality from that which has been reported in Europe and America. Our study of the current fitness status of wheelchair patients in Moscow made it possible to state that definite changes in physical activity are necessary to improve functional status and physi-cal performance even if subjects participate in recreational sports events for the disabled [10]. The main idea of the present study was to clarify whether the new training technol-ogy, developed in Russia - dynamic self-resistance gymnastics (see below) ensures im-

Current Research in Sports Sciences, edited by Rogozkin and Maughan.
Plenum Press, New York, 1996

provement of physical performance of untrained spinal cord injured patients. This type of gymnastics has previously been shown to increase muscle strength and flexibility [5].

METHODS

Subjects were 10 adult males (aged 22–35 yr) with spinal injury (2–8 yr post injury involving levels T6 - L2) who gave their informed consent to participate in the study. All subjects were patients at Moscow City Clinic Number 6.

Experimental Design

The day before the experiment ultrasound scanning of the heart was made to evaluate functional status of the heart and to avoid cardiac disturbances. On the first experimental day each subject performed graded exercise on a treadmill with increasing work load until maximal oxygen consumption (VO2 max) was reached. Total work time (endurance time), VO2, respiratory minute volume (VE), heart rate (HR), blood lactate and acid-base balance were recorded during the VO2max test [1,2]. Following the aerobic power capacity test, patients had three week training involving 30-min bouts of exercise daily. Heart rate was registered every 5 min during the first and the last training session. Before and after the training period the width of muscle and subcutaneous tissue on the front and back surface of the upper arm were determined by an ultrasound technique [3]. After the training period was over another VO2max test was carried out.

Training Procedure

Each training session involved 12 elements of self-resistance dynamic gymnastics involving upper limb muscles and upper trunk muscles. The main principles of self-resistance gymnastics are as follows:

   a. one muscle group produces tension while resisting the force generated by another muscle group. Muscles of the opposing groups contract simultaneously in dynamic and practically isotonic regimen;
   b. exercise intensity is chosen by subject himself according to his subjective feeling of fatigue. Thus muscle tension may vary through the wide range of physiological values and may reach maximal voluntary contractions;
   c. to achieve maximal training effect maximal amplitude of movements and maximal tension should be maintained.

Data Analysis

The Wilcoxon Z-test was used to determine if significance was present between pre- and post-training values for the same subject. Statistical significance was defined as $p < 0.05$. All values are reported as means±SE.

RESULTS

Ultrasound scanning of the heart revealed decreased myocardial perfusion and low functional reserves. No changes in heart morphology were observed. Heart rate values ob-

Table 1. Heart rate changes during the first and the last session of
the 3-week training period

Time periods of HR registration, min	HR during 1st training session, b/min	HR during last training session, b/min
1	92.0 ± 6.1	89.9 ± 7.5
5	117.5 ± 10.0	103.5 ± 8.1
10	114.5 ± 8.6	104.7 ± 5.5 *
15	118.2 ± 7.7	111.2 ± 7.5 *
20	127.5 ± 9.9	119.8 ± 9.9 *
25	137.5 ± 10.0	120.8 ± 10.2 *
30	98.3 ± 9.8	94.7 ± 7.3

Values are means ± SE. HR, heart rate. *Significantly different from initial value at the beginning of training period ($p<0.05$).

tained during the self-resistance training session showed a pronounced decrease by the end of training period (Table 1).

As a result of the 3-wk training period, the width of muscle tissue both on the front and back surfaces of the upper arm increased from 35.2 ± 3.3 to 37.0 ± 40 mm and from 41.0 ± 2.5 to 44.0 ± 5.1 mm respectively ($p<0.05$). Training had no significant effect on changes in width of subcutaneous tissue which varied from 8.9 ± 1.3 to 8.2 ± 2.4 mm and from 12.3 ± 3.4 to 11.1 ± 4.0 mm on front and back surfaces of the upper arm respectively. Training effected maximal cardiorespiratory responses (Table 2) and enhanced wheelchair performance. Total work time increased from 13.5 ± 1.3 to 19.0 ± 2.3 min ($p<0.05$). There was a significant increase in maximal velocity achieved during the VO2max-test after training period: 1.5 ± 0.3m/s before training and 2.1 ± 0.3 m/s after training ($p<0.05$).

Resting blood lactate and acid-base balance values were within the normal range for healthy subjects. Self-resistance training produced significant effects on lactate responses to exercise. After three weeks of this regimen maximal blood lactate accumulation obtained in the VO2max-test increased from 7.50 ± 2.0 to 9.33 ± 3.0 mmol/l ($p<0.05$). Significant differences in submaximal blood lactate levels were observed at the 3rd, 9th, and 12th min of test exercise. Blood lactate concentrations after training were significantly lower than before training. Exercise-induced pH and base excess (BE) decreases appeared to be more pronounced after the training period than before: 7.28 ± 0.02 and 7.30 ± 0.01 ($p<0.05$) for pH values, and 12.0 ± 1.5 and 10.1 ± 2.0 meq/l for BE values.

Table 2. Cardiorespiratory responses to a 3-wk training period
in wheelchair-disabled subjects

	Before training	After training
HRmax, b/min	195.0 ± 8.8	196.1 ± 9.5
VO2max, L/min	1.32 ± 0.30	1.81 ± 0.91 *
VO2max, ml/kg min	21.0 ± 3.4	25.8 ± 4.0 *
VEmax, L/min	58.0 ± 11.0	60.1 ± 13.2

Values are means ± SE. HRmax, peak heart rate, VO2max, peak oxygen intake, VEmax, peak respiratory minute volume. *Significance of differences from pretraining value $p<0.05$.

DISCUSSION

Our data on peak oxygen intake, respiratory minute volume, blood lactate concentrations and acid-base balance changes in spinally injured subjects completely agree with that obtained by other authors [6,7,11]. Self-resistance training proved to be effective enough to induce marked improvement of wheelchair aerobic performance. As a result of this rather short training period, peak oxygen intake increased substantially and thus fitness level of our subjects was changed from "below average" to "average" according to normative values for maximum oxygen intake in the wheelchair disabled data published by R.J.Shephard [11]. Peak heart rate was high initially and remained practically unaffected (Table 2). If we assume that in our investigation self-resistance gymnastics affected muscle strength as was reported to occur in the study on cosmonauts which had a similar training programme after a period of experimental inactivity [5], then training induced enhancement of wheelchair performance can be connected with an increase in upper limb and trunk muscle strength. Our present data on muscle tissue width increase is in favour of this assumption. Moreover, it was shown that muscle development is important to aerobic performance [8] and oxygen intake is fairly closely correlated with local muscle strength [11]. Thus, in spite of the obvious restrictions of the present study the training-induced enhancement of wheelchair performance can be the basis for speculation on the effectiveness of self-resistance gymnastics for conditioning of wheelchair-confined subjects.

REFERENCES

1. Astrand, P.-O., and Rodahl, K., 1977, Textbook of Work Physiology (2nd ed), New York, McGraw-Hill.
2. Bergmeyer, H. U., 1974, Methods of Enzymatic Analysis, NewYork, Academic.
3. Bravaya, D. U., 1984, Physiological analysis of different musclestrength training methods, Diss, Moscow.
4. Dmitriev, V. S., Popov, S. N., Sonkin, V. D., and Suzdalnitsky, R. S., 1994, Physical rehabilitation as part of health promotion system, All-Russia conference on problems of physical fitness of children andthe young, Moscow.
5. Fohtin, V. G., 1991, Athletic gymnastics without apparatus,Moscow, Physical Culture and Sport.
6. Glaser, R. M., Sawka, M. N., Brune, M. F. and Wilde, S. W., 1980, Physiological responses to maximal effort wheelchair and arm crank ergometry, J. Appl.Physiol. 48:1060–1064.
7. Hjeltnes, N., 1988, Physical exercise and physiological exercisetesting in patients with spinal cord injuries - a short review,Scand.J.Sports Sci. 10(2–3):55–59.
8. Kay, C., and Shephard, R. J., 1969, On muscle strength and thethreshold of anaerobic work, International Zeitschrift fur AngewandtePhysiologie 27:311–328.
9. Ozolina, E. V., 1994, Cardiorespiratory, endocrine, and metabolicresponse to maximal exercise in spinal cord injured patients,Rehabilitation of patients with spinal cord injury and lesion of joints, Moscow :94–96.
10. Ozolina, E. V., and Dmitriev, V. S., 1994, Fitness status of wheelchair disabled, Rehabilitation of patients with spinal cord injury and lesion of joints, Moscow :102.
11. Shephard, R. J., 1990, Fitness in Special Population, Human Kinetics Books, Champaign, Illinois.

46

NUTRITION, MODERATE EXERCISE, AND HEALTH

Adrianne E. Hardman

Department of Physical Education, Sports Science and Recreation
    Management
Loughborough University
Loughborough, Leicestershire LE11 3TU, England

INTRODUCTION

People who are physically active tend to be healthier than their sedentary peers. Habitual vigorous exercise, however, is pursued by only a minority, particularly of women (1). From the public health view it is important to assess the potential of moderate, socially-acceptable exercise to confer health gains because programmes of high intensity exercise are associated with poor adherence (2) and high risk of orthopaemic injury (3). The short-term increase in cardiovascular risk associated with each bout of exercise is lower for moderate than for vigorous exercise (4), altering favourably the balance between risk and benefit. A growing body of epidemiological evidence shows relationships between moderate-intensity exercise and mortality from coronary heart disease (CHD), stroke, adult-onset diabetes, hypertension, stroke, and hip fracture.

CORONARY HEART DISEASE

This evidence is particularly strong for coronary heart disease (CHD) where more than 40 epidemiological studies have now compared the risk of CHD in physically active men with that exhibited by their sedentary counterparts. Collectively, they suggest that physical inactivity confers a risk of CHD 90% higher than that exhibited by active men (5,6). Moreover research carried out in many countries strongly supports the suggestion that regular exercise reduces the risk of CHD in men and that its influence is strong, independent and graded (7).

Specifically, several studies have shown that moderate exercise is associated with a reduction in CHD risk; in British civil servants, men who reported that their usual speed of walking was fast (Figure 1) experienced fewer clinical episodes as did those who reported doing considerable amounts of cycling (8). Men in the British Regional Heart

Current Research in Sports Sciences, edited by Rogozkin and Maughan.
Plenum Press, New York, 1996

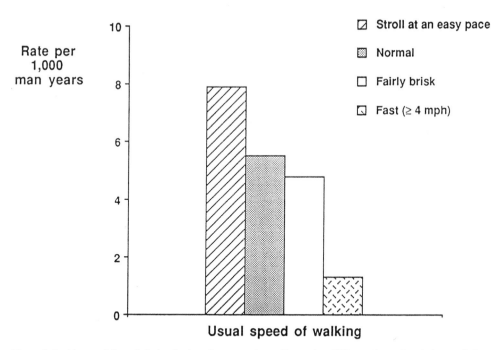

Figure 1. Incidence of first clinical episodes of coronary heart disease in middle-aged men in relation to their re-ported usual speed of walking. Adapted from ref. 8.

Study who reported moderate or moderately vigorous activity experienced a risk 40% lower than that of inactive men ⁻ and, in fact, lower than those who reported vigorous exercise (9). Among Harvard alumni, men who habitually walked more than 9 miles per week had a 21% lower risk than those who walked fewer than 3 miles per week (10).

There are indications in the epidemiological literature of close links between such cardioprotective physical activity and nutrition. For example comparison of food intake records of men who subsequently suffered fatal attacks with the records of survivors suggests a consistently higher energy intakes in the latter group (Table 1). Such observations are not surprising because an increase in exercise level usually results in an increased intake of food energy (12), so that the transition from a sedentary to an active state is associated with a higher energy turnover. This has important implications for the transport, storage and utilization of metabolic fuels, all of which are altered in the trained state such that such individuals experience a lower risk of what has been called "metabolic, hypertensive cardiovascular disease". Higher energy turnover may also be associated with improved weight regulation because food intake appears to be more closely coupled to energy expenditure at higher levels of physical activity (13).

LIPID AND LIPOPROTEIN METABOLISM

Changes in the metabolic milieu probably contribute to the favourable CHD risk profiles commonly seen in physically active individuals. For example, endurance trained individuals are characterised by high serum concentrations of high density lipoprotein (HDL) cholesterol and low concentrations of triacylglycerol (14). The results from longi-

Table 1. Energy intake (kcal/day) and future risk of coronary heart disease

Community studies	Heart disease victims	Survivors
English banking and London bus workers	2,656	2,869
Framingham, Massachusetts, USA	2,369	2,622
Puerto Rico	2,223	2,395
Honolulu, Hawai	2,149	2,319

Adapted from ref. 11.

tudinal studies are less consistent but many show an increase in HDL cholesterol with increased levels of physical activity. It is not clear how much exercise is needed to modify HDL cholesterol but a modest programme of jogging in overweight men (average over one year of 18.9 km or 11.7 miles per week) resulted in a 10.4% increase in HDL cholesterol in relation to controls (15).

These changes may be causally related to reduced levels of body fatness. When an equivalent energy deficit was created either by decreasing energy intake or by increasing exercise levels similar, increases in HDL cholesterol are observed (15) so the inverse relationship between adiposity and HDL cholesterol levels may underly some of the increased risk of CHD observed with obesity in men (16) and in women (17).

Many individuals seek to influence unfavourable lipoprotein profiles by altering dietary habits rather than exercise. Current recommendations for modest, achievable targets are approximately 55 % of total energy from carbohydrate, 30 % from fat, with saturated fat contributing no more than 10 % (18). Whilst a diet following these guidelines is often effective in reducing low density lipoprotein (LDL) cholesterol, it can also reduce HDL cholesterol and, speculatively, may reduce the anticipated beneficial effect of lower LDL cholesterol levels. The potential of exercise-induced increases in HDL cholesterol to offset decreases due to increasing the contribution of carbohydrate to energy intake was examined in a controlled trial of overweight men and women, all of whom followed a "weight-reducing diet" (19). Results differed somewhat in men and women: in men, diet plus exercise produced a greater increase in HDL cholesterol than did diet alone: in women, diet alone decreased HDL cholesterol but the exercise plus dietary intervention prevented this decrease.

Changes in lipoproteins (such as those described above) measured in the blood of fasted individuals, although interesting, may undervalue the role of regular physical activity in reducing the risk of CHD. Atherogenesis, so intimately involved in the underlying pathology of this disease, has been referred to as a postprandial phenomenon (20). Experimental evidence to support this view comes from observations that human cultured arterial smooth muscle cells take up cholesterol from chylomicron remnants even more effectively than from low density lipoproteins (21). In addition, the appearance of chylomicron triacylglycerol (TAG) in the circulation after a meal also increases the persistence in the circulation of very low density lipoproteins (VLDL), augmenting the transfer of endogenous cholesterol to LDL (22). Recent observations of a heightened lipaemic response to consumption of a test meal in patients with known coronary artery disease (22,24) provide support for this view and one report of an association between the progression of atherosclerosis and the persistence in the circulation of small chylomicron remnants is particularly persuasive (25). As man spends most of his life in the postprandial state interventions which reduce the magnitude and duration of lipaemia during the postprandial phase may play an important role in slowing the progression of atherogenesis.

    Both nutrition and exercise influence the lipaemic response to dietary fat. The effect
of exercise on postprandial lipaemia are, however, probably quantitatively greater than
can be achieved by manipulating the type of fat ingested. The high HDL cholesterol levels
evident in physically active men are, arguably, dependent on their high metabolic capacity
for TAG (23): a fall in the concentration of TAG-rich lipoproteins decreases transfer of
cholesteryl esters from HDL to TAG-rich lipoproteins in exchange for TAG. As a consequence, the value measured as HDL cholesterol increases. Thus, elevated TAG levels can
be regarded as the driving force for low HDL cholesterol values. The value of quantifica-
tion of postprandial lipaemia is therefore that it constitutes an integrated measure of TAG
metabolic capacity.

    In line with this thinking, the lipaemic response following a high fat test meal is sig-
nificantly lower in trained individuals than in comparable sedentary controls (26). Recent
work from our laboratory shows that this is probably, at least in part, attributable to the residual effects of the last bout of exercise (27). The lipaemic response to a high fat meal
was determined for twelve normolipidaemic young adults (6 females, 6 males) on two occasions. On one occasion consumption of the meal was preceded by a day of rest. On the
other 2 hours of walking (30% of maximal oxygen uptake, $\dot{V}O_{2max}$) was performed 15
hours prior to ingestion. The total lipaemic response (area under the TAG/time curve) was
31 (sem 7) % lower after the exercise day than after the rest day (Figure 2).

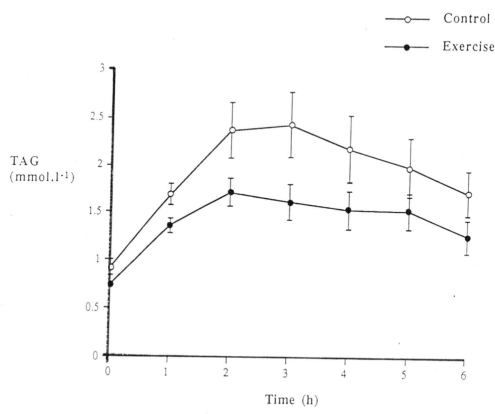

Figure 2. Effect of one bout of walking on the serum triglyceride response to a fatty meal consumed 12 hours after
exercise in young adults (ref. 27). (Mean±SEM, n=12).

Lower concentrations of plasma TAG during the period of recovery from exercise could theoretically be due to either decreased entry into the circulation of TAG-rich lipoproteins or to an increased tissue uptake. In the study described above, however, when lipaemia was determined many hours after exercise there is little reason to suppose that rates of appearance would be lower during the exercise trial. On the other hand the observations are consistent with an increase in the activity of lipoprotein lipase, the enzyme on the capillary endothelium which hydrolyses TAG-rich lipoproteins, during recovery from exercise (28). This mechanism may constitute a directed storage of dietary fat into previously-exercised muscle rather than into adipose tissue. In absolute terms, fat oxidation is probably higher during moderate exercise than during exercise which is either vigorous or of low intensity (29) and there may be a relationship between the amount of muscle TAG utilised during the exercise and the degree of the subsequent mitigation of lipaemia.

The enhanced rate of tissue uptake of TAG during the post-exercise period appears to be associated with increased oxidation of fat. Using a model similar to that described earlier we have recently examined the influence of moderate exercise on the metabolic response to a meal of fat and carbohydrate: when a 90 minute bout of exercise at 60% of VO_{2max} was performed approximately 15 hours earlier, the amount of fat oxidised during the fasted state and throughout a 6-hour postprandial period was increased (30). Thus one bout of moderate exercise (walking) appears to result in lower lipaemia and an increase in lipid oxidation during recovery in both fasting and postprandial states.

Endurance-trained athletes, because they possess a large muscle mass with a well-developed microcirculation, may benefit from effects on postprandial lipaemia which are independent of the last exercise session. However, moderate intensity exercise training – at least when engaged in for a period of 12 weeks (31) – does not necessarily confer a long-lasting decrease in lipaemic response. This finding emphasises the importance of the regularity and frequency of exercise in maintaining this aspect of atherosclerotic risk when the intensity is moderate. One might speculate that the interactions of frequency (and possibly duration) with intensity may be different for moderate and vigorous exercise.

INSULIN/GLUCOSE DYNAMICS

One of the features of metabolic hypertensive cardiovascular disease—alternatively Reaven's syndrome (32)—is hyperinsulinaemia; this develops when the body's cells become insensitive to insulin such that higher concentrations are required to stimulate a given biological activity. Insulin resistance has a pivotal role in the development of the characteristic lipid abnormalities of this syndrome (33); it drives serum TAG up (and hence HDL down) when, because insensitivity of adipocytes to insulin fails to suppress the release of non-esterified fatty acids, hepatic synthesis of very low density lipoproteins is increased.

There is evidence that exercise may increase insulin sensitivity in the body's largest insulin-sensitive tissue, skeletal muscle. It has been known for twenty years that middle-aged, well-trained men exhibit low fasting plasma insulin concentrations and improved insulin sensitivity (34). Both regular physical activity (ie training) and an acute bout of exercise increase insulin sensitivity, whereas deconditioning is associated with the development of insulin resistance (35).

During exercise of moderate intensity the main carbohydrate fuel is blood glucose; glucose uptake in the active muscle is stimulated, despite a fall in plasma insulin concentration, indicating that glucose tolerance is improved, probably as a direct effect of contractile activity (36). However, the sensitivity of the muscle glucose transport process to insulin is also increased. Both the direct and insulin-mediated effects of contraction are sustained during recovery; the direct effect lasts for a few hours (37) but the increase in insulin sensitivity persists for 1 to 2 days (36). For this reason the repeated effects of each bout may cumulatively be important in preventing the development of insulin resistance and ultimately non-insulin-dependent (NIDDM, Type II) diabetes.

Whether chronic, long-term adaptive effects of exercise occur remains a matter of debate but this is possible theoretically because skeletal muscle is quantitatively the most important site for glucose transport and metabolism: adipose tissue mass usually falls with regular exercise, whilst muscle mass can increase; both adaptations are important because overweight and obesity (particularly central) are important features of insulin resistance and NIDDM. Improved vascularisation of skeletal muscle with training may also be advantageous in view of the evidence of extensive insulin binding to the vascular endothelium (38).

Recently, an inverse relation between energy expenditure in leisure-time activity and the subsequent development of NIDDM has been shown amongst ex-students of the University of Pennsylvania (39): overall energy expended in exercise was the important factor in prevention of NIDDM, incidence of NIDDM decreasing by some 6 percent for each 2.1 MJ expended per week in physical activity. Although vigorous activities (eg sports, running, swimming) were most protective, moderate activities were also associated with decreased risk. Of course moderate exercise can be sustained for longer periods than more vigorous activities and is, arguably, likely to be more regularly engaged in because of the lower risk of injury; it may therefore make the major contribution for most people to overall exercise energy expenditure in the long term.

The evidence that physical activity may decrease the risk of developing NIDDM is not restricted to men. Women who take part in regular, vigorous exercise have only 67% of the risk of developing NIDDM compared with women who did not exercise every week, independent of body mass index (40). Thus exercise may be effective in the primary prevention of NIDDM as well as in the management of individuals with known disease.

Because of its influence on insulin sensitivity exercise may be an effective adjunct to weight reduction in the treatment of NIDDM. Walking (pace–normal to brisk), which typifies moderate exercise, elicits 50 to 60% of $\dot{V}O_{2max}$—the level of which maintenance of stable blood glucose concentration is most likely. For the many older patients (with lower maximal oxygen uptake values) even slow walking could have the desired effect. Moderate exercise is more likely to be tolerated than harder exercise in this group and a worthwhile contribution to management of the disease will only be achieved if the activity is engaged in regularly and over a long period. One specific problem with walking is microcirculatory damage which often gives patients trouble with their feet: in this case good alternatives would be cycling (stationary if necessary), swimming or dancing.

One view of the metabolic advantage of enhanced insulin sensitivity in physically active individuals has been the decreased load on the ß cells of the pancreas (41). Recent work, however, challenges this view (42). These workers, recognising that athletes have a higher habitual intake of food energy (43), compared the responses to glucose loads between trained and untrained subjects in two different ways: on one occasion both groups received the same absolute glucose load whereas on the other the load provided the same fraction of predetermined daily carbohydrate intake. In addition, blood insulin and glucose levels were monitored during a normal day. Insulin and glucose responses were, as expected, lower in athletes than in untrained men when a standard glucose load was administered. The novel findings were that these responses were actually very similar when the

carbohydrate challenge reflected daily living habits. One interpretation is that these adaptations allow the necessary increase in food intake without the development of potentially
harmful hyperglycaemia. These observations demonstrate how important it is to consider
nutritional implications of changes in levels of physical activity; without such "real-life"
studies at a whole body level our understanding of the integration of adaptive changes will
be incomplete.

HYPERTENSION

During moderate exercise there is marked vasodilation in active skeletal muscle
which persists after exercise. When the body's large muscles are involved this undoubtedly contributes to the sustained lowering of peripheral resistance and blood pressure
which is associated with regular dynamic aerobic exercise.

In normotensive men and women walking at 50% of VO_{2max} (1 hour, 5 day/week,
for 4 weeks) lowered systolic blood pressure by 2 to 3 mm mercury, compared with a period of sedentary living (44). In this study the change was not attributable to weight loss.
In a longer study, hypertensive patients aged 60–69, walked for one hour, three times a
week at an estimated 50% VO_{2max}; after nine months, blood pressure was lower by 20/12
mm mercury, again independent of weight loss and significantly more than the 1/2 mm
mercury fall in controls (45) (Figure 3). Overall, moderate to moderately vigorous exercise—at 40–70% VO_{2max}—appears to be as effective in reducing blood pressure as more
vigorous exercise approaching 85% of VO_{2max} (46).

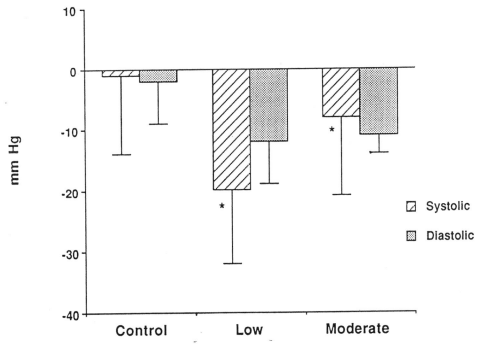

Figure 3. Change in blood pressure in elderly hypertensives after a 9-month programme of either low or moderate
intensity exercise. Mean and SD. Adapted from ref. 45.

In line with these experimental data, epidemological study of middle aged men has shown is a lower occurrence of hypertension among active/fit men relative to their sedentary peers, which is independent of age, smoking, body weight or family history (47, 48).

SKELETAL HEALTH

Osteoporosis, a decrease in bone mass and strength leading to an increase in spontaneous fractures, is a major public health problem in the UK and elsewhere. Current treatment does not significantly restore lost bone, but only decreases the rate of loss. Consequently, prevention is the strategy of choice and is concerned with increasing peak bone mass at skeletal maturity and/or reducing the rate of bone loss after the menopause.

The rate of bone remodelling is subject to several influences, including mechanical loading. Consequently, bed rest or weightlessness result in bone loss (49) whereas increased mechanical loading increases bone mineral content (50). Bone mass in some skeletal sites is markedly higher in athletes than in comparable controls (51) but evidence that professional tennis players show higher bone mineral density in their dominant arm than in their non-playing arm (52) shows that this cannot be attributed solely to constitutional factors. The obvious interpretation is that bone responds locally to loading in man, in a manner consistent with the animal studies. Athletes, however, invariably consume higher intakes of nutrients and it is possible – even probable – that adequate availability of nutrients may be a pre-requisite for optimal exercise-induced influence on bone quality and density.

Although less research has focused on attainment of peak bone mass than on loss in later life, recent findings point to the importance of exercise during the period of skeletal growth. Prospective study for 5 years of 156 healthy, college-aged women showed that physical activity and dietary calcium both exert positive – and quantitatively similar – effects on bone gain (53). None of these women were representative athletes, those with the highest levels of activity participating regularly in recreational rather than competitive sports. Longer follow-up (11 years) of Finnish males and females aged 9 to 18 years at entry found that exercise was an important predictor of bone mineral density – at the hip for both sexes and at the lumbar spine for men (54); no analysis was presented on intensity of activity, only frequency of participation. These studies show the importance of both adequate physical activity and calcium intake during the years of bone growth (up to and into the third decade). No specific information is available on the effect of exercise intensity but theoretically exercise mode is probably more important than (cardiovascular) intensity; because mechanical loading is the stimulus to bone formation, swimming – a popular moderate intensity exercise – will likely be ineffective in this regard.

Theoretically several interventions have the potential to offset the age-related loss of bone mineral in postmenopausal women at risk of, eg exercise, calcium supplementation and hormone replacement therapy, and these are likely to interact. Some studies (eg 55) have found calcium supplementation to be effective in slowing bone loss in postmenopausal women. However, in women with low bone density a reduced rate of loss was seen only when exercise or hormone replacement therapy (HRT) was added to the calcium supplementation (56). Although HRT was the more effective adjunct to the exercise regimen it also caused more side effects. These side effects, with the need for medical supervision of hormone replacement regimens suggest that an exercise and calcium regimen may be the therapy of choice for the many women with intermediate (as opposed to low) bone density values.

What is uncertain is whether a modest amount of less intense physical activity is likely to be effective in the prevention of osteoporosis. The most effective exercise regimen for stimulating an osteogenic response probably involves high loading during short periods of weight bearing activity but experimental evidence does not always confirm this. For example, performing low intensity strength training in addition to aerobic exercise (walking, jogging, dance routines) did not provoke an increase in bone mass greater than that demonstrated by a comparable group of women who undertook aerobic exercise only (57). In contrast, women with a history of osteoporotic fractures who performed simple resistance exercises achieved a 3.8% increase in bone mineral density, compared with a 1.9% loss in a matched control group over the same five month period (58). Consequently there may be some scope for therapeutic exercise to thicken and stabilise the remaining trabeculae, even in established osteoporosis.

There may be a dose-response relationship between levels of physical activity and indices of bone density: in a group of 83 women aged between 30 and 85 those with high levels of activity (evaluated by questionnaire) had higher values for bone mineral content, corrected for skeletal size, than did the moderate and low activity groups (59). From the public health point of view, however, it is important rigorously to examine the efficacy of moderate amounts of physical activity in adult women. Whilst a programme likely to be maximally effective at a number of skeletal sites may be desirable, it may be difficult to implement on any scale. The most easily accessible form of weight bearing exercise is walking and this activity is widely recommended as both a therapeutic and a preventive measure but whether it is sufficient to confer a clinically worthwhile effect is not known.

The frequency of walking uphill has been associated with a 40% reduction in the risk of hip fracture in Asian women (60), an observation concordant with a report of higher bone density of the legs and trunk in postmenopausal women who habitually walked more than seven and a half miles per week compared with those who walked less than 1 mile per week (61). In another cross-sectional study, however, premenopausal women who walked 6 km per working day carrying mail had bone densities at the femoral neck and lumbar spine which did not differ from those of comparable office workers (62).

The number of hours of walking per day has been found to be correlated with lumbar and femoral neck densities (63) but prospective studies have so far yielded conflicting results. White and colleagues studied the influence of walking (progressing to 2 miles on 4 days per week) on the bones of recently menopausal women. Over six months walkers lost bone mineral content at the same rate as did controls (64). In one of the few randomised trials reported, postmenopausal women walking about 7 miles per week over 3 years exhibited a loss of cortical bone density which did not differ from that shown by controls (34). Both these studies relied on measurements of the radius and may therefore have failed to reveal adaptive changes due to local loading.

More recently, broadband ultrasonic attenuation (BUA) has been used to study the effects of brisk walking on bone condition. In women aged 30 to 61 years (mean 44) who walked briskly for 16 to 18 km per week BUA measured in the calcaneus was increased after one year by 12% whereas it decreased by 4% for the controls (65). The site of measurement, the lumbar spine, may be one reason why a recent study employing a similar brisk walking regimen failed to find a beneficial effect on bone density (66). However, BUA of the calcaneus is reported to be a very good predictor of lumbar vertebral bone density (67) so it seems unlikely that the measurement site alone can account for the inconsistency between these two reports. Further work is clearly needed to determine the amount and type of physical activity needed to influence skeletal health.

The detrimental effects on the skeleton of extreme levels of intense exercise can also be invoked to demonstrate the advisability of moderate exercise. Low bone mineral densities have been reported for young women athletes who experience "athletic amenorrha" (68). Loss of oestrogen, one of the factors stimulating bone formation, (possibly secondary to extremely low levels of body fat) results in an excess of bone resorption over bone formation so that density falls and the propensity to fracture is increased. In many of these individuals poor dietary habits with a low intake of calcium and energy (69) may exacerbate the problem. Decreasing the volume of training has been reported to reverse bone loss somewhat (70) but long term effects of the repeated spells of amenorrha experienced by a majority of young women athletes (71) are unknown.

Functional Capacities in the Elderly

For most normal adults less than optimal levels of functional capacity do not place limitations on the activities of daily living. This is not so for the elderly whose functional capacities are reduced to the point where they fall below thresholds which may limit the quality of life. For example, the timing of traffic signals at pedestrian crossings and junctions commonly assumes a walking speed of about 2.5 mph. This is faster than the 2 mph which a typical, healthy, 75 to 80 year old woman finds comfortable (72) so she will feel hurried and insecure.

Whilst some of the decline in capacities is attributable to the aging process per se, a decrease in daily physical activity probably contribute: older men who maintain their activity levels experience a rate of decrease in $\dot{V}O_{2max}$ which is about half that found in sedentary individuals, ie about 5% per decade (73). Moderate exercise is especially suitable for the elderly. As at younger ages, aerobic training increases $\dot{V}O_{2max}$ (74) and improvements in the oxidative capacity of muscle, which determine the individual's stamina, are as great in the elderly as in young adults (75). Again, walking is obviously a sensible form of exercise with little risk of orthopaedic injury; for example, during 13 weeks of training by walking only one injury occurred amongst 57 healthy men and women in their seventies (76).

Summary

Physical inactivity is now recognised as a risk factor for coronary heart disease and, arguably, NIDDM and hypertension. Data on the incidence of hip fracture (60), stroke (77), upper respiratory tract infections (78) and the risk of developing colon cancer (79) suggest additional health gains in physically active persons. Whilst the relative risk is what matters to the individual, the population attributable risk—depending also on its prevalence—is of greater concern for the development of public health policy. Physical inactivity is widespread in Western societies and so the potential to reduce morbidity by stimulating an increase in the physical activity levels of the presently sedentary majority is great. High intensity exercise programmes are poorly adhered to and carry a higher risk of orthopaedic injury and cardiac arrest; the greatest net public health gain is likely to derive from an increase in population levels of moderate intensity exercise.

REFERENCES

1. Allied Dunbar National Fitness Survey: Main Findings. London: Sports Council and Health Education Authority, 1992.

2. Martin JE, Dubbert PM. Exercise applications and promotion in behavioural medicine: current status and future directions. J Consulting Clin Psychol 50:1004–1017, 1982.

3. Pollock ML, Gettman LR, Milesis CA, Bah MD, Durstine L, Johnson RB. Effects of frequency and duration of training on attrition and incidence of injury. Med Sci Sports Exerc 9:31–36, 1977.

4. Siscovick DS, Weiss NS, Fletcher RH, Lasky T. The incidence of primary cardiac arrest during vigorous exercise. N Engl J Med 311:874–877, 1984.

5. Powell KE, Thompson PD, Casperson CJ, Kendrick JS. Physical activity and the incidence of coronary heart disease. Ann Rev Public Health, No. 8:253–287, 1987.

6. Berlin JA, Colditz A. A meta-analysis of physical activity in the prevention of coronary heart disease. Am J Epidemiol 132:612–627, 1990.

7. Blair SN. Physical activity, fitness, and coronary heart disease. In Physical activity, fitness and health. ed Bouchard C, Shephard RJ, Stephens T. Human Kinetics, Champaign, Illinois: 1994, 579–590.

8. Morris JN, Clayton DG, Everitt MG, Semmence AM, Burgess EH. Exercise in leisure-time: coronary attack and death rates. Br Heart J 63:325–334, 1990.

9. Shaper AG, Wannamethee G. Physical activity and ischaemic heart disease in middle-aged British men. Br Heart J 66:384–394, 1991.

10. Paffenbarger RS, Hyde RT, Wing AL, Hsieh C-C. Physical activity, all-cause mortality, and longevity of college alumni. New Engl J Med 314:605–613, 1986.

11. Wood PD. Exercise, plasma lipids, weight regulation. In Exercise, heart, health. London: Coronary Prevention Group, 1987.

12. Wood PD, Terry RB, Haskell WL. Metabolism of substrates: diet, lipoprotein metabolism, and exercise. Fed Proc 44:358–363, 1985.

13. Saris WHM. Physiological aspects of weight cycling. Am J Clin Nutr 49:99–104, 1989.

14. Haskell WL. The influence of exercise training on plasma lipids and lipoproteins in health and disease. In: Physical activity in health and disease, eds Astrand P-O & Grimby G. Acta Med Scand, Suppl. 711: 25–37, 1986.

15. Wood PD, Stefanick ML, Dreon DM et al. Changes in plasma lipids and lipoproteins in overweight men during weight loss through dieting as compared with exercise. N Engl J Med 319:1173–1179, 1988.

16. Hubert HB, Feinleib M, McNamara PM, Castelli WP. Obesity as an independent risk factor for cardiovascular disease: a 26-year follow-up of participants in the Framingham Heart Study. Circulation 67:968–977, 1983.

17. Manson JE, Colditz GA, Stampfer MJ et al. A prospective study of obesity and risk of coronary heart disease in women. N Engl J Med 322:882–889, 1990.

18. National Cholesterol Education Program. Report of the expert panel on detection, evaluation and treatment of high blood cholesterol in adults. Bethesda, Maryland. National Institutes of Health. DHHS publication no. (NIH) 8902925, 1989.

19. Wood PD, Stefanick ML, Williams PT, Haskell WL. The effects on plasma lipoproteins of a prudent weight-reducing diet, with or without exercise, in overweight men and women. N Engl J Med 325:461–466, 1991.

20. Zilversmit DB. Atherogenesis: A postprandial phenomenon. Circulation 60:473–485, 1979.

21. Florén C-H, Albers JJ, Bierman EL. Uptake of chylomicron remants causes cholesterol accumulation in cultured human arterial smooth muscle cells. Biochem Biophys Acta 663:336–349, 1981.

22. Schneeman BO, Kotite L, Todd KM, Havel RJ. Relationships between the responses of triglyceride-rich lipoproteins in blood plasma containing apolipoproteins B-48 and B-100 to a fat-containing meal in normolipidemic humans. Proc Natl Acad Sci USA 90:2069–73, 1993.

23. Patsch JR, Miesenböck G, Hopferwieser T et al. Relation of triglyceride metabolism and coronary artery disease. Studies in the postprandial state. Arteriosclerosis Thrombosis 12:1336–1345, 1992.

24. Groot PHE, van Stiphout WAHJ, Krauss XH et al. Postprandial lipoprotein metabolism in normolipidemic men with and without coronary artery disease. Arteriosclerosis Thrombosis 11:653–662, 1991.

25. Karpe F, Steiner G, Uffelman K, Olivecrona T, Hamsten A. Postprandial lipoproteins and progression of coronary atherosclerosis. Atherosclerosis 106:83–97, 1994.

26. Merrill JR, Holly RG, Anderson RL et al. Hyperlipemic response of young trained and untrained men after a high fat meal. Arteriosclerosis 9:217–223, 1989.

27. Aldred HE, Perry IC, Hardman AE. The effect of a single bout of brisk walking on postprandial lipemia in normolipidemic young adults. Metabolism 43:836–841.

28. Keins B, Lithell H, Mikines KJ, Richter EA. Effects of insulin and exercise on muscle lipoprotein lipase activity in man and its relation to insulin action. J Clin Invest 84:1124–1129, 1989.

29. Romijn JA, Coyle EF, Sidossis LS, Gastal Delli A, Horowitz JF, Endert E, Wolfe RR. Regulations of endogenous fat and carbohydrate metabolism in relation to exercise intensity and duration. Am J Physiol 265:E380-E391, 1993.

30. Tsetsonis NV, Roche DM, Hardman AE. The influence of treadmill walking on the lipaemic and metabolic responses to a high-fat meal in women aged 35 to 50 years. Proc Nutr Soc, in the press.

31. Aldred HE, Hardman AE, Taylor S. The influence of 12 weeks of training by brisk walking on postprandial lipemia and insulinemia in sedentary, middle-aged women. Metabolism 44:390–397, 1995.

32. Reaven GM. Role of insulin resistance in human disease. Diabetes 37:1595–1607.

33. Frayn KN. Insulin resistance and lipid metabolism. Curr Opin Lipidol 4:194–204, 1993.

34. Björntorp P, Fahlén M, Grimby G, Gustafson A, Holm J, Renström P and Scherstén T. Carbohydrate and lipid metabolism in middle-aged, physically well-trained men. Metabolism 21:1037–1044, 1972.

35. Horton ES. Exercise and physical training: effects on insulin sensitivity and glucose metabolism. Diab Metab Rev 2:1–17, 1986.

36. Wallberg-Henriksson H, Constable SH, Young DA, Holloszy JO. Glucose transport into rat skeletal muscle: interaction between exercise and insulin. J Appl Physiol 65:909–913, 1988.

37. Young DA, Wallberg-Henriksson H, Sleeper MD, Holloszy JO. Reversal of the exercise-induced increase in muscle permeability to glucose. Am J Physiol 253:E331-E335, 1987.

38. Holmäng A, Björntorp P, Rippe B. Tissue uptake of insulin and inulin in red and white skeletal muscle in vivo. Am J Physiol 263:H1-H7, 1992.

39. Helmrich SP, Ragland DR, Leung RW, Paffenbarger RS. Physical activity and reduced occurrence of non-insulin-dependent diabetes mellitus. N Engl J Med 325:147–152, 1991.

40. Manson JE, Rimm EB, Stanpfer MJ et al. Physical activity and incidence of non-insulin-dependent diabetes mellitus in women. Lancet 338: 774–778, 1991.

41. Consensus Development Panel: Consensus development conference on diet and exercise in NIDDM. Diabetes Care 10:639–644, 1987.

42. Dela F, Mikines KJ, von Linstow M, Galbo H. Does training spare insulin secretion and diminish glucose levels in real life? Diabetes Care 15, suppl 4:1712–1715, 1992.

43. Walker M. Dietary planning for performance. in Nutrition in Sport. Milton Keynes, Shaklee, 1986.

44. Kingwell BA, Jennings GL. Effects of walking and other exercise programs upon blood pressure in normal subjects. Med J Australia 158:234–238, 1993.

45. Hagberg JM, Montain SJ, Martin WH, Ehsani AA. Effect of exercise training in 60- to 69-year-old persons with essential hypertension. Am J Cardiol 64:348–353, 1989.

46. Arroll B, Beaglehole R. Does physical activity lower blood pressure: a critical review of the clinical trials. J Clin Epidemiol. 45:439–447, 1992.

47. Paffenbarger RS, Wing AL, Hyde RT, Jung DL. Physical activity and incidence of hypertension in college alumni. Am J Epidemiol 117:245–257, 1983.

48. Blair SN, Goodyear NN, Gibbons LW, Cooper KH. Physical fitness and incidence of hypertension in healthy normotensive men and women. J Am Med Assoc 252:487–490, 1984.

49. Krølner B & Toft B. Vertebral bone loss: an unheeded side effect of therapeutic bedrest. ClinSci 64:537–540, 1983.

50. Rubin CT & Lanyon LE. Regulation of bone formation by applied dynamic loads. J BoneJoint Surg, 66A:397–402, 1984.

51. Lane NE, Bloch DA, Jones HH et al. Long distance running, bone density and osteoarthritis. J Am MedAssoc 255:1147–1151, 1986.

52. Pirnay F, Bodeux M, Criwelaard JM, Franchimont P Bone mineral content and physical activity. Int J Sports Med 8:331–335, 1987.

53. Recker RR, Davies KM, Hinders SM et al. Bone gain in young adult women. J Am Med Assoc 268:2403–2408, 1992.

54. Välimäki Mj, Kärkkäinen M, Lamberg-Allardt C et al. Exercise, smoking, and cal

55. Reid IR, Ames RW, Evans MC, Gamble GD, Sharpe SJ. Effect of calcium supplementation on bone loss in postmemopausal women. N Engl J Med 328:460–464, 1993.

56. Prince RL, Smith M, Dick IM et al. Prevention of postmenopausal osteoporosis. A comparative study of exercise, calcium supplementation, and hormone-replacement therapy. N Engl J Med. 325:1189–1195, 1991.

57. Chow R, Harrison JE & Notarius C. Effect of two randomised exercise programmes on bone mass in healthy postmenopausal women. Br MedJ 295:1441–1444, 1987.

58. Simkin A, Ayalon J, Leichter I. Increased trabecular bone density due to bone-loading exercises in postmenopausal osteoporotic women. Calcif Tiss Int 40:59–63, 1987.

59. Zylstra S, Hopkins A, Erk M, Hreshchyshyn MM, Anbar M Effect of physical activity on lumbar spine and femoral neck bone densities. IntJSports Med 10, 181–186, 1989.
60. Lau E, Donnan S, Barker DJP, Cooper C. Physical activity and calcium intake in fracture of the proximal femur in Hong Kong. Br Med J 297:441–443, 1988.
61. Krall EA, Dawson-Huges B. Walking is related to bone density and rates of bone loss. Am J Med 96:20–26, 1994.
62. Uusi-Rasi K, Nygård C-H, Oja P, Pasanen M, Sievänen, Vuori I. Walking at work and bone mineral density of postmenopausal women. Osteoporosis Int., in the press.
63. White MK, Martin RB, Yeater RA, Butcher RL, Radin EL. The effects of exercise on the bones of postmenopausal women. Int Orthop 7:209–214, 1984.
64. Cauley JA, Sandler R, LaPort RE et al. Physical activity and postmenopausal bone loss: results of a three year randomised clinical trial. Am J Epidemiol 24:525, 1986.
65. Jones PRM, Hardman AE, Hudson A, Norgan NG. Influence of brisk walking on the broadband ultrasonic attenuation of the calcaneus in previously sedentary women aged 30 to 61 years. Calcif Tiss Int 49:112–115, 1991.
66. Cavanaugh DJ, Cann, CE. Brisk walking does not stop bone loss in postmenopausal women. Bone 9:201–204, 1988.
67. Baran DT, Kelly AM, Karellas A et al. Ultrasonic attenuation of the os calcis in women with osteoporosis and hip fractures. Calcif Tiss Int 43:138–142, 1988.
68. Cann CE, Martin MC, Genant HK, Jaffe RB. Decreased spinal mineral content in amenorrheic women. J Am Med Assoc 251:626–629, 1984.
69. Nelson ME, Fisher EC, Catsos PD et al. Diet and bone status in amenorrheic runners. Am J Clin Nutr 43:910–916, 1986.
70. Drinkwater BL, Nilson K, Ott S, Chesnut CH. Bone mineral density after resumption of menses in amenorrheic athletes. J Am Med Assoc 256:380–382, 1986.
71. Drinkwater BL, Bruemner B, Chesnut CH. Menstrual history as a determinant of current bone density in young athletes. J Am Med Assoc 263:545–548, 1990.
72. Danneskiold-Samsøe B, Kofod V, Muner J et al. Muscle strength and functional capacity in 78–81-year-old men adn women. Europ J Appl Physiol. 52:310–314, 1984.
73. Hagberg JM. Effect of training on the decline of VO_{2max} with aging. Fed Proc 46:1830–1833, 1987.
74. Blumenthal JA, Emergy CF, Madden DJ et al. Cardiovascular and behavioural effects of aerobic exercise in healthy older men and women. J Gerontol 44:M147-M157, 1989.
75. Meredith CN, Frontera WR, Fisher EC et al. Peripheral effects of endurance training in young and old subjects. J Appl Physiol 66:2844–2849, 1989.
76. Pollock MJ, Carroll JF, Graves JE et al. Injuries and adherence to walk/jog and resistance training programs in the elderly. Med Sci Sports Exerc 23:1194–200, 1991.
77. Wannamethee G, Shaper AG. Physical activity and stroke in British middle-aged men. BMJ 304:597–601, 1992.
78. Nieman DC. Exercise, upper respiratory tract infection, and the immune system. Med Sci Sports Exerc 26:128–139, 1994.
79. Lee I-Min, Paffenbarger RS, Hsieh C-C. Physical activity and risk of developing colorectal cancer among college alumni. J Nat Cancer Inst 83:1324–1329, 1991.

47

METABOLIC AND VENTILATORY EFFECTS OF CAFFEINE DURING LIGHT INTENSITY EXERCISE IN TRAINED AND SEDENTARY LOW HABITUAL CAFFEINE USERS

H.-J. Engels,[1] J.C. Wirth,[1] and E.M. Haymes[2]

[1] Division of HPR-Exercise Science
Wayne State University, Detroit, Michigan 48202
[2] Department of Nutrition, Food and Movement Sciences
Florida State University
Tallahassee, Florida 32306

SUMMARY

This study examined selected physiological effects of caffeine during constant load, light intensity exercise in aerobically trained (TM) and sedentary (SM) low habitual caffeine users (<150 mg·day^{-1}). Twelve healthy nonsmoking males (6 TM, 6 SM), following a 12-hr caffeine and food abstinence, participated in two separate 45-min trials consisting of constant load treadmill exercise at 30% of $\dot{V}O_{2max}$ after caffeine (5 mg·kg^{-1}) or placebo administered 60 min prior to experimental data acquisition. It was observed that pre-trial caffeine intake lowered respiratory exchange ratio (RER) and increased oxygen uptake (l·min^{-1}) and metabolic rate (kJ·min^{-1}) during light intensity exercise ($p<0.05$). Minute ventilation (l·min^{-1}) and tidal volume (ml·min^{-1}) was elevated in caffeine when compared to placebo trials ($p<0.05$). On the other hand, caffeine had no effect on frequency of breathing (breath·min^{-1}) and carbon dioxide output via the lungs (l·min^{-1}). There was no statistically significant difference in physiological response to caffeine between the SM and TM subject groups for any of the variables studied. It is concluded that caffeine enhances fat oxidation, augments exercise metabolic rate, and alters ventilatory dynamics during light intensity exercise. The present findings indicate that these effects are similar in aerobically trained and sedentary low habitual caffeine consumers.

INTRODUCTION

Caffeine is a methylated xanthine found in a large number of beverages and foods throughout the world. Like the closely related compounds theophylline and theobromine,

Current Research in Sports Sciences, edited by Rogozkin and Maughan.
Plenum Press, New York, 1996

it is classified as an alkaloid and has a variety of pharmacological properties which are commonly used for therapeutic purposes (Rall 1985). Many scientific studies have examined the effects of caffeine and its health consequences in man (Curatolo and Robertson 1983). The findings have shown that caffeine influences virtually every organ system in the human body and that these effects may be quite variable from person to person (Leonard, Watson and Mohs 1987). Although some researchers have expressed concern about high levels of caffeine consumption, a proposed link between established health risks and caffeine is far from being established (Curatolo and Robertson 1983).

Caffeine is among the various substances that have been hypothesized to affect exercise performance. In the past, a primary interest has been the study of caffeine's ability to influence metabolism and alter substrate utilization during prolonged exercise. While some studies were designed to examine the potential of caffeine to enhance the endurance capacity of elite athletes during submaximal exercise of high aerobic intensity (Costill et al. 1978; Graham and Spriet 1991; Tarnopolsky et al. 1989), other investigations employed prolonged exercise using more moderate work loads ($\leq$ 55% VO_{2max}) and untrained, sedentary study populations (Chad and Quigley 1989; Donelly and McNaughton 1992; Engels and Haymes 1992). In general, past metabolic findings from exercise studies of caffeine have shown considerable variation and are often difficult to interpret. Although the inconsistency in findings has been attributed to a number of factors, according to Bucci (1993), some of the research variables in human experimental studies of caffeine that are difficult to control may greatly affect results.

Among the potential confounding variables of caffeine research, the influence of habituation and the development of tolerance to caffeine on exercise responses have recently received increased research attention (Bangsbo et al. 1992; Brown et al. 1991; Fisher et al. 1986; Tarnopolsky et al. 1989; Van Soeren et al. 1993). In contrast, the influence of aerobic training status on caffeine's physiological effects during exercise has not been thoroughly examined (Knapik et al., 1983; Toner et al., 1982). It is known that aerobic training results in important biological adaptations that alters the functions of many organ systems during exercise (Åstrand and Rohdal 1986). Conceivably, some of these changes may also affect the physiological response to caffeine ingestion. Therefore, the primary purpose of this study was to compare the effects of prior caffeine intake on metabolic responses and ventilatory adjustments during prolonged submaximal exercise in aerobically trained and sedentary individuals. A secondary objective was the attempt to reproduce findings from a previous exercise study (Engels and Haymes 1992) obtained in a different group of sedentary males with similar caffeine consumption habits. Light intensity exercise was chosen to enhance the limited existing knowledge of caffeine's effects during constant load exercise at an exertion level that is typical of many work, leisure, and household activities. Since at the present time the available research on caffeine habituation and tolerance is incomplete, only individuals with similar low caffeine consumption habits were selected.

METHODS

Subjects

Twelve nonsmoking males between the ages of 20 to 32 years of age volunteered as subjects for this study. They were selected on the basis of their aerobic capacity and low habitual caffeine intake levels (<150 mg·day^{-1}). Prior to participation in the study, all volunteers were given a complete explanation of procedures and possible risks and discomforts involved in the study and informed consent was obtained. The experimental protocol was approved by the institutional review board regarding the use of human subjects.

Preexperimental Procedures

Two preliminary laboratory visits preceded the two experimental testing sessions. In the first meeting, subjects were questioned about their health and medical history, nonsmoking status, and regular habitual physical activity level. In addition, they completed a detailed questionnaire (coffee, softdrinks, tea, medications, etc.) regarding their habitual caffeine consumption habits without the knowledge of the 150 mg·day^{-1} limit for inclusion in the study. If volunteers met the first criteria for participation in the study, they then performed a standard progressive treadmill ergometer test until voluntary exhaustion to determine VO_{2max}. If the results of the VO_{2max} test indicated that the subjects were either in the ≥90th percentile (TM group) or in the ≤50th percentile (SM group) of their age and sex specific aerobic power (ml·kg^{-1}·min^{-1}) (Pollock and Wilmore 1990), they were included in the study. During the second preliminary laboratory visit subjects were introduced to the experimental procedures and a submaximal work load corresponding to 30% of VO_{2max} was established which was subsequently employed during the two experimental trials. In addition, body composition was determined using a standard underwater weighing technique (Pollock and Wilmore 1990).

Experimental Procedures

All experimental tests were performed in the morning following a 12-hr overnight fast and caffeine abstinence. The protocol for each of the tests was identical with the exception of the contents of the test drink (caffeine vs. placebo). As soon as the subjects reported to the laboratory they were given either the caffeine (CAF) or placebo (CON) solution. A 5 mg·kg^{-1} caffeine dosage was chosen based on previous research studies. The subjects then rested in a chair in the laboratory for 60 min before the start of exercise and were only allowed to leave the laboratory if they had to use the bathroom.

During each exercise trial, respiratory gases and airflow were measured continuously by open circuit spirometry indirect calorimetry and collected at 5-min intervals. Expired air was analyzed using a model S-3A oxygen analyzer and a model CD-3A electrochemical carbon dioxide analyzer (Applied Electrochemistry, Pittsburgh, PA). Both gas analyzers continuously sampled expired air at a flow rate of 100 ml·min^{-1} through Drierite anhydrous $CaSO_4$ (W.A. Hammond, Xenia, OH) from a Vacuumed gas mixing chamber. Before each test, gas analyzers were calibrated with gases of known concentration using a primary standard medical gas mixture. Subjects breathed through a low resistance, large 2-way Hans Rudolph breathing valve (Model # 2700, Hans Rudolph, Kansas City, MO). Inspired air flow was measured with a turbine volume meter (Applied Electrochemistry, Pittsburgh, PA). Exercise metabolic rate (kJ·min^{-1}), including percent fat oxidation, was computed from measured oxygen consumption and the caloric equivalent per liter oxygen at a given respiratory exchange ratio (Zuntz 1901). Tidal volume (TV, ml·min^{-1}) was calculated from minute ventilation VE, l·min^{-1}) and the frequency of breathing (f_b, breath·min^{-1}).

Anhydrous caffeine (Firma Caesar & Loretz, Hilden, Germany) was used as the caffeine source. Prior to oral administration, 5 mg·kg^{-1} body weight anhydrous caffeine was dissolved in 240 ml of distilled water which was sweetened with an artificial sweetener. A placebo solution consisted of 240 ml of distilled water and the same artificial sweetener only. The addition of the sweetener served to disguise the taste of caffeine and to assure that subjects did not know whether they had received caffeine or not. The study was performed using a placebo controlled, double blind design. The two experimental trials were spaced one week apart and their order was randomized and counterbalanced.

Statistical Methods

Differences across training groups were identified using a one-way analysis of variance. All experimental data were analyzed using a three factor mixed design ANOVA (group, treatment, time) with repeated measures on two factors (treatment, time). Duncan's multiple range test was used to determine the significance of interactions and differences across exercise time. A probability of rejecting the null hypothesis of less than 0.05 was considered a statistically significant difference. All data are presented as mean values ± standard error of the mean (M ± SEM).

RESULTS

Subjects

The physical characteristics of the SM and TM subject groups are presented in Table 1. In addition to a significantly higher VO_{2max} ($ml \cdot kg^{-1} \cdot min^{-1}$) the TM group had a lower body fat percentage ($p < 0.05$); however, both groups were similar in age, weight, and height ($p > 0.05$). All subjects regularly consumed some caffeine but there was no significant difference between the two groups in view of their absolute ($mg \cdot day^{-1}$) or relative ($mg \cdot kg^{-1} \cdot day^{-1}$) low habitual caffeine consumption levels ($p > 0.05$). The estimated daily caffeine intake level of the SM subjects was 77.1 ± 21.5 $mg \cdot day^{-1}$ (1.02 ± 0.3 $mg \cdot kg^{-1} \cdot day^{-1}$) as compared to 65.0 ± 12.7 $mg \cdot day^{-1}$ (0.96 ± 0.2 $mg \cdot kg^{-1} \cdot day^{-1}$) for the TM group.

Caffeine Compared to Placebo

Caffeine intake 60-min prior to exercise caused a significant rise in exercise VO_2 ($l \cdot min^{-1}$) and reduction in RER in sedentary and trained subjects (Fig. 1).

Changes in RER suggest that a 14.3% increase in fat usage had occurred in caffeine (CAF: $54.3 \pm 1.9\%$ fat oxidation) compared to control (CON: $40.0 \pm 1.7\%$ fat oxidation) trials ($p < 0.05$). Together, the caffeine mediated alterations in VO_2 and RER transposed into a 0.46 $kJ \cdot min^{-1}$ increase in metabolic rate from 20.87 ± 0.3 $kJ \cdot min^{-1}$ to 21.33 ± 0.3 $kJ \cdot min^{-1}$ during light intensity exercise ($p < 0.05$).

Table 1. Basic characteristics of subjects

	TM (n=6)	SM (n=6)
Age (yrs)	25.7 ± 1.8	24.2 ± 1.1
Height (cm)	177.3 ± 1.4	178.7 ± 3.5
Weight (kg)	69.1 ± 2.1	79.7 ± 5.3
Body fat (%)	10.2 ± 0.9	18.3 ± 1.9
VO_{2max} ($ml \cdot kg^{-1} \cdot min^{-1}$)	56.6 ± 1.4	40.2 ± 0.7

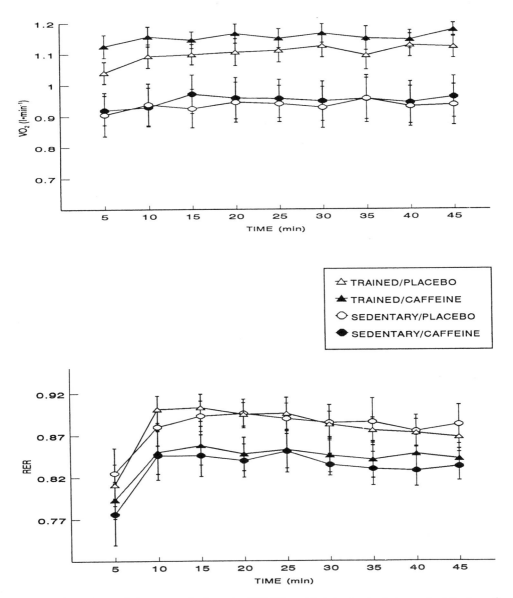

Figure 1. VO_2 and RER during constant load exercise (30% VO_{2max}) in trained and sedentary males following caffeine or placebo intake.

Significantly higher minute ventilation volumes ($l \cdot min^{-1}$) (Fig. 2) in caffeine trials were achieved through an enlargement of tidal volumes. Mean tidal volumes in CAF trials were 1260.9 ± 20.2 $ml \cdot min^{-1}$ as compared to 1192.4 ± 18.4 $ml \cdot min^{-1}$ in CON trials ($p < 0.05$). Caffeine had no effect on breathing frequency (CON: 19.2 ± 0.3 $breath \cdot min^{-1}$, CAF: 18.9 ± 0.2 $breath \cdot min^{-1}$) and did not affect carbon dioxide output via the lungs (Fig. 2). No further interactive effects between group and treatment factors were noted.

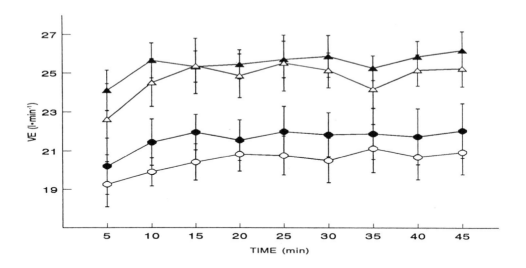

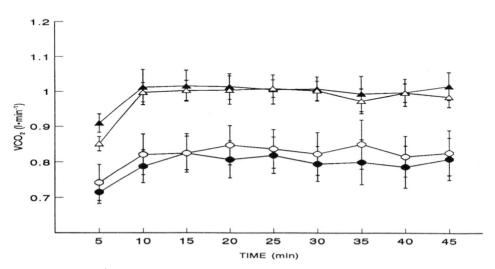

Figure 2. $\dot{V}E$ and $\dot{V}CO_2$ responses during light intensity exercise in trained and sedentary males following caffeine or placebo intake.

SM Compared to TM Subjects

As expected, with and without caffeine, TM subjects exhibited a significantly higher exercise O_2 uptake ($l \cdot min^{-1}$), CO_2 production ($l \cdot min^{-1}$) and metabolic rate ($kJ \cdot min^{-1}$) compared to the SM group while exercising at the same relative exercise intensity (30% VO_{2max}). Mean constant load energy expenditure rates were 23.01 ± 0.2 $kJ \cdot min^{-1}$ and 19.19 ± 0.3 $kJ \cdot min^{-1}$ in TM and SM subjects, respectively. There was no difference in the percentage contribution of fat (TM $46.3 \pm 1.8\%$ fat oxidation, SM $47.9 \pm 2.0\%$ fat oxidation) as measured by RER between the two subject groups ($p > 0.05$).

It was observed that a significantly higher exercise minute ventilation volume ($l \cdot min^{-1}$) in TM subjects was achieved primarily through a larger tidal volume ($ml \cdot min^{-1}$) ($p < 0.05$); in contrast, breathing frequency ($breath \cdot min^{-1}$) was similar compared to the SM group. Experimental findings on exercise ventilatory dynamics in trained and sedentary males with and without prior caffeine intake are summarized in Figure 3.

Changes across Exercise Time

Small time dependent shifts in constant load exercise responses were noted across the 45-min trial duration. These drifts were mainly apparent during the initial phase of the exercise bout and are generally consistent with the existing research knowledge of physiological changes during steady rate prolonged submaximal exercise (Ekelund 1967). There was no significant interaction present between the time factor and either the group and/or treatment factor.

DISCUSSION

The primary purpose of this study was to compare the effects of caffeine on metabolic responses and ventilatory adjustments in aerobically trained and sedentary males during prolonged light intensity exercise. It was observed that pre-trial caffeine intake resulted in an elevation of oxygen uptake during subsequent constant load exercise. Although the mean increase in VO_2 following caffeine was on the average higher in aerobically trained compared to sedentary subjects, no statistically significant group difference was observed ($p = 0.15$). Caffeine's effect on oxygen consumption was paralleled by a reduction in respiratory exchange ratio during light intensity exercise in both trained and sedentary subjects. The observed changes in oxygen uptake and RER are in agreement with several recent studies that examined untrained females during prolonged moderate exercise (55% VO_{2max}) (Chad and Quigley 1989; Donelly and McNaughton 1992) and, collectively, suggest that the rise in VO_2 following caffeine is related to a shift in metabolic substrate usage.

Previous resting studies have shown that training status may be an important variable to determine the metabolic response to caffeine (LeBlanc et al. 1985; Poehlman et al. 1985). However, questions remain whether caffeine exerts a more pronounced metabolic effect in aerobically trained (LeBlanc et al. 1985) or in sedentary (Poehlman et al. 1985) resting individuals. Two previous exercise studies that directly compared trained and untrained subjects both failed to observe a caffeine effect on substrate use (as measured by RER) during prolonged constant load exercise at 60% of VO_{2max} (Knapik et al. 1983) and during an incremental exercise test (Toner et al. 1982). Toner et al. (1982) also reported no influence of caffeine on exercise oxygen uptake in either trained and untrained sub-

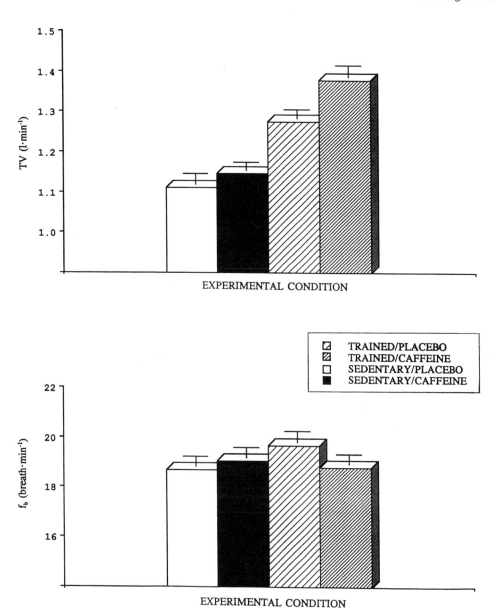

Figure 3. Ventilatory dynamics in trained and sedentary subjects following caffeine or placebo intake. Columns represent the integrated response for the total 45-min constant load exercise period.

jects, whereas Knapik et al. (1983) noted a training dependent difference to caffeine in serum energy substrate levels, but did not report oxygen uptake.

Examination of other pertinent studies suggests that exercise RER values were not reduced following caffeine in trained (Casal and Leon 1985; Graham and Spriet 1991; Tarnopolsky et al. 1989) and untrained subject populations (Engels and Haymes 1992; Van Soeren et al. 1993) during prolonged submaximal exercise. Similarly, oxygen uptake has been shown not to be systematically changed by caffeine in a number of studies that employed either trained athletes (Brown et al. 1991; Casal and Leon 1985; Costill et al. 1978; Tarnopolsky et al. 1989) or sedentary individuals (Engels and Haymes 1992). The available evidence does not yet provide a conclusive answer on the importance of training status to alter caffeine's metabolic effects both at rest and during exercise.

The caffeine mediated changes in $\dot{V}O_2$ and RER in the present study indicate a modest 0.46 kJ·min^{-1} increase (+2.2%) in metabolic rate during light intensity exercise. Moreover, findings by Donelly and McNaughton (1992) in untrained women with similarly low habitual caffeine intake levels indicate that this effect can be expected to continue for some time into the postexercise recovery period. However, from an applied physiological viewpoint this transient caffeine-induced rise in thermogenesis is probably too small for sedentary individuals to be considered a viable means to increase the energy cost of light intensity exercise for control of body weight. This is particularly true considering that tolerance to caffeine's physiological effects during exercise has been observed to develop in habitual users (Fisher et al. 1986). For aerobically trained subjects, increasing fat oxidation is a desirable outcome to enhance endurance capacity; however, a caffeine induced rise in energy expenditure while exercising at the same submaximal work load theoretically could impair performance ability for prolonged activity. Whether the observed small thermogenic effect of caffeine has any true ergolytic consequences must be seriously questioned though. From a quantitative perspective, the present findings suggest that pretrial caffeine resulted in only a modest 5 kcal increase in total caloric expenditure over the 45 min exercise duration in study participants.

An increase in minute ventilation following caffeine has been reported in several prolonged exercise studies at both high (Casal and Leon 1985; Spriet et al. 1992) and low constant work loads (Engels and Haymes 1992). Findings from the present investigation suggest that the increase in pulmonary ventilation was primarily achieved by an increase in the depth of breathing. Using a smaller caffeine dose (3.3 mg·kg^{-1}), Brown et al. (1991) also observed elevated tidal volumes following caffeine during prolonged constant load exercise at 50% of $\dot{V}O_{2max}$ in both low and high caffeine consuming subjects. However, these increases were opposed by equal reductions in breathing frequency so that minute ventilation volumes were not changed. In contrast, Powers et al. (1986) in a single-blind study noted a rise in minute ventilation following ingestion of 7 mg·kg^{-1} caffeine to be due to increases in breathing frequency with no change in tidal volume. D'Urzo et al. (1990), following 650 mg caffeine, observed both increases in tidal volume and frequency of breathing to elevate minute volume. Both of these investigations employed a short-term exercise protocol and prescribed workload below the ventilatory/anaerobic threshold. Although other factors may also be important, it appears that the ability of caffeine to alter pulmonary ventilation mechanics during exercise varies depending on dosage and may not necessarily involve a change in minute volume (Brown et al. 1991). According to D'Urzo et al. (1990) both central and peripheral mechanisms may be involved in mediating the respirogenic action of caffeine in humans. An evaluation of the dynamics of pulmonary ventilation changes following caffeine intake remains an area of study which requires additional clarification.

Although rarely reported in prolonged exercise studies of caffeine, a measurement of carbon dioxide production is important not only due to caffeine's known respiratory effect but also because of its potential to affect acid-base balance, both of which can result in an increase in CO_2 output. This would cause an inflation of the $\dot{V}CO_2/\dot{V}O_2$ ratio and, hence, invalidate the use of RER to estimate substrate utilization. In agreement with two previous submaximal exercise studies (Casal and Leon 1985; Engels and Haymes 1992), the present findings indicate no systematic rise in $\dot{V}CO_2$ despite consistent increases in VE following 5 $mg \cdot kg^{-1}$ caffeine. Two other prolonged exercise studies also reported no change in CO_2 homeostasis following either administration of 3.3 $mg \cdot kg^{-1}$ (Brown et al. 1991) or 6 $mg \cdot kg^{-1}$ (Tarnopolsky et al. 1989) caffeine prior to exercise. However, both of these investigations also failed to observe increased minute volumes in caffeine trials. Furthermore, Tarnopolsky et al. (1989) reported similar plasma lactate levels in caffeine and control trials. It is important to note that the work by Powers et al. (1986) and D'Urzo et al. (1990) indicates an increase in minute volume is associated with a rise in $\dot{V}CO_2$ during short term, constant load exercise below the ventilatory/anaerobic threshold when somewhat higher caffeine dosages (≥ 7 $mg \cdot kg^{-1}$) are used. It would be helpful in future exercise studies of caffeine to routinely report pulmonary CO_2 outflow results as this is important for the validity of estimating substrate usage based on the respiratory exchange ratio. Although the evidence is limited, this seems to be particularly important when high caffeine dosages are used in the research protocol.

An ancillary, but nonetheless important objective of this study was the attempt to reproduce selected findings from a previous investigation conducted in a different group of male subjects who had similar physical and caffeine intake characteristics (Engels and Haymes 1992). While caffeine induced changes in minute ventilation and carbon dioxide production were successfully replicated in the present examination, statistical analysis of the oxygen uptake and RER responses lead to different study conclusions. The variation in the upper threshold of habitual caffeine consumption level (300 vs. 150 $mg \cdot day^{-1}$) and the small difference in pre-trial caffeine abstinence periods (12 vs. 36 hrs) for admission to the two studies do not seem to offer suitable explanations for the difference in the metabolic results. In both investigations, the pre-trial caffeine withdrawal period served to assure that essentially all dietary caffeine had been eliminated at the time of the test (Blanchard and Sawers 1983). On the other hand, findings by other investigators provide some insight why it may be difficult to reproduce the precise metabolic effects of caffeine. Bangsbo et al. (1992) recently noted that caffeine's most important mechanism of action to alter metabolism is probably an epinephrine response. In agreement with this hypothesis, caffeine's metabolic effects may be quite variable because epinephrine has the ability to stimulate both lipolysis and glycogenolysis in various tissues (Bangsbo et al. 1992). According to van Soeren et al. (1993), the activity of some of the breakdown products of caffeine metabolism may play an important part in determining the effects of caffeine intake. In addition, pharmacological research studies indicate a substantial biological variability to caffeine ingestion in humans (Blanchard and Sawers 1983; Grant et al. 1983) which may contribute to variable results. Although these observations do not offer a definite answer, they might help explain the inconsistency among metabolic findings that is seen in both resting and exercise studies of caffeine and that was noted between the present and earlier work (Engels and Haymes 1992) under similar experimental control conditions.

CONCLUSIONS

In view of the present findings the following conclusions can be drawn from this study: (1) caffeine intake one hour prior to exercise results in an enhanced fractional contribution of fat for energy and exerts a mild thermogenic effect during constant load light intensity exercise; (2) the respiratory stimulant action of caffeine may result in an increase of the exercise minute ventilation volume without a necessary rise in carbon dioxide output via the lungs; (3) the caffeine mediated increase in $\dot{V}E$ is primarily achieved through an enhancement of the depth of breathing; and (4) the effects of caffeine on metabolism and on the mechanics of pulmonary ventilation are similar in aerobically trained and sedentary low habitual caffeine users. It is also suggested that, despite similar experimental controls, caffeine's precise effects during exercise may be difficult to reproduce.

RECOMMENDATIONS FOR FUTURE RESEARCH

Only a few studies have directly compared the effects of caffeine in aerobically trained and sedentary/untrained subjects either at rest or during exercise and so far all of them have been cross-sectional in nature. It would be interesting to conduct a longitudinal research study in which the same individuals are followed and their response to caffeine is tested before and again after successfully completing an aerobic training program.

Energy substrate utilization is commonly determined from respiratory gas exchange measurements using the relationship between CO_2 production and O_2 consumption. Since an effect of caffeine on carbon dioxide output via the lungs may invalidate the use of the respiratory exchange ratio to estimate substrate oxidation mix, it is important that future metabolic studies of caffeine evaluate and report $\dot{V}CO_2$ findings.

In contrast to early caffeine studies, more recent experimental research protocols have mostly employed a relative, body weight based, caffeine dosage (5–10 $mg \cdot kg^{-1}$). The rationale behind this approach is to account for the volume of distribution between subjects of different body weights when examining the acute physiological effects of caffeine. However, caffeine intake habits of study participants are routinely still estimated on an absolute daily basis (e.g. 105 $mg \cdot day^{-1}$ as compared to 1.5 $mg \cdot kg^{-1} \cdot day^{-1}$ for a 70 kg individual). In order to examine and/or better control for the potential confounding effects of habitual caffeine intake level and its associated tolerance it would seem judicious for future studies to recruit subjects on the basis of their relative ($mg \cdot kg^{-1} \cdot day^{-1}$) rather than absolute ($mg \cdot day^{-1}$) caffeine consumption levels.

REFERENCES

Åstrand PO, Rohdal K (1986) Textbook of Work Physiology. McGraw-Hill, New York, NY

Bangsbo J, Jacobsen K, Nordberg N, Christensen NJ, Graham T (1992) Acute and habitual caffeine ingestion and metabolic responses to steady state exercise. J Appl Physiol 72:1297–1303

Blanchard J, Sawers SJA (1983) The absolute bioavailability in caffeine in man. Eur J Clin Pharmacol 24:93–98

Brown DD, Knowlton RG, Sullivan JJ, Sanjabi PB (1991) Effect of caffeine ingestion on alveolar ventilation during moderate exercise. Aviat Space Environ Med 62:860–864

Bucci L (1993) Nutrients as ergogenic aids for sports and exercise. CRC Press, Boca Raton, FL

Casal DC, Leon AS (1985) Failure of caffeine to affect substrate utilization during prolonged running. Med Sci Sports Exerc 17:174–179

Chad K, Quigley B (1989) The effects of substrate utilization, manipulated by caffeine, on post-exercise oxygen consumption in untrained female subjects. Eur J Appl Physiol-59:48–54

Costill DL, Dalsky GP, Fink WJ (1978) Effects of caffeine ingestion on metabolism and exercise performance. Med Sci Sports 10:155–158

Curatolo PW, Robertson D (1983) The health consequences of caffeine. Ann Intern Med 98:641–653

Donelly K, McNaughton L (1992) The effects of two levels of caffeine ingestion on excess post-exercise oxygen consumption in untrained women. Eur J Appl Physiol 65:459–463

D'Urzo AD, Jhirad R, Jenne H, Avendano MA, Rubenstein I, D'Costa M, Goldstein RS (1990) Effect of caffeine on ventilatory responses to hypercapnia, hypoxia, and exercise in humans. J Appl Physiol 68:322–328

Ekelund LG (1967) Circulatory and respiratory adaptations during prolonged exercise. Acta Physiol Scand 70: Suppl 292

Engels HJ, Haymes EM (1992) Effects of caffeine ingestion on metabolic responses to prolonged walking in sedentary males.Int J Sport Nutr 2:386–396

Fisher SM, McMurray RG, Berry M, Mar MH, Forsythe WA (1986)Influence of caffeine on exercise performance in habitual caffeine users. Int J Sports Med 7:276–280

Graham TE, Spriet LL (1991) Performance and metabolic responses to a high caffeine dose during prolonged exercise. J Appl Physiol71: 2292–2298

Grant DM, Tang BK, Kalow W (1983) Variability in caffeinemetabolism. Clin Pharmacol Ther 33:591–602

Knapik JJ, Jones BH, Toner MM, Daniels WL, Evans WJ (1983) Influences of caffeine on serum substrate changes during running in trained and untrained individuals. Biochem Exerc-13:514–519

LeBlanc J, Jobin M, Cote J, Samson P, Labrie, A (1985) Enhanced metabolic response to caffeine in exercise-trained human subjects. J Appl Physiol 59:832–837

Leonard TK, Watson RR, Mohs ME (1987). The effects of caffeine on various body systems. J Am Diet Assoc 87:1048–1053

Poehlman ET, Despres JP, Bessette H, Fontaine E, Tremblay A, Bouchard C (1985) Influence of caffeine on the resting metabolic rate of exercise-trained and inactive subjects. Med Sci Sports Exerc 17:689–694

Pollock ML, Wilmore JH (1990) Exercise in health and disease (2nd edn.). WB Saunders, Philadelphia, PA

Powers SK, Dodd S, Woodyard J, Mangum M (1986) Caffeine alters ventilatory and gas exchange kinetics during exercise. Med Sci Sports Exerc 18:101–106

Rall TW (1985) The methylxanthines. In: Goodman Gilman A, Goodman LS, Rall TW, Murad F (eds) The pharmacological basis of therapeutics (7th edn.). Mac Millan Publishing Company, New York, NY

Spriet LL, MacLean DA, Dyck DJ, Hultman E, Cederblad G, Graham TE (1992) Caffeine ingestion and muscle metabolism during prolonged exercise in humans. Am J Physiol 262:E891-E898

Tarnopolsky MA, Atkinson SA, MacDougall JD, Sale DG, Sutton JR (1989) Physiological responses to caffeine during endurance running in habitual caffeine users. Med Sci Sport Exerc 21:418- 424

Toner MM, Kirkendall DT, Delio DJ, Chase JM, Cleary PA, Fox EL (1982) Metabolic and cardiovascular responses to exercise with caffeine. Ergonomics 25:1185–1196

Van Soeren MH, Sathasivam P, Spriet LL, Graham TE (1993) Caffeine metabolism and epinephrine responses during exercise in users and nonusers. J Appl Physiol 75:805–812

Zuntz, N (1901) Über die Bedeutung der verschiedenen Nährstoffe als Erzeuger der Muskelkraft. Pflügers Arch 83:557–571

48

EXERCISE TRAINING AND HIGH CARBOHYDRATE DIET

Effects of Vitamin C and B6

O. V. Karpus and V. A. Rogozkin

Research Institute of Physical Culture
Dynamo Ave. 2, 197042, St. Petersburg, Russia

INTRODUCTION

Proper nutrition is critical for optimal physical performance. Carbohydrate is the most important nutrient in an athletes diet because it is the only fuel that can power intense exercise over prolonged periods, yet its stores within the body are relatively small (Coyle, 1991). However, the systematic use of large amounts of carbohydrate provokes hyperglycamia and creates the conditions for the development of non-enzymatic glycation of proteins. Over the last few years a relationship between hyperglycamia and an increase in content of glycated proteins of tissues leading to the development of diabetes and the acceleration of aging has been discovered (Brownlee et al., 1984; VanBoekal, 1991; Brownlee, 1994). A high content of blood glycated proteins in athletes practisicing winter and summer sports is to be found (Rogozkin et al., 1991). This is comparable to data obtained from those who suffer from diabetes and therefore, shows some disturbance of the regulation of carbohydrate metabolism in this group of athletes. Subsequent experiments using animal models showed that with an abundant administration of carbohydrate food together with aerobic exercise created a higher concentration of glucose which is not only used as a source of energy, but can also lead to the modification of many proteins in the reaction of non-enzymatic glycation (Karpus,1993; Karpus et al., 1993, 1994; Rogozkin et al., 1993). This encouraged us to the search for the substance which competed with glucose for binding to protein amino groups. In the present paper we investigated the content of glycated proteins in blood under standard conditions and with different kinds of physical exercise, as well as with various concentrations of carbohydrates and vitamin C and B6 in normal diets.

METHODS

The investigation were carried out on 146 white male rats with a body mass 260–320g. All rats received two isocaloric diets: a standard diet (18% of energy in the

Current Research in Sports Sciences, edited by Rogozkin and Maughan.
Plenum Press, New York, 1996

333

form of protein, 27% fat, 55% carbohydrate) and a carbohydrate diet (18% of energy in the form of protein, 12% fat, 70% carbohydrate). To study the influence of dietary components on the level of rats blood glycated proteins we investigated 32 animals at rest on weeks 1, 2, 3 and 4 of the carbohydrate diet and at weeks 2 and 4 of the standard diet. In order to study the influence of systematic physical exercise on various metabolic aspects of the rats blood glycation 62 animals received either a standard or carbohydrate diet. The animals were divided into three groups with different physical activities: aerobic exercise consisted of uninterrupted swimming with a load equivalent to 3% of body mass attached and swimming duration was gradually increased from 3 to 40 min; anaerobic exercise consisted of swimming for a 5 min period with a gradual increase in load from 3 to 8% of body mass; the control group of animals undertook no exercise. In order to investigate the influence of the administration of vitamins on the level of protein glycation, 52 rats on the carbohydrate diet and who performed aerobic exercise for a month were administrated 10 mg/kg vitamin C or 0.4 mg/kg vitamin B6 in 1 ml of 6% solution of glucose immediately after training. In order to study the rate of protein exchange, a radioactive cocktail containing 14C-tyrosine and 3H-glucose was administrated inside the peritoneum 2 hours before slaughter of the rats . The inclusion of 14C-tyrosine and 3H-glucose in albumin was isolated by means of electrophoresis on PAAG (Karpus et al., 1993). The animals were investigated after exercise and after 1, 4 and 24 hours of rest. The rats were killed by means of decapitation and blood samples were collected. Fibrinogen was obtained by recalcification of citrated plasma (Sheicher & Wieland, 1986). Glycated serum proteins (GSP), glycated fibrinogen (GF), and glycated haemoglobin (GH) were assayed using affinity chromatography on phenylboronic acid modified sepharose columns (Schmid & Vormbrock, 1984). The concentration of glucose in serum was determined by "Reflotron" (Boehringer Mannheim GMBH). The concentrations of vitamin C (Jagota & Dani, 1982) and vitamin B6 (Pyatniska & Gvozdova, 1987) in blood were also determined.

RESULTS

The high carbohydrate diet evoked an increase in the degree of serum protein glycation up to $6.1 \pm 0.2\%$ (compared to $5.5 \pm 0.1\%$ in the animals on the standard diet) two weeks after the beginning of the experiment. This index increased to $6.3 \pm 0.1\%$ and $6.5 \pm 0.2\%$ three and four weeks later. The glycation of rapidly exchanging fibrinogen increased up to $2.8 \pm 0.2\%$ (compared to $2.4 \pm 0.2\%$ in the rats on the standard diet) one week after the beginning of the experiment, then make up to $3.2 \pm 0.2\%$ after two weeks and then remained at that level. The degree of slow exchanging haemoglobin did not change within the period of study (fig.1).

It should be noted that in spite of the fact that during a month, the consumption of the carbohydrate diet lead to a significant increase in the level of glycated serum proteins but remain within the physiological range. However, the levels of GSP in rats with experimental diabetes were 2.5 times higher (Day et al.,1980). Four weeks aerobic exercise training to an increase in the level of rats blood protein glycation (Table 1).

The results are depended on the diet content. Aerobic exercise together with the carbohydrate diet lead to the maximum increase of GSP [25% (p<0.01)] and GF [13% (p<0.05)]. The same exercise with the standard diet gave an increased in the level of GSP by 18% (p<0.05) and GF by 9% (p<0.05). In this case the indexes remained stable during the 24 hour recovery period. As shown in Table 1 there were no differences between the level of glycated haemoglobin in the groups of rats with different diets and physical activ-

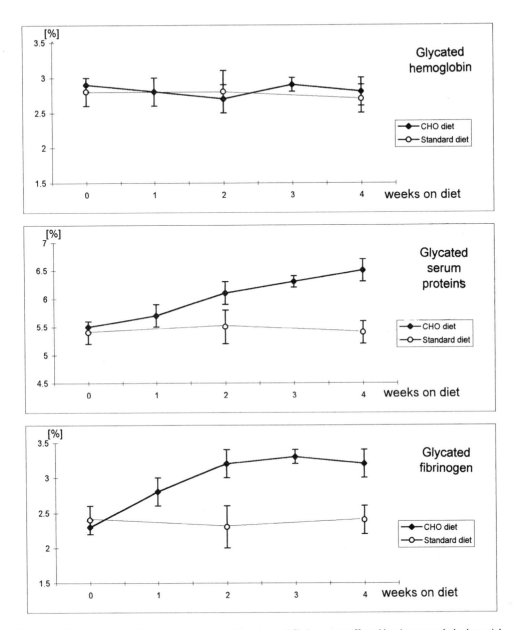

Figure 1. Blood haemoglobin, glycated serum proteins, glycated fibrinogen as affected by time on carbohydrate-rich and standard diets.

ity. The four-week training with aerobic exercise did not influence the level of blood protein glycation irrespective of diet. In order to investigate the mechanisms that are responsible for the increased level of glycated proteins under the influence of aerobic exercise, we investigated some indexes of carbohydrate and protein metabolism at various times during the period of recovery after. By analysing the content of 14C-tyrosine and 3H-glucose in serum albumin it was shown that neither physical exercise or diet had any influ-

Table 1. Effect of systematic exercise and carbohydrate diet on the levels
of glycated proteins in the blood of rats

Group of animals (n = 11)	Glycated fibrinogen, %	Glycated serum proteins, %	Glycated haemoglobin, %
Standard diet			
Aerobic exercise	2.8 ± 0.2**	6.1 ± 0.1**	2.9 ± 0.2
Anaerobic exercise	2.4 ± 0.2	5.5 ± 0.2	2.8 ± 0.3
Sedentary	2.5 ± 0.1	5.4 ± 0.2	2.9 ± 0.1
CHO diet			
Aerobic exercise	3.8 ± 0.2**	7.7 ± 0.1*	2.7 ± 0.3
Anaerobic exercise	3.5 ± 0.1	6.5 ± 0.2	2.8 ± 0.2
Sedentary	3.4 ± 0.2	6.5 ± 0.1	2.9 ± 0.2

*P< 0.001. **P< 0.01.

ence on the rate of protein exchange. Therefore, this can not influence the level of blood proteins glycation. The study of carbohydrate metabolism showed that the concentration of blood glucose was 13 - 15% higher in all the groups that received the carbohydrate diet. If in the control groups of the animals, as well as in the groups that were investigated immediately after exercise, because 1 hour after exercise this is seen only as a tendency, then in the groups of the rats that were investigated after 4 and 24 hours of rest, these differences were significant ($p<0.05$).

We investigated the influence of vitamins C and B6 on protein glycation by performing aerobic exercise together with a carbohydrate diet.

Table 2 gives the data on the level of vitamin C, vitamin B6, glucose and insulin in blood immediately after physical exercise, as well as after 4 or 24 hours of rest. As shown in Table 2 the administration of vitamin C does not lead to an increase in the level of vitamin C in blood. This is probably because the rats synthesize vitamin C under normal physiological conditions anyway and this will only change under conditions that stimulate or oppress biosynthesis of ascorbic acid. Also, the concentration of vitamin C in blood was measured 4 and 24 hours after its last administration, and the surplus of this vitamin is removed from the organism during the first 2–3 hours. The administration of vitamin B6 gave a significant increase in this parameter (70%) from 23.0 ± 3.1 mmol/l to 38.5 ± 4.8 mmol/l ($p<0.001$) immediately after exercise and was maintained at these levels for the whole recovery period, in both the control groups and in the groups that received vitamin B6. It is interesting that with a high carbohydrate diet and by performing aerobic exercises, the inclusion into the diet of both vitamin C and vitamin B6 led to change in carbohydrate metabolism which gave a significant change in the levels of carbohydrate in blood. In an experiment on men who did not perform any muscular activity, the administration of analogous doses of vitamin C did not exert any influence on the concentration of insulin and glycaemia (Stolba et al., 1987; Davie et al., 1992). As shown in Table 2 the level of glucose was higher in all control groups in comparison to the groups that received vitamin C or vitamin B6, both immediately after exercise and in the period of recovery. In most cases these differences were significant. The exception is in the group which received vitamin C and which was investigated 4 and 24 hours after aerobic training where there was only a tendency for this to occur. It should be noted that the levels of glucose in the groups that received vitamin C and vitamin B6 did not differ significantly. The same was also observed for blood insulin. There was a negative correlation between this and the level of glycaemia. As shown in Table 2, the concentration of insulin was lower in all con-

Table 2. Effect of vitamin C and B6 on the levels of glucose and insulin in the blood of rats

Group of animals (n = 6)	Glucose mmol/l	Insulin ng/ml	Vitamin C mmol/l	Vitamin B6 nmol/l
0h of recovery				
Control	9.2 ± 0.3	0.5 ± 0.1	90.3 ± 5.7	23.0 ± 3.1
Vitamin C	7.5 ± 0.4**	1.0 ± 0.1**	97.7 ± 7.4	25.1 ± 4.2
Vitamin B6	7.9 ± 0.4**	0.8 ± 0.1**	89.2 ± 6.8***	38.5 ± 4.8
4h of recovery				
Control	8.4 ± 0.3	0.5 ± 0.1	86.9 ± 6.2	24.5 ± 3.8
Vitamin C	7.8 ± 0.3	0.8 ± 0.1	93.2 ± 8.5	21.6 ± 2.3
Vitamin B6	7.2 ± 0.2*	1.3 ± 0.3**	91.5 ± 4.6	45.7 ± 2.5***
24h of recovery				
Control	8.1 ± 0.2	0.6 ± 0.1	92.6 ± 5.7	22.2 ± 2.6
Vitamin C	7.4 ± 0.3	0.9 ± 0.1	99.4 ± 9.6	26.6 ± 3.4
Vitamin B6	6.9 ± 0.3*	1.4 ± 0.2**	95.0 ± 7.9	40.9 ± 2.9***

*$p<0.05$. **$p<0.01$. ***$p<0.001$.

trol groups in comparison with the groups that received the vitamins. Like the level of glucose, these differences were statistically significant for most groups with the exception of those which received vitamin C and that were investigated 4 and 24 hours after physical exercise. It is likely that vitamin C did not exert a considerable influence on carbohydrate metabolism, or at least, it caused a lesser increase than with vitamin B6 in insulin concentration 24 hours after exercise ($p<0.05$). It is possible to conclude that the administration of the selected doses of vitamins for a month lead to visible changes in carbohydrate metabolism, and in particular to a reduction in blood glucose concentration and a arise in the level of metabolites competing with glucose for binding to proteins.

The diet with vitamin C and B6 did not influence the levels of GH. While the level of GSP was reduced from 7.3 ± 0.2% in the control group to 6.6 ± 0.1% in group that received vitamin C and to 6.5 ± 0.2% in the group that received vitamin B6 immediately after exercise and remained at this level. After physical exercise the level of GF was reduced from 3.7 ± 0.1% to 3.2 ± 0.2% in the groups that received vitamin C and to 3.1 ±

Table 3. Effect of vitamin C and vitamin B6 on the levels of glycated proteins in the blood of rats

Group of animals (n=6)	Glycated fibrinogen, %	Glycated serum proteins, %	Glycated haemoglobin, %
0h of recovery			
Control	3.7 ± 0.1	7.3 ± 0.2	2.8 ± 0.2
Vitamin C	3.1 ± 0.2	6.5 ± 0.2	2.7 ± 0.2
Vitamin B6	3.2 ± 0.2	6.6 ± 0.1	2.9 ± 0.3
4h of recovery			
Control	3.8 ± 0.3	7.4 ± 0.2	2.8 ± 0.1
Vitamin C	3.3 ± 0.1	6.4 ± 0.3	2.7 ± 0.2
Vitamin B6	3.1 ± 0.2	6.3 ± 0.2	2.9 ± 0.3
24h of recovery			
Control	3.6 ± 0.1	7.4 ± 0.1	2.7 ± 0.2
Vitamin C	3.0 ± 0.3	6.6 ± 0.3	2.8 ± 0.3
Vitamin B6	3.1 ± 0.2	6.5 ± 0.2	2.9 ± 0.2

0.2% in the groups that received vitamin B6, and these values remained stable during the period of recovery. In this case the levels of GF and GSP decreased to values which are typical of the groups of animals that received the standard diet and performed physical exercise. Also, it should be noted that the administration of vitamin B6 and vitamin C reduced the given parameters to the same degree. This is probably associated with the doses administered which took into account the levels of metabolites in the blood of the rats.

CONCLUSION

The penetration of carbohydrates into the body is a critical factor for ensuring that athletes have a maximum work-capacity. At the same time, abundant intake of carbohydrates into the body, together with intense aerobic exercises leads to the disturbance of carbohydrate metabolism and increases the content of glycated proteins in tissues. The inclusion into the carbohydrate diet of natural metabolites, as well as vitamins C and B6 competing with glucose for binding to proteins leads to the standardization of the level of glycated proteins. Therefore, to organize a diet for the athletes it is necessary to make up a balanced diet which would contain vitamin C and B6.

REFERENCES

Brownlee, M., 1994, Glycation and diabetic complications, Diabetes 43:836–841.
Brownlee, M., Vlassara, H., Cerami, A., 1984, Nonenzymatic glycosylation and the pathogenesis of diabetic complications, Ann. Int. Med. 101:527–537.
Coyle, E. F., 1991, Timing and method of increased carbohydrate intake to cope with heavy training, competition and recovery, J. Sports sciences 9:29–52.
Davie, S. J., Gould, B. S., Yudkin, J. S., 1992, Effect of vitamin C on glycosylation of proteins, Diabetes 41:167–173.
Day, J. F., Ingelbretsen, C. G., Ingelbretsen, W. R. et al., 1980, Nonenzymatic glycosylation of serum proteins and hemoglobin: response to changes in blood glucose levels in diabetic rats, Diabetes 29:524–527.
Jagota, S. K., Dani, H. M., 1982, A new colometric technique for the estimation of vit C using Folin Phenol reagent, Anal. Biochem. 127(1):178–182.
Karpus, O. V., 1993, Influence of systematic physical exercise and dietary carbohydrate intake on levels of glycated proteins in blood or liver of rats, Int. J. Sports Med. 14(5):296.
Karpus, O., Rogozkin, V., Feldkoren, B., 1994, Exercise, blood glycated proteins and carbohydrate supplementation, Clin. Sci. 87(supplement):67.
Karpus, O., Rogozkin, V., Melgunova, E., Feldkoren, B., 1993, Effect of different metabolic type of physical training and carbohydrate diet on levels of glycated proteins in blood of rats, Physiol. J. 79(12):39–43.
Pyatniska, I. N., Gvozdova, L. G., 1987, Vitamin B6, Scien. papers of Nutrition Research Institute :109–120.
Rogozkin, V. A., 1993, Physical exercise and carbohydrates, Physical exercise performance and nutrition, St.Petersburg :12–26.
Rogozkin, V., Karpus, O., Melgunova, E., Feldkoren, B., 1993, Effect of food carbohydrates, anaerobic and aerobic training on metabolism in rats, XV International Congress of Nutrition, Adelaide, Abstract 1:369.
Rogozkin, V., Osipova, E., Feldkoren, B., 1991, Bloodglycated proteins as an index of carbohydrates metabolism in athletes, Second IOG World Congress on Sport' Sciences, Barcelona:238
Scheicher, E., Wieland, O., 1986, Kinetic analysis of glycation as a tool for assessing the half-life of proteins, Biochim. Biophys. Acta 884:199–205.
Schmid, G., Vormbrock, R., 1984, Determination of glycosylated hemoglobin by affinity chromatography, Anal. Chem.317:703–704.
Stolba, P., Hatle, K., Krnakova, V., et al. 1987, Effects of ascorbic acid on non-enzymatic glycation of serum proteins in vitro and in vivo, Diabetologia 30:585A.
Van Boekal, M. A., 1991, The role of glycation in aging and diabetes mellitus, Mol. Biol. Rep. 15:57–64.

49

RADIOISOTOPIC INVESTIGATION OF GASTRIC EMPTYING AND SMALL INTESTINE FUNCTION AT DIFFERENT EXERCISE LEVELS

A. P. Kuznetsov, V. I. Kozhevnikov, and A. V. Rechkalov

State Teacher Training Institute
Sovetskaja st. 63, 640000 Kurgan, Russia

INTRODUCTION

The digestive tract of a person in the process of adaptation to physical loading undergoes specific changes which are aimed at meeting the power and plastic needs of the organism fully and in a short period of time. Factors regulating the ingestion of food and all the stages of the process of digestion mechanical treatment and movement of chyme in the intestine, separation of substances according to the components suitable for absorption and so on are described (Chesta et al, 1990; Kaufman et al, 1990; Mistae net al, 1990). In the present study, we pay attention to the evacuation of the stomach contents and the transit of food through the intestine of people with different levels of everyday activity. It is well known that motor activity is a powerful factor influencing all the body systems, including the gastroenteric tract (Kuznetsov, 1985). There has, however, been little information about this question in the published literature.

METHODS

The evacuation of food from the stomach and its transit through the small intestine were studied in 11 people. Six of them were in the control group who participated in sport as far as the program of higher educational institutions requires. 5 of the subjects were in the experimental group. All of them were highly skilled sportsmen. The study of the evacuation of food from the stomach and its transit through the intestine was performed by radio-isotope scanning, which allows a quantitative evaluation of the results. According to the results of Ischmuhametov (1979), " the organ's contours stand exposed clearly and according to them the size of the stomach in the scanning plane coincides with the anatomical data of the healthy man." After the first scan, we could determine the position of the stomach within the abdominal cavity and mark the points above which further external de-

Current Research in Sports Sciences, edited by Rogozkin and Maughan.
Plenum Press, New York, 1996

tection would be conducted. It was very important to use a preparation during the experiment which is poorly absorbed by the intestine. Otherwise it is not possible to get clear contours of the stomach. Scanning was done on the apparatus "Plannescanning system" KE-32 (Italy, The Netherlands). It was conducted in the laboratory of radiology in the Russian Scientific Center "Restorative Traumatology and Orthopaedics." People under testing were given 200 g of 10% semolina on an empty stomach (12–14 hours after the last meal). Semolina was marked by 0.5 mCi of a colloidal solution of 99 mTc (preparation TCK-17, France). The radiation load on the critical organ did not exceed 0.13 mSv. Those undergoing testing were fed directly under the apparatus in a sitting position. Before this the plan of the stomach region and of the ileocecal sphincter was marked. After taking the porridge, the subjects lay down and the scanning was done over 8–10 minutes. According to the results of this initial scan, 3 points were marked on the plan of stomach: X1- cardial section of the stomach, X2- the body of stomach, X3 - the pyloric canal. The radiometric control was held every 15 minutes after ingestion of the test meal until the mark appeared in the region of the ileocecal valve. The control was always done when subjects were lying. Between the times when the control measurement was made they completed a standard walk. According to the data received the penetration of nuclide into blood was not detectable. Repeated research was held after an interval of two weeks and the same methods were used. But before taking the test, subjects exercised on a cycle ergometer for 30 minutes at an intensity of 75% of maximum oxygen consumption. The time of gastric emptying (T), the speed of gastric emptying in 15 minute intervals, the time of transit of the trial breakfast to the ileocecal valve were calculated. With the help of the method of electric gastrography the following qualitative characteristics of the process of gastric emptying were studied: amplitude, the period of contraction, the factor of unbalance of contractile waves, and the quantity of evacuative and mixed waves. With this aim the subjects ingested 200 g of 10% semolina and after this the electrical activity of the stomach was recorded. All the studies were done in the morning hours and the conditions were equal. To study the morphofunctional connections of endocrine cells of the stomach and the motor-evacuating function of the alimentary canal, the volunteers were biopsied by gastroendoscopy with the biopsy collected from the body and the entering section of the stomach. The endocrine cells (G-, Ec- and Ecl-cells) were identified according to the scheme of Uspensky & Golofeevsky (1980). The endocrine cells were stained and counted by means of an ocular micrometer.

RESULTS

The preliminary investigation shows that the difference in the level of everyday motor activities influences the gastric motor evacuating function. The half time for gastric emptying in the experimental and control groups was equal to 49 ± 4.4 min and 54 ± 5.4 min ($p > 0.05$) respectively. The time for full evacuation from the stomach for the trained subjects was 171 ± 14.7 min, and the corresponding value for the less active group was 145 ± 12.6 min. In the experimental group, the most rapid rate of evacuation from the stomach occurred from the 15th to the 45th minute after eating the test meal. In the control group the maximum speed of emptying occurred from the 45th to the 75th minute after eating the test breakfast (fig.1).

In the laboratory of sports gastroenterology it has been shown by gastric aspiration methods that sportsmen who regularly subject themselves to dehydration (wrestlers, boxers, weight-lifters) have a reduced rate of gastric secretion. We might suppose that the re-

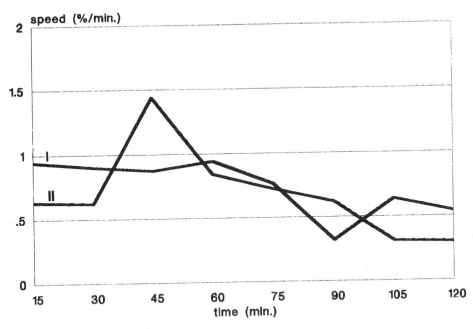

Figure 1. Speed of gastric emptying at rest.

duction of HCl in the gastric secretion of these sportsmen induces a rapid emptying of the contents of the stomach into the small intestine. However, after 45 minutes of taking food the speed of gastric emptying in the experimental group begins to reduce and in the period of 75 - 90 minutes it achieves the minimum value, even though only 60 - 65% of the meal has been emptied by that time. In the control group after the 75th minute we also observed a much more rapid reduction of the speed of emptying. We are apt to think that this reduction is conditioned by the fact that by the 75 - 90th minutes after the test breakfast most of the chyme (80–90%) had already been emptied into the small intestine. There were no substantial differences between the experimental and control groups in the electrogastrogram. The results in the experimental group are the following: the amplitude of contraction was 0.132 ± 0.01 mV, the average frequency of peristaltic waves was 2.9 ± 0.26 contractions per minute. The coefficient of asymmetry was 0.987, and the percentage proportion of the evacuating and mixing waves was 40.9% and 59.1% respectively. The data of the control group are the following: the average amplitude of the contractions was 0.164 ± 0.01mV, the average frequency of peristaltic waves was 2.4 ± 0.15 contractions per minute, and the coefficient of asymmetry was 0.996, the percentage proportion of the evacuating and mixing waves was 49.2 and 50.8%.

For revealing the mechanism of regulation of the motor function of the alimentary canal of subjects with different training levels by means of gastrofibrescopy with biopsy and followed by morphometry, the number of G-, Ec- and Ecl-cells in the body and antral part of the stomach is useful. It is known that G-cells produce serotonin and Ecl-cells are supposed to produce gystaminum. According to the results the control group and the sportsmen have some differences in the number of G- and Ecl- cells in the body of the stomach. In subjects with low exercise levels the number of G-cells in the body of the

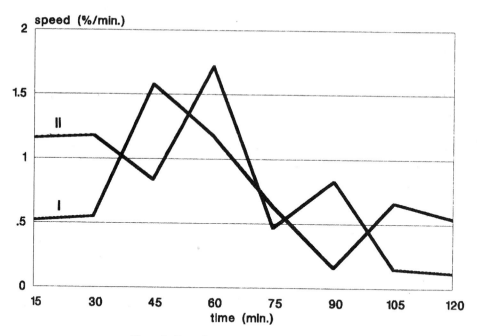

Figure 2. Rate of gastric emptying after exercise.

stomach was 7.8 ± 1.2 and the number of Ecl-cells was 44.5 ± 2.8; for the athletes these figures were respectively 5.5 ± 1.1 and 38.4 ± 3.2 in 1 mm.sq. In the antral part of the stomach of the control group and athletes there were no significant differences in the quantity of G- and Ecl-cells. Comparing the results of the research of the endocrine cells of the stomach and motor function of the alimentary canal one may suppose that people with different exercise levels the differences in the speed of evacuation and the transit of food in the quantity of G- and Ecl-cells in the mucous membrane of the body of stomach. The transit time of chyme from the stomach up to the ileocecal valve varies between 70 - 220 minutes. It was considered before that such differences can not be explained by physiological indices. Analysis of the transit times of the test breakfast up to the ileocecal sphincter gives the following results: in the experimental group, transit time was 78 ± 7.3 min, and in the control group it was 251 ± 26.5min (p<0.01).

In the second stage of the experiment after the cycle exercise, the control and experimental groups again showed differences in the transit time of food up to the ileocecal valve. The transit time in the experimental group was 75 ± 15 min, and in the control group 210 ± 24.3 min (p<0.01). The rapid intestinal transit in the trained subjects at rest and after exercise appears to be an adaptational change correlated with the level of daily physical activity. After exercise, the trained subjects showed an increased half emptying time (61 ± 8.7 min), while the time to complete emptying did not change significantly (173 ± 22.5 min). In the control group on the contrary, there was a tendency for a reduction in the half emptying time at rest (43 ± 3.7 min) and also of the full emptying (110 ± 13.2 min). The dynamics of stomach emptying after muscular exercise were also changed substantially. In athletes the maximal speed of evacuation occurred from the 30th to 60th minutes, and the slowest rate from the 75th to 90th minutes after the test meal. In the control group high rates of emptying were observed immediately after ingestion of the meal and continued to the 60th min: the lowest rates occurred from the 90th min onward (Fig 2).

Gastrointestinal hormones of the proximal segment of the small intestine (gastrin, motilin, CCK-PZ) stimulate the secretory function of the stomach, and the hormones of the distal segment of the small intestine release neurotensin and enteroglucagon which act to suppress secretion. A similar effect may have occurred in the trained subjects, when the first portions of chyme, having reached the ileocecal region, suppressed the secretion and motoric activity of the stomach, so that the time to full emptying of the stomach turned out to be so prolonged.

CONCLUSION

Many changes in work of the digestive system are the result of adaptational changes which make its work more economical and effective. Apparently, the rapid transit of food along the small intestine in the experimental group is connected with adaptational restructuring as a reaction to the intensive training, connected with the specific character of the given kind of sport. Thus, judging by the preliminary data, it is possible to conclude that persons with different levels of everyday physical activity in a state of rest and also after exercise show considerable differences in the length of the transit time for a test carbohydrate breakfast to the ileocecal sphincter and insignificant differences in the process of emptying the stomach.

REFERENCES

Chesta, J., Debnam, E. S., Srai, S. K., Epstein, O., 1990, Delayed stomah to caecum transit time in the diabetic rat. Possible role of hyperglucagonaemia, Gut. 31:660–662.

Ischmuhametov, A. I., 1979, Radioisotopnaja diagnostica organov pischevorenija, Medisina:36.

Kaufman, P. N., Richter, J. E., Chilton, H. M., et. al., 1990, Effects of liqued versus solid diet on colonic transit scintigraphy, Gastroenterology 98:73–81.

Kuznetsov, A. P., 1985, Vozrastnaja sportivnaja gastroenterologija, Cheljabinsk:100.

Mistaen, W., Van Hee, R., Bloskx, P., Hubens, A., 1990, Gastric emptying for solids in patients with duodenal ulcer before and after highly selective vagotomy, Digest. Dis. Sci.35: 310–316.

Uspensky, V. M., Golofeevsky, V. U., 1980, K metodike himicheskoj identifikasii endokrinnih kletok slizistoj obolochki jeludka i dvenadsatiperstnoj kischki, Arhiv patologii 42:81–84.

INDEX